FORENSIC GENETICS

FORENSIC GENETICS

Volume 2

Marly Cardoso
Eduardo Penna

Kruger Brentt
Publishers

2025

Kruger Brentt Publishers UK. LTD.
Company Number 9728962

Regd. Office: 68 St Margarets Road, Edgware, Middlesex HA8 9UU

ISBN: 978-1-78715-431-5 (Vol. 2)

For information on all our publications visit our website at http://krugerbrentt.com/

PREFACE

"Forensic Genetics" delves into the fascinating intersection of genetics and forensic science, offering readers a comprehensive exploration of the principles, techniques, and applications that define the field of forensic genetics. This book serves as an essential resource for students, researchers, forensic scientists, law enforcement professionals, and legal practitioners seeking to understand the role of genetics in criminal investigations, identification of individuals, and the administration of justice.

Forensic genetics utilizes genetic information, such as DNA analysis, to solve crimes, identify victims, exonerate the innocent, and establish biological relationships in legal proceedings. With advances in technology and the increasing availability of genetic data, forensic genetics has become an indispensable tool in criminal justice systems worldwide, revolutionizing the way investigations are conducted and evidence is analyzed.

The primary objective of this book is to provide readers with a comprehensive overview of forensic genetics, covering key topics such as DNA extraction and profiling, forensic DNA databases, kinship analysis, forensic genealogy, molecular autopsy, and ethical considerations in genetic testing. Each chapter offers insights, case studies, and practical examples to help readers understand the fundamental principles and applications of forensic genetics in real-world scenarios.

Contributions from leading experts in the field explore various aspects of forensic genetics, including the use of DNA evidence in criminal investigations, mass disaster victim identification, missing persons cases, paternity testing, and the challenges of interpreting complex genetic data in court. Special emphasis is placed on emerging technologies and methodologies, such as next-generation sequencing, single nucleotide polymorphism (SNP) analysis, and probabilistic genotyping algorithms.

As forensic genetics continues to evolve and expand its capabilities, it is essential for practitioners and stakeholders to stay abreast of the latest developments, ethical considerations, and legal implications in the field. By fostering collaboration, transparency, and interdisciplinary approaches, we can ensure that forensic genetics remains a reliable and powerful tool for achieving justice and truth in the criminal justice system.

We extend our sincere gratitude to the contributors whose expertise and dedication have enriched this book. It is our hope that "Forensic Genetics" will serve as a valuable resource and catalyst for dialogue, innovation, and collaboration in advancing the science and practice of forensic genetics worldwide.

Marly Cardoso

Eduardo Penna

CONTENTS

11 ASSESSMENT OF THE FORENSIC SUITABILITY OF BIOSTATISTICAL TECHNIQUES FOR DETERMINING ANCESTRY BASED ON AUTOSOMAL GENETIC MARKERS

1. INTRODUCTION: AN OVERVIEW

The increased availability of whole human genome sequences in the public domain is an invaluable data resource in many genomics, biomedical, and anthropological research areas. In particular, the data repositories focusing on the genomic diversity among human populations (e.g., HapMap Project [**1**], 1000 Genomes Project [**2**], Simons Genome Diversity Project [**3**], and more recently curated in the Genome Aggregation Database Project, gnomAD [**4**]) have contributed to the understanding of human evolution, migration histories, waves, and patterns.

Typically, genetic samples from populations with different geographical locations, cultural backgrounds, tribal memberships, and linguistic groups constitute the data used to identify genomic differences among the derived populations. These genomic differences are the result of a mixture of causes: mutations, recombination, genetic drift, selection, and migration [**5**]. The variations in the human genome can be observed both in tandemly repeated DNA sequences (e.g., short tandem repeats, STRs) and in single-nucleotide polymorphisms (SNPs) but are also manifested in sequence variations in lineage markers (e.g., Y-chromosome and mitochondrial DNA) and structural variations (e.g., copy-number variation) [**6**].

The geography of human populations (ancestry) and genetic polymorphisms are closely related [**7**,**8**,**9**,**10**,**11**,**12**]. Hence, identifying ancestry informative markers (AIMs) can be accomplished by analysing the publicly available genome sequences. Several measures of informativeness has been derived in order to rank candidate markers [**13**,**14**].

In forensic genetics, STRs are presently the standard markers used for identification and relationship testing. Furthermore, variation in allele frequencies may be used to infer an individual's ancestry [**15,16**]. Other types of AIMs are microhaplotypes [**17**], which are groups of closely located SNPs (often positioned within 500 bp), and insertion–deletion polymorphisms (indels) [**18,19**]. However, SNPs are more commonly used for ancestry analysis [**20,21**], where several commercial assays have been developed for ancestry investigations. The selection of SNPs to be included in commercial kits depends both on their biochemical properties (e.g., allele balances and primer site location) and their informativeness regarding specific continental and regional populations. Hence, the companies focused on specific population structures (e.g., intracontinental differences) may result in poor resolution in other regions and among populations of close proximity. Consequently, all results obtained from the genotyping and subsequent analysis are conditioned on the specific SNPs included in the analysis. This is important to bear in mind when interpreting the results of an analysis, as general patterns and expectations to population variability may not be contained on a narrowly selected set of markers.

2. BIOSTATISTICAL METHODS

The study of population structure from genomic data has a long history within the genetic and statistical literature. The increasing complexity and availability of data from ancestry informative markers on multiple populations have required the development of appropriate methodologies to model and capture the subtle genetic structures in the data: from the first studies of a few hundred markers and individuals to more recent studies with thousands of individuals genotyped on highly dense assays or from whole genome sequences.

This progress, driven by biotechnological advances, has been challenging not only from a methodological point of view but increasingly so from a computational perspective. Where independence between the markers was ensured by their genetic distance, the closer proximity of genetic markers results in statistical dependence (e.g., due to linkage disequilibrium, LD). This increases the complexity of the analysis since such dependencies must be adequately modelled and their dependence structure learned from the data. Some methods incorporate the structure into their modelling framework, either by identifying stretches supposedly inherited as a complete block (e.g., [**22**]) or attempts to learn the association structure using graphical models [**23**]. Others operate by removing the dependence among markers by pruning or thinning the data based on LD (e.g., [**24**]) or using the residual signal after successively regressing the markers on each other (e.g., [**25**]).

The most widely used methodologies for analysing genetic structure can broadly be divided into four different groups: principal components analysis, model-based clustering, classification and likelihood-based, and hypothesis test-based methods. In addition, tree-based and distance/dissimilarity clustering methods are used but not included in this review. Most of the development in the study of population structures is driven by interests in medical genetics or population genetics, where controlling for population stratification may be vital for some applications or where the study of population dynamics on its own is the main purpose. Hence, few methods have been developed with the applications of forensic genetics in mind. In the subsequent sections, the various methodologies are discussed with references to key publications in their domain.

To demonstrate and visualise the results from the various methods, data from [26] are used in the following sections. The data consist of samples from six reference metapopulations (comprised by a total of $n$ref=3453 samples) and $n$test=517 samples all genotyped on the 164 ancestry informative SNPs included in the Precision ID Ancestry Panel (excluding marker rs10954737 due to a high degree of missingness). The reference samples are known to originate from the indicated metapopulation, are of high sequencing quality, and used to estimate the metapopulation's allele frequencies. The 517 test samples were harvested online, where their alleged metapopulation of origin was derived from the sampling location and other available meta information [26]. Both the reference and test samples originated from six regional areas, where the regional specific $n$ref/$n$testref/test counts are: Sub-Saharan Africa (668/37668/37), North Africa (283/49283/49), Europe (1014/1731014/173), Middle East (377/52377/52), South/Central Asia (489/85489/85), and East Asia (622/121622/121).

2.1. Principal Components Analysis

Principal components analysis (PCA) has a long history of application in the study of population structure. The usage of PCA in the analysis of population structure was pioneered by [**27**], as an efficient way to capture the underlying structure for visualisation purposes. PCA is able to capture continuous admixture between populations, implying that admixed populations typically falls on the *lines* between its *parental* populations. More recent PCA-based methods for inference include SMARTPCA [**25**] and EIGENSTRAT [**28**]. These methodologies provide further insight as to how many PCs are required to capture detectable population structures by the use of hypothesis testing on the magnitude of the PCA's eigenvalues. For the reference samples (cf. above) from [**26**], there are four significant PCs according to EIGENSTRAT. The first three PCs are plotted in **Figure 1** (plots including the fourth PC do not visually separate the metapopulations). The often detected *triangle* spanned by Sub-Saharan Africa, Europe, and East Asia is cleary visible in the plot of PC1 and PC2.

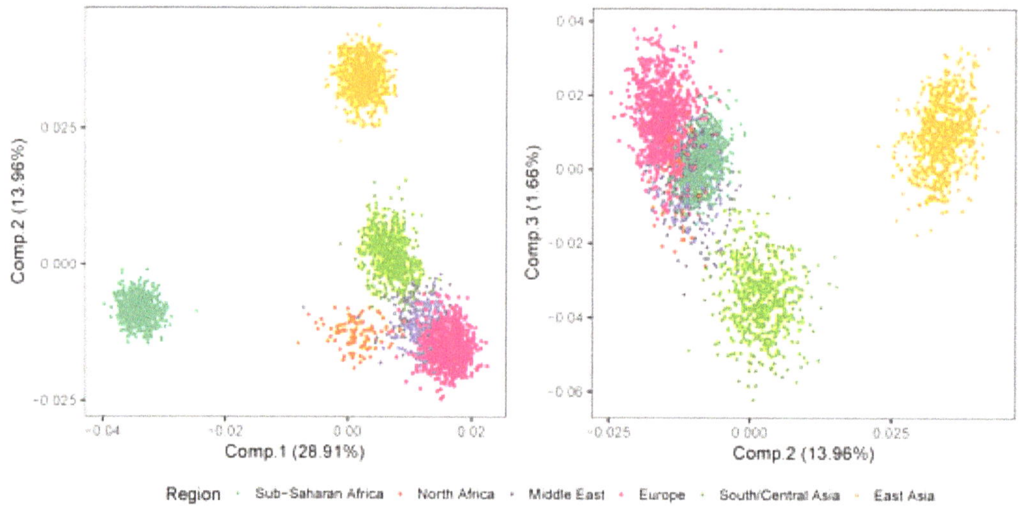

Figure 1. Plot of the first three PCs from the PCA of the reference samples.

Besides being an efficient tool for visualising high-dimensional (genetic) data, PCA also yields the best linear low-dimensional approximation of the data in terms of Euclidean distance. This, together with the orthogonality of the principal components (PCs), implies that population substructures can be detected by the discriminant analysis of the PCs, DAPC [**29**]. Primarily relying on linear algebra, DAPC is computationally much faster than STRUCTURE and gives comparable results. However, several papers discuss some of the known pitfalls when using PCA [**30,31,32,33**]. The most important issue to be aware of is PCA's sensitivity to the sampling of individuals and populations. An unbalanced sampling of some populations forces the PCs to account for the variation caused by the majority groups. Since most models and methods try to minimise a *loss function*, this implies that in the estimation of the unknown quantities (e.g., model parameters, here, the PCs) emphasis is given to the majority groups in the data [**30,31**]. The loss function is typically proportional to the deviation between the observed data and the model's predictions (or expectation), e.g., measured by the model's likelihood or squared distance of the residuals. For PCA, the loss function is inversely proportional to the fraction of explained variance of the first PCs, with DAPC's loss also depending on the intracluster variances as a function of the numbers of cluster K (measured by a Bayesian Information Criterion, BIC). Another property shared by several methods and statistical models is that they provide a *compression* of the data, in the sense that the methods are trying to retrieve as much information from the data as possible and discard a nonsignal as noise. Taken a bit further, one may interpret the model output and parameter estimates as *data summaries* (some more informative, advanced, and interpretable than others). Consequently, some of these summaries (e.g., the PCs and STRUCTURE as discussed next) are similar or even identical for different population genealogies [**31**].

2.2. Model-Based Clustering Methods

The seminal STRUCTURE paper [**34**] introduced a population genetics methodology based on a statistical finite mixture model. Using a Bayesian approach, the posterior probabilities for cluster memberships and cluster-specific allele frequencies are estimated using an MCMC-algorithm (MCMC: Markov chain Monte Carlo). STRUCTURE has been used in numerous studies and has, according to Google Scholar, more than 30,000 citations, which indicates its enormous influence in the field of population genetics [**35**]. Forensic applications were mentioned as one of motivations for STRUCTURE ([**34**], p. 945) in order to identify cryptic population structures and assess their influence on the detection of immigrants and calculation of match probabilities [**36,37,38**]. In essence, the initial STRUCTURE model is a simple Hardy–Weinberg model with the relaxation that subpopulations have different allele frequencies and individuals may inherit alleles from several subpopulations, i.e., being admixed [**35**]. Following the initial publication, several modifications both on the population genetic model (allowing for, e.g., linked markers and null alleles) [**22,39**] and computational aspects (faster and more efficient algorithmic schemes) [**40,41,42**] have been suggested. See [**35,43**] for further references and remarks.

Despite the many successful applications of STRUCTURE (including variants and similar methodologies, e.g., FRAPPE [**40**] and ADMIXTURE [**41**]), to population genetic data and research questions, no model is better or more general than its underlying assumptions. Similar stories apply to simple linear regression models, where the assessment and evaluation of the residuals is part of any analysis. However, because of STRUCTURE's complexity, it is harder to conduct the assessment of the model fit. A recently published instructive tutorial [**44**] provides a critical view on how to assess the outcome of STRUCTURE analysis using the tools badMIXTURE [**45**], GLOBETROTTER [**46**], fineSTRUCTURE, and CHROMOPAINTER [**47**]. In particular, the authors highlighted the fact that different population scenarios (e.g., recent admixture, bottleneck, and admixture contribution from untyped populations) result in similar STRUCTURE barplots [**7,48**] (the typical data summary from a STRUCTURE analysis). Supplementing the STRUCTURE analysis by residual plots based on badMIXTURE, patterns were detected enabling these scenarios to be distinguished [**44**]. Similar to PCA, STRUCTURE is sensitive toward biased sampling and population sample sizes. Specifically, imbalances between the analysed populations may influence how and which of the sampled populations that exhibit patterns of population admixture ([**44**], Case Study 3, pp. 5–8). However, choosing the appropriate (or in a sense, *correct*) value for K, the number of population clusters, is not a well-defined problem, i.e., only heuristic methods exist for guiding the specification of K [**34,35,43,44**]. Different choices of K may result in rather different results and interpretations of the population stratification and history.

In **Figure 2**, the estimated STRUCTURE admixture components, $\boldsymbol{q}$=($q^{(1)}$,⋯,$q^{(K)}$, for the reference samples (cf. above) are shown for K=4. There is a clear visual distinction in the distribution of $\boldsymbol{q}i$ across the 3453 reference samples.

Figure 2. Barplots of the STRUCTURE admixture components (K=4) for the reference samples from six metapopulations. There is a clear visual difference between the admixture components for the six metapopulations.

The four clusters of **Figure 2**, K=4, may be associated with Sub-Saharan African, European, South/Central Asian, and East Asian components. The samples from North African and Middle Eastern regions show the strongest admixture of these four components.

2.3. Classification and Likelihood-Based Methods

The statistical problem solved by STRUCTURE [**34**] and the related methodologies (e.g., ADMIXTURE [**41**]) is a rather complex one: Assign individuals to populations while estimating the unknown populations' unknown allele frequencies. As discussed above (**Section 2.2**), this is performed by finite mixture models, where the assignment of individuals and updating of population allele frequencies are conducted iteratively (either by Bayesian MCMC methods [**34**] or using efficient variants of expectation–maximisation algorithms [**41**]). However, if analysis of population structure is not the purpose and the allele frequencies are known for a set of populations, the assignment of individuals is simpler.

In this case, the assignment of nonadmixed individuals can be conducted by a Bayes classifier, where the individual is assigned to the most probable population among the ones included in the reference database. As such, the population assignment problem can be solved by several types of classification algorithms known from the machine learning literature (e.g., classification trees, random forest, support vector machines, multinomial regression, partial least squares, discriminant analysis, nearest neighbours, gradient boosting, and other ensemble learners). Common to most of these methods is their ability to account for interaction effects between the explanatory variables on the outcome (here, the marker interactions when predicting the population of origin). Furthermore, similar to the methods of **Section 2.1** and **Section 2.2** (e.g., PCA and STRUCTURE), sample sizes and imbalanced training data typically influence the tuning and therefore performance of the machine learning methods. When training a classification algorithm on a specific dataset, the objective is to drive the overall loss function downwards. Typically, the loss function in classification contexts decreases with more samples being accurately classified (e.g., the misclassification rate or smooth functions, thereof). However, if some populations have low proportions in the sample, the classifier gains little from classifying these correct compared to the frequently occurring populations. Hence, the relative composition of the populations in the data may impact the trained classifier quite substantially. In the context of forensic genetics, the risk of incomplete AIMs profiles obtained from, e.g., crime scenes, makes some methods less applicable to samples with some or more untyped markers (e.g., due to low signals or marker drop-out). Some methods (e.g., naïve Bayes classifiers and classification trees using *surrogate splits*) may readily deal with partial profiles, whereas in particular regression, models are susceptible to missing data.

From a forensic perspective evaluating the weight of the evidence is typically conducted using the likelihood ratio principle. For multiple propositions (in this case, multiple populations), this approach generalises to several ratios assessing the pairwise agreement between data and the alleged populations of origin. The predicted population of origin is the population where the profile is most probable. The likelihood function typically assumes a specific population genetic model, e.g., within each population assuming Hardy–Weinberg equilibrium and independent genetic markers. This corresponds to a naïve Bayes classifier assuming a Binomial distribution for each marker given the population specific allele frequencies. Two forensically relevant implementations including several reference populations and AIMs panels are freely available online (FROG-kb (**http://frog.med.yale.edu/FrogKB/**, accessed on 13 December 2021) [**49,50**] and Snipper (**http://mathgene.usc.es/snipper**, accessed on 13 December 2021) [**51**]). Other approaches toward classification have been considered and compared (e.g., multinomial regression and genetic distances) [**52**]. These methods are in particular relevant in the case of admixed individuals, where the single population likelihoods do not suffice [**53,54**].

2.4. Hypothesis Test-Based Methods

An underlying and important assumption for both clustering and classification is that of *exhaustive* populations. This means that for clustering the admixture components *must* sum to one, indicating that *all* the genetic admixture of an individual can be explained by contributions from the *K* clusters. However, the *Ghost admixture* example of ([**44**], p. 3) shows that the noninclusion of important branches of a genealogy may result in misleading results. In that case, the contribution from the unsampled reference population to the admixture was replaced by the reference population most similar to the unsampled one. This scenario may easily occur in the forensic setting, where only major populations are included in the reference material, but where conclusions are wanted on a finer (sub)population scale.

For classifications, an underlying assumption is that an observation belongs to *exactly* one class (i.e., one and only one). Hence, this *forced* classification means that even when all the possible classes are (highly) unlikely, the DNA profile under classification *must* be assigned to precisely one of the populations (typically the *least unlikely* population).

This issue was the main motivation for the derivation of the GenoGeographer methodology [**55,56**]. Rather than forcing an observation to be classified to any of the prespecified populations in the reference material, it was assessed whether the sample could originate from each of the reference populations. Logically, this implied that the sample was tested for being an outlier in each of the possible populations, and this was formally conducted by testing the hypothesis of the DNA profile sample and database sample originated from the same underlying population or not [**55**]. This outlier test is equivalent to a Fisher's exact test and thus enjoys many of the same statistical properties, e.g., increasing power with increasing sample size and robustness toward missing observations. Other hypothesis-based methodologies include a likelihood ratio test for recent admixture [**57**].

For each reference sample i=1,⋯,3453STRUCTURE estimates the admixture components, $\boldsymbol{q}_i = (q_i^{(1)}, \ldots, q_i^{(K)})$, where $\sum_{k=1}^{K} q_i^{(k)} = 1$ by definition. From these, the average admixture component is calculated for each of the reference metapopulations, by the average over the admixture components for reference samples belonging to this metapopulation:

$$\bar{q}_j = (\bar{q}_j^{(1)}, \ldots, \bar{q}_j^{(K)}), \quad \text{with} \quad \bar{q}_j^{(k)} = n_j^{-1} \sum_{i \in R_j} q_i^{(k)}, \quad k = 1, \ldots, K,$$

where R_j is the set of the n_j samples from metapopulation j.

As for the training samples STRUCTURE admixture components q_l was calculated for the test samples, l=517,⋯,1. The closer q_l $\bar{q}_j$ for some test sample l and referece metapopulation j, the more likely it is to assume l to originate from metapopulation j. However, how (dis)similar should these admixture components be to declare l (not) to originate from j? Moreover, how should this similarity be measured? One possible measure of proximity could be to use a Brier-like score, B_{lj}, which is the sum of squared admixture components differences between test sample l and reference metapopulation j:

$$B_{lj} = \frac{1}{K} \sum_{k=1}^{K} (q_l^{(k)} - \bar{q}_j^{(k)})^2.$$

The closer B_{lj} is 0, the more similar is sample l to metapopulation j in terms of STRUCTURE admixture components (Supplementary materials of [**26**]). The maximal value of B_{lj} is $2/K2/$, which happens if $q(k)l$=1 and $q(k')j$=1=1 for $k≠k'$. In **Figure 3**, the boxplots of the Brier scores for each of six metapopulations are grouped according to the test samples' status inferred by the GenoGeographer methodology (in colours) [**55**]. Added to plot in grey boxplots are the Brier scores for the reference samples, which are the samples used to compute $\bar{q}_j$. Thus, the grey boxplots are expected to show lower Brier scores compared to those of the coloured boxplots.

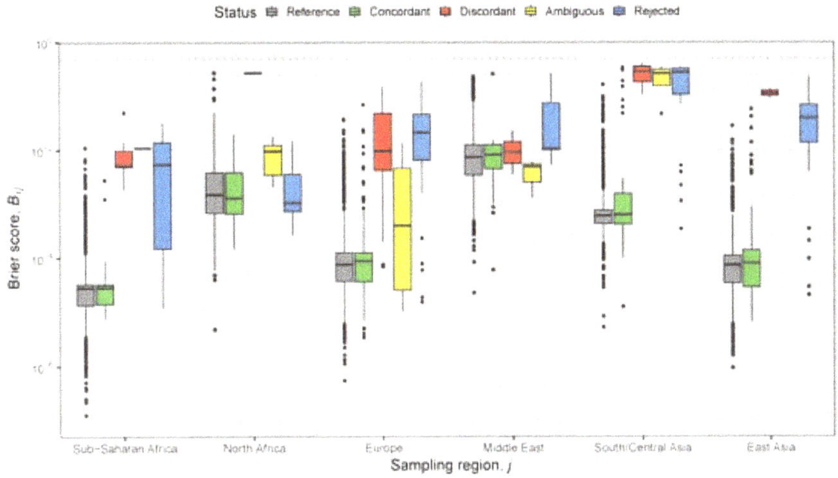

Figure 3. Boxplots of the Brier scores, *Bij*, with *K*=4 in the STRUCTURE analysis and *j* running through the metapopulations of [58]. *Reference:* **Samples constituting the metapopulations (and used to calculate $q^{-}j$).** *Concordant:* **A sample is accepted in the metapopulation coinciding with its sampling region (with a likelihood value significantly larger than any other metapopulation).** *Discordant:* **A sample is accepted in a metapopulation different than its sampling region (with a likelihood value significantly larger than any other metapopulation).** *Ambiguous:* **A sample's likelihood value is not significantly different in two or more of the accepted metapopulations.** *Rejected:* **A sample is rejected in all metapopulations. The concordant test samples (green) have similar Brier scores as those of the reference samples (grey). Discordant samples (red) tend to have the largest Brier scores, followed by the rejected samples (blue) with the ambiguous (yellow) in between (see text for definitions). The horizontal dashed and dotted lines in the top of the plot indicates *Bij*=0.5 and 0.250.25, respectively ([26], inspired by Supplementary Figure S8).**

The Brier scores, $B_{ij,}$ in **Figure 3**, tend to be larger for discordant, ambiguous, and rejected samples (see definitions in caption of **Figure 3**). However, from the visual inspection of the STRUCTURE barplots (data not shown), it is often rather hard to judge when a sample deviates substantially from a given metapopulation to declare it extreme compared to the other samples. The advantage of GenoGeographer is that such visual inspections or *ad hoc* defined thresholds for dissimilarity (e.g., using a

Brier score) are not required. The conclusions are based solely on a hypothesis test and a predefined significance level.

The model formulation of STRUCTURE implies that any spurious and complex type of admixture can be modelled. For the GenoGeographer framework, however, only specific admixtures can be assessed. Currently, first-order admixtures can be accounted for [**56**], but the admixture approach can be generalised to handle outlier tests for higher-order admixtures (e.g., second-order where the parents themselves may be admixed). However, the ancestry SNPs may not be sufficiently informative to distinguish between certain admixture configurations (e.g., two pedigrees with different founder populations may be equally likely).

One solution to this may be to use linked markers as these typically are more informative for pedigree analyses [**59**]. In order to handle linked markers in the framework of hypothesis tests, e.g., microhaplotypes [**17**], the interactions between the markers was modelled using decomposable graphical models, where association structures between the SNPs could be different among the reference populations [**23**] (e.g., the linkage between SNPs was allowed to be different between continental regions). For a microhaplotype with K biallelic SNPs, there exist $2K2$ different alleles, implying that many samples are needed to obtain accurate allele frequency estimates for moderate values of K. By exploiting the conditional independence structures, the framework of graphical models provides a more data efficient modelling of the microhaplotype frequencies and thus requires fewer samples.

3. DISCUSSION

STRUCTURE remains to be a valuable methodology and approach for analysing population structure in forensic genetics. There exist several guides and tutorials for how to prepare the population data and choosing parameter settings for STRUCTURE (e.g., [**60,61**], with some emphasis on forensic applications). However, as warned by [**44**], the resulting barplots should not be overinterpreted. The badMIXTURE approach [**45**] provides the means to reduce the risk of being mislead by the summaries provided by STRUCTURE and supporting software (e.g., [**7,48**]).

However, STRUCTURE's popularity in the field of population genetics to study population structure is well deserved. It complemented PCA with a quantitative method that assigns sample specific weights to each of the K populations specified in the study. This is a powerful way to summarise the data in terms of cluster membership probabilities and the estimated allele frequencies for the identified populations. Both PCA and STRUCTURE are valuable for exploratory analysis, where encoding errors (e.g., of missing data) and warnings about misspecification of origin may be detected by visual inspections of the results. From a forensic point of view, the results from both PCA and STRUCTURE are hard to report in terms of a weight of evidence calculation. The similarity (or dissimilarity) between the sample and reference materials can be reported but typically only in terms of their visual appearance. By contrast, classification and likelihood-based methods are able to give weight to the evidence. This may be in terms of a posterior probability or likelihood ratio, where the assignment would be to the most probable population. However, none of these methods take into account the risk of assigning the profile to the *least unlikely* population in the case where none of the populations are representative for the true population of origin.

Inferring a DNA profile's geographic region or population of origin has been an active field of research in forensic genetics for several years. The vast collections of publicly available databases of whole-genome sequences provide a unique resource for researchers. Rather than spending time and consumables on collecting samples, these DNA profiles can now be harvested online. However, in doing so, one relies on the quality of the data provided by others, which includes the genetic typing and base calling but also the metadata regarding sampling location and ethnic information.

In the case of forensic genetics, the finer resolutions are often of interest. In order to increase the specificity, it may be necessary to supplement the selected AIMs with more locally informative markers, specific to separating the local and often related populations of interest. In the search for such markers, the publicly available datasets are of immense importance as they can be used to screen for candidate AIMs. In particular, the samples from regions close to the specific populations of interest are essential for selecting potential markers.

Combining the flexibility of STRUCTURE with the appealing visual features of PCA (e.g., using EIGENSTRAT to determine the number of significant components) is essential in the exploratory phase of ancestry inference. However, the forensic questions related to ancestry is typically different from those of population genetics, where the focus may be on *population* specific patterns (e.g., bottleneck and expansion events or migration). The typical forensic use case focuses on a specific *individual* (or human remains), for which the most likely population of origin needs to be identified. In such cases, likelihood ratio or classification methods can be used. However, the fundamental assumption of the existence of an appropriate population in the reference material may very often be violated. Hence, DNA profiles are assigned to the population that is most similar to the sample, which due to human evolution and history may geographically and culturally be very far away from the true population. The GenoGeographer methodology attempts to overcomes this logical problem by using statistically based outlier tests.

4. AT A GLANCE

Many forensic genetic laboratories now offer services that involve inferring ancestry. The motivation behind studying ancestry may be to uncover valuable clues or determine the geographical roots of unidentifiable individuals who are reported missing. Several biostatistical methods have been developed specifically for studying population structure in the field of population genetics. Nevertheless, the challenges and inquiries in the field of forensic genetics vary slightly from those in the study of population histories and genealogies, as the focus is primarily on determining the source of a particular sample. This paper provides a comprehensive assessment of the approaches used for modeling population mixing and inferring ancestral populations. The focus is on evaluating their strengths and limitations specifically in the context of ancestry inference in forensic investigations.

REFERENCES

1. International HapMap Consortium. The International HapMap Project. *Nature* **2003**, *426*, 789–796. **[Google Scholar] [CrossRef] [PubMed] [Green Version]**
2. The 1000 Genomes Project Consortium. A global reference for human genetic variation. *Nature* **2015**, *526*, 68–74. **[Google Scholar] [CrossRef] [PubMed] [Green Version]**
3. Mallick, S.; Li, H.; Lipson, M.; Mathieson, I.; Gymrek, M.; Racimo, F.; Zhao, M.; Chennagiri, N.; Nordenfelt, S.; Tandon, A.; et al. The Simons Genome Diversity Project: 300 genomes from 142 diverse populations. *Nature* **2016**, *538*, 201–206. **[Google Scholar] [CrossRef] [PubMed]**
4. Karczewski, K.J.; Francioli, L.C.; Tiao, G.; Cummings, B.B.; Alföldi, J.; Wang, Q.; Collins, R.L.; Laricchia, K.M.; Ganna, A.; Birnbaum, D.P.; et al. The mutational constraint spectrum quantified from variation in 141,456 humans. *Nature* **2020**, *581*, 434–443. **[Google Scholar] [CrossRef]**
5. Cavalli-Sforza, L.L.; Menozzi, P.; Piazza, A. *The History and Geography of Human Genes*; Princeton Universily Press: Princeton, NJ, USA, 1994. **[Google Scholar]**
6. Jobling, M.A.; Hollox, E.; Hurles, M.; Kivisild, T.; Tyler-Smith, C. *Human Evolutionary Genetics*, 2nd ed.; Garland Science Taylor & Francis Group: New York, NY, USA, 2014. **[Google Scholar]**
7. Rosenberg, N.A.; Pritchard, J.K.; Weber, J.L.; Cann, H.M.; Kidd, K.K.; Zhivotovsky, L.A.; Feldman, M.W. Genetic structure of human populations. *Science* **2002**, *298*, 2381–2384. **[Google Scholar] [CrossRef] [PubMed] [Green Version]**
8. Cavalli-Sforza, L.L.; Feldman, M.W. The application of molecular genetic approaches to the study of human evolution. *Nat. Genet. (Suppl.)* **2003**, *33*, 266–275. **[Google Scholar] [CrossRef]**
9. Serre, D.; Pääbo, S. Evidence for gradients of human genetic diversity within and among continents. *Genome Res.* **2004**, *14*, 1679–1685. **[Google Scholar] [CrossRef] [Green Version]**

10. Manica, A.; Prugnolle, F.; Balloux, F. Geography is a better determinant of genetic differentiation than ethnicity. *Hum. Genet.* **2005**, *118*, 366–371. **[Google Scholar] [CrossRef] [Green Version]**
11. Novembre, J.; Johnson, T.; Bryc, K.; Kutalik, Z.; Boyko, A.R.; Auton, A.; Indap, A.; King, K.S.; Bergmann, S.; Nelson, M.R.; et al. Genes mirror geography within Europe. *Nature* **2008**, *456*, 98–101. **[Google Scholar] [CrossRef] [Green Version]**
12. Wang, C.; Zöllner, S.; Rosenberg, N. A quantitative comparison of the similarity between genes and geography in worldwide human populations. *PLoS Genet.* **2012**, *8*, e1002886. **[Google Scholar] [CrossRef] [Green Version]**
13. Rosenberg, N.; Li, L.; Ward, R.; Pritchard, J. Informativeness of genetic markers for inference of ancestry. *Am. J. Hum. Genet.* **2003**, *73*, 1402–1422. **[Google Scholar] [CrossRef] [Green Version]**
14. Rosenberg, N.A. Algorithms for Selecting Informative Marker Panels for Population Assignment. *J. Comput. Biol.* **2005**, *12*, 1183–1201. **[Google Scholar] [CrossRef] [PubMed] [Green Version]**
15. Brinkmann, B.; Junge, A.; Meyer, E.; Wiegand, P. Population Genetic Diversity in Relation to Microsatellite Heterogeneity. *Hum. Mut.* **1998**, *11*, 135–144. **[Google Scholar] [CrossRef]**
16. Alladio, E.; Della Rocca, C.; Barni, F.; Dugoujon, J.M.; Garofano, P.; Semino, O.; Berti, A.; Novelletto, A.; Vincenti, M.; Cruciani, F.; et al. A multivariate statistical approach for the estimation of the ethnic origin of unknown genetic profiles in forensic genetics. *Forensic Sci. Int. Genet.* **2020**, *45*, 102209. **[Google Scholar] [CrossRef]**
17. Oldoni, F.; Kidd, K.K.; Podini, D. Microhaplotypes in forensic genetics. *Forensic Sci. Int. Genet.* **2019**, *38*, 54–69. **[Google Scholar] [CrossRef]**
18. Yang, N.; Li, H.; Criswell, L.A.; Gregersen, P.K.; Alarcon-Riquelme, M.E.; Kittles, R.; Shigeta, R.; Silva, G.; Patel, P.I.; Belmont, J.W.; et al. Examination of ancestry and ethnic affiliation using highly informative diallelic DNA markers: Application to diverse and admixed populations and implications for clinical epidemiology and forensic medicine. *Hum. Genet.* **2005**, *118*, 382–392. **[Google Scholar] [CrossRef]**
19. Moriot, A.; Santos, C.; Freire-Aradas, A.; Phillips, C.; Hall, D. Inferring biogeographic ancestry with compound markers of slow and fast evolving polymorphisms. *Eur. J. Hum. Genet.* **2018**, *26*, 1697–1707. **[Google Scholar] [CrossRef]**
20. Phillips, C. Forensic genetic analysis of bio-geographical ancestry. *Forensic Sci. Int. Genet.* **2015**, *18*, 49–65. **[Google Scholar] [CrossRef] [PubMed]**

21. Phillips, C.; Santos, C.; Fondevila, M.; Carracedo, Á.; Lareu, M.V. Inference of Ancestry in Forensic Analysis I: Autosomal Ancestry-Informative Marker Sets. In *Forensic DNA Typing Protocols*; Methods in Molecular Biology; Goodwin, W., Ed.; Springer: Berlin/Heidelberg, Germany, 2016; Volume 1420, pp. 234–253. **[Google Scholar]**

22. Falush, D.; Stephens, M.; Pritchard, J.K. Inference of Population Structure Using Multilocus Genotype Data: Linked Loci and Correlated Allele Frequencies. *Genetics* **2003**, *164*, 1567–1587. **[Google Scholar] [CrossRef]**

23. Lindskou, M.; Eriksen, P.S.; Tvedebrink, T. Outlier detection in contingency tables using decomposable graphical models. *Scand. J. Stat.* **2020**, *47*, 347–360. **[Google Scholar] [CrossRef]**

24. Purcell, S.; Neale, B.; Todd-Brown, K.; Thomas, L.; Ferreira, M.A.; Bender, D.; Maller, J.; Sklar, P.; de Bakker, P.I.; Daly, M.J.; et al. PLINK: A Tool Set for Whole-Genome Association and Population-Based Linkage Analyses. *Am. J. Hum. Genet.* **2007**, *81*, 559–575. **[Google Scholar] [CrossRef] [Green Version]**

25. Patterson, N.; Price, A.L.; Reich, D. Population Structure and Eigenanalysis. *PLoS Genet.* **2006**, *2*, e190. **[Google Scholar] [CrossRef]**

26. Mogensen, H.S.; Tvedebrink, T.; Børsting, C.; Pereira, V.; Morling, N. Ancestry prediction efficiency of the software GenoGeographer using a z-score method and the ancestry informative markers in the Precision ID Ancestry Panel. *Forensic Sci. Int. Genet.* **2020**, *44*, 102154. **[Google Scholar] [CrossRef]**

27. Menozzi, P.; Piazza, A.; Cavalli-Sforza, L.L. Synthetic maps of human gene frequencies in Europeans. *Science* **1978**, *201*, 786–792. **[Google Scholar] [CrossRef] [PubMed]**

28. Price, A.L.; Patterson, N.J.; Plenge, R.M.; Weinblatt, M.E.; Shadick, N.A.; Reich, D. Principal components analysis corrects for stratification in genome-wide association studies. *Nat. Genet.* **2006**, *38*, 904–909. **[Google Scholar] [CrossRef] [PubMed]**

29. Jombart, T.; Devillard, S.; Balloux, F. Discriminant analysis of principal components: A new method for the analysis of genetically structured populations. *BMC Genet.* **2010**, *11*, 94. **[Google Scholar] [CrossRef] [PubMed] [Green Version]**

30. Novembre, J.; Stephens, M. Interpreting principal component analyses of spatial population genetic variation. *Nat. Genet.* **2008**, *40*, 646–649. **[Google Scholar] [CrossRef]**

31. McVean, G. A Genealogical Interpretation of Principal Components Analysis. *PLoS Genet.* **2009**, *5*, 1–10. **[Google Scholar] [CrossRef] [Green Version]**

32. Wangkumhang, P.; Hellenthal, G. Statistical methods for detecting admixture. *Curr. Opin. Genet. Dev.* **2018**, *53*, 121–127. **[Google Scholar] [CrossRef] [Green Version]**
33. Miller, J.M.; Cullingham, C.I.; Peery, R.M. The influence of a priori grouping on inference of genetic clusters: Simulation study and literature review of the DAPC method. *Heredity* **2020**, *125*, 269–280. **[Google Scholar] [CrossRef]**
34. Pritchard, J.K.; Stephens, M.; Donnelly, P.J. Inference of population structure using multilocus genotype data. *Genetics* **2000**, *155*, 945–959. **[Google Scholar] [CrossRef] [PubMed]**
35. Novembre, J. Pritchard, Stephens, and Donnelly on Population Structure. *Genetics* **2016**, *204*, 391–393. **[Google Scholar] [CrossRef] [PubMed]**
36. Rannala, B.; Mountain, J.L. Detecting immigration by using multilocus genotypes. *Proc. Natl. Acad. Sci. USA* **1997**, *94*, 9197–9201. **[Google Scholar] [CrossRef] [PubMed] [Green Version]**
37. Foreman, L.; Smith, A.; Evett, I. Bayesian analysis of DNA profiling data in forensic identification applications. *J. R. Stat. Soc. A* **1997**, *160*, 429–469. **[Google Scholar] [CrossRef]**
38. Roeder, K.; Escobar, M.; Kadane, J.B.; Balazs, I. Measuring heterogeneity in forensic databases using hierarchical Bayes models. *Biometrika* **1998**, *85*, 269–287. **[Google Scholar] [CrossRef]**
39. Falush, D.; Stephens, M.; Pritchard, J.K. Inference of population structure using multilocus genotype data: Dominant markers and null alleles. *Mol. Ecol. Notes* **2007**, *7*, 574–578. **[Google Scholar] [CrossRef]**
40. Tang, H.; Peng, J.; Wang, P.; Risch, N. Estimation of Individual Admixture: Analytical and Study Design Considerations. *Genet. Epidemiol.* **2005**, *28*, 289–301. **[Google Scholar] [CrossRef] [Green Version]**
41. Alexander, D.H.; Novembre, J.; Lange, K. Fast model-based estimation of ancestry in unrelated individuals. *Genome Res.* **2009**, *19*, 1655–1664. **[Google Scholar] [CrossRef] [Green Version]**
42. Raj, A.; Stephens, M.; Pritchard, J.K. fastSTRUCTURE: Variational Inference of Population Structure in Large SNP Data Sets. *Genetics* **2014**, *197*, 573–589. **[Google Scholar] [CrossRef] [Green Version]**
43. Novembre, J. Variations on a Common STRUCTURE: New Algorithms for a Valuable Model. *Genetics* **2014**, *197*, 809–811. **[Google Scholar] [CrossRef] [Green Version]**
44. Lawson, D.J.; van Dorp, L.; Falush, D. A tutorial on how not to over-interpret STRUCTURE and ADMIXTURE bar plots. *Nat. Commun.* **2018**, *9*, 3258. **[Google Scholar] [CrossRef] [Green Version]**

45. Lawson, D. badMIXTURE: Validating Structure With Chromosome Painting; R Package Version 0.0.0.9000. 2018. Available online: **https://github.com/danjlawson/badMIXTURE** (accessed on 13 December 2021).

46. Hellenthal, G.; Busby, G.B.J.; Band, G.; Wilson, J.F.; Capelli, C.; Falush, D.; Myers, S. A Genetic Atlas of Human Admixture History. *Science* **2014**, *343*, 747–751. **[Google Scholar] [CrossRef] [Green Version]**

47. Lawson, D.J.; Hellenthal, G.; Myers, S.; Falush, D. Inference of Population Structure using Dense Haplotype Data. *PLoS Genet.* **2012**, *8*, 1–16. **[Google Scholar] [CrossRef] [PubMed] [Green Version]**

48. Jakobsson, M.; Rosenberg, N.A. CLUMPP: A cluster matching and permutation program for dealing with label switching and multimodality in analysis of population structure. *Bioinformatics* **2007**, *23*, 1801–1806. **[Google Scholar] [CrossRef] [Green Version]**

49. Cheung, K.H.; Osier, M.V.; Kidd, J.R.; Pakstis, A.J.; Miller, P.L.; Kidd, K.K. ALFRED: An allele frequency database for diverse populations and DNA polymorphisms. *Nucleic Acids Res.* **2000**, *28*, 361–363. **[Google Scholar] [CrossRef] [Green Version]**

50. Pakstis, A.; Kang, L.; Liu, L.; Zhang, Z.; Jin, T.; Grigorenko, E.L.; Wendt, F.R.; Budowle, B.; Hadi, S.; Qahtanif, M.S.A.; et al. Increasing the reference populations for the 55 AISNP panel: The need and benefits. *Int. J. Legal Med.* **2017**, *131*, 913–917. **[Google Scholar] [CrossRef] [Green Version]**

51. Phillips, C.; Prieto, L.; Fondevila, M.; Salas, A.; Gomez-Tato, A.; Alvarez-Dios, J.; Alonso, A.; Blanco-Verea, A.; Brion, M.; Montesino, M.; et al. Ancestry analysis in the 11-M Madrid bomb attack investigation. *PLoS ONE* **2009**, *4*, e6583. **[Google Scholar] [CrossRef] [PubMed] [Green Version]**

52. Cheung, E.Y.; Gahan, M.E.; McNevin, D. Prediction of biogeographical ancestry from genotype: A comparison of classifiers. *Int. J. Legal Med.* **2017**, *131*, 901–912. **[Google Scholar] [CrossRef]**

53. McNevina, D.; Santos, C.; Gómez-Tato, A.; Álvarez-Dios, J.; Casares de Cal, M.; Daniel, R.; Phillips, C.; Lareu, M.V. Anassessment of Bayesian and multinomial logistic regression classification systems to analyse admixed individuals. *Forensic Sci. Int. Genet. Suppl. Ser.* **2013**, *4*, e63–e64. **[Google Scholar] [CrossRef]**

54. Cheung, E.Y.Y.; Gahan, M.E.; McNevin, D. Prediction of biogeographical ancestry in admixed individuals. *Forensic Sci. Int. Genet.* **2018**, *36*, 104–111. **[Google Scholar] [CrossRef]**

55. Tvedebrink, T.; Eriksen, P.S.; Mogensen, H.S.; Morling, N. Weight of the Evidence of Genetic Investigations of Ancestry Informative Markers. *Theor. Popul. Biol.* **2018**, *120*, 1–10. **[Google Scholar] [CrossRef]**

56. Tvedebrink, T.; Eriksen, P.S. Inference of admixed ancestry with Ancestry Informative Markers. *Forensic Sci. Int. Genet.* **2019**, *42*, 147–153. **[Google Scholar]** **[CrossRef]**

57. Pfaffelhuber, P.; Sester-Huss, E.; Baumdicker, F.; Naue, J.; Lutz-Bonengel, S.; Staubach, F. Inference of recent admixture using genotype data. *Forensic Sci. Int. Genet.* **2021**, *56*, 102593. **[Google Scholar]** **[CrossRef]**

58. Tvedebrink, T.; Eriksen, P.S.; Mogensen, H.S.; Morling, N. GenoGeographer—A tool for genogeographic inference. *Forensic Sci. Int. Genet. Suppl. Ser.* **2017**, *6*, e463–e465. **[Google Scholar]** **[CrossRef]** **[Green Version]**

59. Kling, D.; Tillmar, A.; Egeland, T.; Mostad, P. A general model for likelihood computations of genetic marker data accounting for linkage, linkage disequilibrium, and mutations. *Int. J. Legal Med.* **2015**, *129*, 943–954. **[Google Scholar]** **[CrossRef]**

60. Porras-Hurtado, L.; Ruiz, Y.; Santos, C.; Phillips, C.; Carracedo, Á.; Lareu, M. An overview of STRUCTURE: Applications, parameter settings, and supporting software. *Front. Genet.* **2013**, *4*, 98. **[Google Scholar]** **[CrossRef]** **[Green Version]**

61. Santos, C.; Phillips, C.; Gomez-Tato, A.; Alvarez-Dios, J.; Carracedo, Á.; Lareu, M.V. Inference of Ancestry in Forensic Analysis II: Analysis of Genetic Data. In *Forensic DNA Typing Protocols*; Methods in Molecular Biology; Goodwin, W., Ed.; Springer: Berlin/Heidelberg, Germany, 2016; Volume 1420, pp. 255–285. **[Google Scholar]**

12 ETHICS AS LIVED PRACTICE: ANTICIPATORY CAPACITY AND ETHICAL DECISION-MAKING IN THE FIELD OF FORENSIC GENETICS

1. INTRODUCTION: AN OVERVIEW

Many forensic geneticists prioritize the use of ethical values in their work to assure the reliability of the tools they develop and use for casework, while also contributing to a fair society. However, the field of forensic genetics, which includes research scientists, laboratory practitioners, journals, industry, professional bodies, and users of forensic genetic analyses in policing and other areas, still undergoes scrutiny of new and emerging forensic genetic technologies. This scrutiny encompasses their social, ethical, and legal implications, as well as their scientific and technical aspects. An examination conducted by the German Government between 2017 and 2019 examined forensic DNA phenotyping and biogeographic ancestry testing. As a result, forensic DNA phenotyping was legalized in 2019, however biogeographic ancestry testing was not due to concerns about racial profiling. [**1**]. The Law Commission of New Zealand reported on new and emergent forensic genetics in 2020, recommending the comprehensive revision of legislation; increased oversight; and greater consideration of concerns surrounding privacy and discrimination [**2**]. The Law Commission of Ontario reviewed Artificial Intelligence (AI) and probabilistic genotyping DNA tools in June 2021, suggesting that inherent biases in these technologies challenge the notion of DNA profiling as a "gold standard" and requires that there is continuous scrutiny as well as new legislation [**3**]. The Scottish Government and the UK House of Lords have commenced parallel enquiries into developments in technologies for policing in 2021 [**4,5**], scoping opportunities, and challenges arising for policing from new technologies.

While these enquiries are undertaken, reports published, and debates held, demand continues for the utility of databases (including genetic ones) to be maximised, and the promise of forensic genetics—often portrayed as the greatest crime-fighting tool since fingerprinting—be realised and capitalised upon. Meanwhile, new techniques are being developed and debated. Despite remaining contentious, internationally, the use of genetic 'phenotyping' in police investigations is growing. The Swiss Parliament is the latest to prepare legislation to allow for forensic DNA phenotyping amidst, both, supportive and critical debate [**6**]. Similarly, since 2018, forays into forensic genetic genealogy have been spreading since the successful detection of the perpetrator with the help of the technique in the 'Golden State Killer' case in the USA [**7**,**8**,**9**,**10**].

In these areas, as well as generally, forensic scientists face institutional pressures, both from casework and their scientific community (including employers/funders), which must be met with scientific professionalism and ethical practice. Yet it was not long ago that scientists themselves conceded that forensic science 'is not sufficiently well developed as a profession' (p. 523, [**11**]). Within the scientific community, there is increased scrutiny regarding the way that data are sourced and shared, informing both forensic research and casework analysis [**12**]. The intensification of external scrutiny, and productivity pressures demand the highest levels of professionalism, which must include a strong, internalized code of ethics that has not yet been proven to be universal [**13**].

In some parts of the world, forensic genetics research and casework have become an important aspect of policing, criminal justice, and victim identification. In multiple ways, forensic genetics contributes to rendering these domains more agile vis-à-vis existing and emerging demands and challenges. Novel technologies are developed in response to scientific insights and practical problems while remaining subject to the same responsibilities that scientists and users have towards their specific institutions, disciplines, and to society as a whole.

Therefore, in this paper, we offer an approach through which the forensic genetics community can reflect on their daily decisions in both scientific research and casework as part of their professional and ethical forensic practice. While we cannot, nor do we aim to, provide technical or legal solutions, we offer a framework for ethical conduct. We consider the problem of 'thin' ethics in forensic genetics and propose understanding ethics as an intrinsic part of the lived practice of a scientist. We aim to support scientists in developing anticipatory capacity [**14**]—to understand the wider operational and societal impacts of their daily decisions around forensic genetics—by making visible the ethical decisions in both their scientific practice and in the application and implementation of forensic technologies. We do so by exploring ethical aspects of decision-making in forensic genetics and illustrating the discussion by way of three case studies of emerging forensic genetics technologies.

2. MATERIALS AND METHODS

This paper builds on literature analysed in three previous reviews of social and ethical aspects of scientific, technological, and operational developments in forensic genetics [**15,16,17**], adding consideration of scholarship published since 2017, and focusing on three technologies currently being debated: forensic DNA phenotyping (FDP), forensic genetic genealogy (FGG), and forensic epigenetics (FEpi). In this endeavour, the concept of the "social life of DNA," proposed by Corinna Kruse [**18**], which captures the different social relationships observed in the processing of DNA as it moves from crime scene to courtroom, resonates. As such, we analyse the different life stages of advanced forensic DNA technologies. While DNA profiling is a relatively widely accepted technology, having stabilized and been implemented across different contexts, the three technologies we focus on are yet to reach such a point of stability. In reviewing these examples, we highlight key ethical concerns to assist forensic practitioners in expanding their anticipatory capacity and thus work towards ethically sound decision making. By anticipatory capacity, we refer to antecedent considerations in science, and in real-world application of science, set within a wider context of liberal-democratic social and cultural norms and values. These three technologies were selected, firstly, due to controversy concerning their utility and reliability/validity specifically, and their ethics more generally [**19,20**]. Secondly, these technologies present opportunities to engage with ethics as lived practice, as they pose ethical dilemmas. Thirdly, debates are ongoing over their use, and measures in place to regulate these technologies differ across countries. Finally, each of these technologies is currently at different life stages. Forensic DNA phenotyping, although still the subject of complex politics of legitimation and contestation [**20,21,22,23**], is currently regulated and applied in some countries such as Slovakia and Germany. Forensic genetic genealogy is under intense scrutiny in many jurisdictions and upcoming regulation can be expected (see for example recent initiatives in the USA and in Australia). Forensic epigenetics is being developed and has not (yet) been considered widely with respect to regulation and/or application.

3. THE LIMITATIONS OF "THIN" ETHICS IN FORENSIC GENETICS

Controversies surrounding advances in forensic genetics could (at least in part) be a consequence of a structural deficiency in comprehensive, systematic understandings of what ethics is—or should be—in forensic genetics and related technological innovations. Let us explain what we mean by a "thin" understanding of ethics.

Firstly, forensic geneticists will encounter ethics as part of an administrative process to be followed to enable research, innovation, and implementation: following codes of practice, completing forms, attending training, and submitting authorizations to some form of the institutional oversight body. Such administrative tasks and tools are clearly essential in ensuring forensic scientists navigate the most basic rules of research and technology use. However, the main thrust of this 'procedural ethics' [**24**] is to comply with administrative and legal procedural requirements, with ethics conceived of as 'passive', lending itself to the interpretation of ethics as a hurdle to research, innovation, and (positive) societal impact that forensic genetics could otherwise make. Consequently, exclusive adherence to this "thin" procedural ethics can extinguish the light that a (pro-) active approach can shine on the complex environment within which forensic work takes place. It also 'disengages' forensic geneticists (both in research and casework) from ethics to the detriment of forensic genetics, and more widely to science in the criminal justice system.

Secondly, within this "thin" understanding of ethics, 'privacy' tends to be perceived as an interpretable legal concept, addressed by compliance with the law. Moreover, 'data protection' is often used as a proxy for 'privacy', whereby observance with principles of data protection will suffice, yet 'privacy' is far more expansive and manifold than mere data protection (pp. 33–50, [**25,26**]). A reductionist understanding of ethics as privacy obscures other vital aspects of ethical practice, including attending to justice and fairness, human dignity (and its cultural values, see, e.g., recommendations by the Law Commission of New Zealand vis-a-vis forensic genetics and the Māori), and the physical and informational integrity of a person and/or a group of persons (community). A focus on privacy also misdirects debate into a cul-de-sac of arguments over balancing 'personal privacy' with 'public security' since this dichotomy is erroneous; privacy and security are symbiotic [**27,28**].

Thirdly, technology may be perceived as a mere tool, process, or method. However, as several years of consolidated scientific research have shown (particularly in social studies of science and technology in health research, and more recently in the social studies of forensic science), technology is a product of social, cultural, and moral facets of society. Technology is always technology-in-practice [**29**] resulting from, and shaping social interactions and the organisation of social life. As Greenhalgh and Swinglehurst (p. 3, [**30**]) suggest, considering technology as social practices "moves us on from studying either people or technologies" towards a rich, inclusive analysis of technology as an aspect of social life, considering technology-in-use as well as the discourses and practices of technology [**31**].

Reducing technology to its instrumental aspects conceals many ethical decisions that shape technology, from its conception to application and maintenance as well as to its oversight. Thus, we consider "thin" or 'procedural ethics' insufficient in forensic research and application, with privacy as just one ethical concern. Nor is technology a mere tool or method but should be viewed as social practice. Thus, we propose an approach to ethics as intrinsic to the lived practice of forensic geneticists, making deliberate decisions about daily practices and procedures, acting in light of these decisions, while continually reflecting upon their broader impacts now and in the future.

4. ETHICS AS LIVED PRACTICE

Ethics as lived practice extends throughout the life cycle of science: from development and implementation to regulation and oversight. Regardless, then, of the role(s) that professionals fulfil—whether as research scientist engaged with the laboratory-based development of forensic applications; as a forensic practitioner working on criminal cases; as an expert advising on casework or giving evidence in court—ethical engagement is inalienably part of professional practice. Such ethical engagement is inescapable because (i) the development of forensic technologies will affect human beings; (ii) those conducting forensic casework, and working with forensic data and information, are, as all other human beings, subjective, open to influence, and fallible; and (iii) forensic casework builds upon and contributes to selecting, categorising, and curating data and data repositories as reference points, thus making choices that will affect future forensic research and casework, and as such, many more humans.

Our understanding of ethics as lived practice stands in contrast to the notion of ethics as a necessary evil and to the "thin" ethics discussed above—where ethics only arise at bounded points in time when the ethics 'hurdle' is jumped. Ethics as lived practice, instead, recognises that decision-making is ongoing, and an intrinsic part of the practice. Continual reflection, then, becomes central: taking time, sometimes briefly, sometimes for a longer period; alone or as a team; discipline; or community, to (re)consider the basis and motivations on which decisions are being taken, as well as their consequences. This also calls for serious engagement with others, including those from outside the laboratory (such as criminal investigators, social scientists, policymakers forging legislation and regulation, as well as the civil society), to discuss research, analysis, and future application, within their wider social and cultural contexts. We call upon the forensic genetics community to build on emerging strengths here by making ethical reflection—as a way of building anticipatory capacity—an integral part of forensic genetics work.

The remarkable potential of forensic science, to not only contribute to the pursuit of justice, but to shape and direct society, is highly valued, but comes with enormous responsibility. Acknowledgment of the wider impact of forensic work means that ethics ought to be reflected in scientists' and practitioners' training, in their daily decision-making, and in contributions to scientific and public debates about forensics. Guillemin and Guillam [**24**] explain that ethics work occurs in researchers' daily practice beyond procedural ethics. Such a practice-related understanding of ethics has been suggested as a way of incorporating principle-based reflections about the impact and implications of scientific design and work into everyday decisions [**32**]. Furthermore, Guillemin and Guillam suggest that researchers may encounter ethically relevant moments, where the approach taken or decision made has important ethical ramifications, but where researchers do not necessarily feel themselves to be on the horns of a dilemma. In fact, in some cases, it might be clear how the researcher should respond or proceed, and yet there is still something ethically important at stake (p. 265, [**24**]). While the scientist or practitioner might not perceive this moment as particularly vital or difficult, the decisions made at this moment can have significant long-term implications and impacts: on the person making the decision and their forensic specialism; on forensic science more widely and its perception among users, policy and publics; on the domains using forensic science services; and, via these domains, on ethical issues such as: social justice; equality; equity; dignity; privacy; integrity, and so forth. In these 'ethical moments' (for a look at ethical moments in forensic genetics, see e.g., [**33**]) different perspectives are brought together, to identify and resolve disagreements, making visible ethical reasoning about scientific practice and the use of scientific knowledge [**20**].

Consider, for example, the forensic scientist tracing a partial DNA match indicative of a familial relationship between a suspect DNA profile and that of a dragnet/mass test volunteer. If permitted by law, the modus operandi is clear, and the practitioner should always follow what the law prescribes. However, the law does not, and cannot, provide for every situation practitioners will come across in their daily practice. If the law has not explicitly provided for a situation, there is a significant decision to be made: should the expert report a partial match and thus a potential familial relationship, even though familial searching is not explicitly regulated in their jurisdiction? In pursuit of a perpetrator, it may be legitimate to report, even if such analyses are not formally accepted practice (e.g., not regulated by law, nor a laboratory accredited for), thus shifting the ultimate ethical decision to others, such as the investigating judge (in predominantly inquisitorial systems) or the police and prosecution (in more adversarial systems). Nonetheless, the forensic geneticist is required to make an ethical decision about their practice at this point, and procedures to support them in doing so—perhaps by calling on them to reflect upon their decision—may often be non-existent or very limited. This example, like many others occurring daily in forensic genetics, clearly demonstrates that ethics cannot be limited to an initial passing of a procedural ethical hurdle, but relies on many decisions around reliability, utility, and legitimacy [**20**] that an individual or their team take in forensic research and casework.

Ethics as lived practice demands more attention, while a "thin" understanding of ethics, anchored, for example, in 'procedural ethics', is insufficient for responsible (and as such also accountable) forensic genetics practice. We argue that ethics as lived practice, therefore, means that forensic researchers and practitioners identify and understand the need to address opportunities as well as limitations of their work set within an institutional and societal context—e.g., in criminal justice, migration management, or disaster victim identification. In the next section, we discuss ethical decision-making in forensic genetics practice using the example of data repositories because their creation, curation, and use are vital to forensic genetics research and can play a significant role in investigative work and law enforcement.

5. BUILDING UPON STRONG FOUNDATIONS: ETHICAL DECISION-MAKING IN GENETIC DATABASING

Central to forensic genetics is the data collected from individuals from diverse groups and curated for use by the scientific community. However, as we show in the example of forensic genetic genealogy below, data collected and curated for non-forensic purposes are also increasingly of interest to forensic geneticists and law enforcement agencies. Data collection and curation in repositories form a key life stage of genetic data and ethical considerations in forensic genetics, as with any human genetics field, therefore begin with the sampling design for DNA data for research and for databases such as EMPOP or YHRD. Since some forensic genetics services depend on population data, genetic data from diverse populations have been widely collected from individuals across the globe for decades. Since the 1970s, the fields of human and medical genetics have developed and adopted a number of ethical standards and procedures, including, most notably, requirements for informed consent and approval by ethics committees [**34**]. Such processes have been continuously refined and upgraded, and yet ethical debates around genetic data continue [**35,36,37,38,39**].

In contrast, the forensic genetics community's leading journal, Forensic Science International: Genetics, introduced statements about these basic ethical standards (informed consent and ethics approval) as a publication requirement as late as 2010 [**40**]. In 2020, guidelines were updated and extended, with a specific focus on population data from minorities or indigenous communities [**38**]. Until 2010, there were no ethical standards for authors to meet when publishing DNA data; and data collected before 2010 without informed consent and ethical approval is still used, shared, and uploaded. Furthermore, forensic genetic databases usually require the publication of data in peer-reviewed journal articles before such data can be uploaded, along with ethical approval and/or evidence of informed consent. However, in some databases, data can be uploaded without meeting these criteria [**41**]. Notably, such databases are not only used for research, but also for investigation by law enforcement authorities. Ethical debates around these databases have only now begun.

In human genetics, the demands of obtaining consent for samples from sufficiently large numbers of individuals requires the building of an ethical regime of institutional and personal infrastructures in order to enable the legal and ethical use of samples and profiles (e.g., [**42**]). Numerous ethical decisions, consciously or not, are inevitably made when recruiting donors: whom to include or exclude, how to frame the research to donors about why samples and data are requested, what to give back to donors and their community, and so forth. In forensic genetics the stakes are higher, as persuading individuals to donate their DNA is challenging. Willingness to donate DNA for forensic purposes used to be considerably lower than for health purposes [**43**]; albeit more recently there has been an increase in voluntary offers of existing genetic profiles or donation of new genetic data to enable the forensic use of ancestry, health, and research databases [**8**]. Such donations place a particular responsibility upon researchers and analysts to be very clear about the potential uses of such data, not only at the time of sample collection but also afterwards, as data moves along different life stages.

For such data repositories to serve law enforcement, data acquisition needs to be as extensive as possible to represent diverse populations. Previous resolutions to this problem have attracted criticism: E.g. some forensic practitioners have argued that overrepresentation of marginalised minorities on databases means they are more likely to be exonerated, but this theory has yet to be supported with evidence while emerging database uses for biogeographic ancestry testing and capacities to de-anonymize profiles increase the riskiness of overrepresentation [**12**]. Another example is that in China, but also in numerous other countries, police and security forces have been involved in collection efforts, nullifying any supposed 'consent' [**1,41,44,45**]. Recent discussions focus upon the reuse of data and samples previously collected under unclear ethical conditions, or without any demonstrable adherence to ethical principles [**1,12**]. Some of these datasets have been obtained from vulnerable populations and shared by, and between researchers from forensic, medical, and population genetics, raising serious questions about whether such data transfers violate general ethical standards.

Even if data have been sampled ethically and safely stored in databases, to be used as training and test data in calibration exercises, or as reference data for criminal investigative application, social and ethical issues may still arise. A key issue is whether the data is representative of the 'target' population [**1**]. Yet rarely discussed are potential problems arising from non-representative data collection. For example, if the training data for an FDP application has been sampled too selectively (e.g., [46]) and hence is very homogeneous, probability calculations might be overly optimistic. Then, in a criminal investigation, the technology will overestimate its own reliability against the backdrop of a local population that is not as homogeneous as the training data. In biogeographic ancestry testing (BGA), so-called "mixed populations" in training and reference data lead to erroneous test results and overly optimistic reliability estimates [47]. However, "mixed ancestries" will most likely be more commonplace in real life but, in a worst-case scenario, the technology will produce erroneously high probability estimates, and yet these are more likely to be trusted than other evidence-based on perceptions of scientific objectivity and DNA's standing as "biological witness"—and thus risk leading investigations astray [45].

It is critical, therefore, that ethics as lived practice commences at the very first life stages of DNA data: when collecting and curating data and establishing repositories. As this DNA data is then utilised by forensic practitioners within the context of the criminal justice system, the next vital life stage of forensic genetic data, further ethical decision-making is required.

6. CHALLENGES OF CONTEXT AND COMMUNICATION: ETHICAL DECISION-MAKING AS FORENSIC EXPERT

Forensic DNA technologies have been framed by metaphorical affirmations of DNA as representing a "truth machine" [48] or the "gold standard" of evidence [49], reflecting high expectations based upon unrivalled (alleged) levels of validity, certainty, reliability, and objectivity. Besides disregarding issues arising during the "social life of DNA" [18], as it moves from crime scene (or database) to court, such metaphors eclipse the challenges of communicating probabilistic DNA evidence to non-experts [50,51]. These challenges occur at many different contact points: in the regulatory process leading up to new or amended laws or regulations; in the training of geneticists and investigators; in public communications during police investigations; during and after a trial; and in the aftermath of a solved case when the efficiency of different investigative tools is compared and highlighted. In addition, as DNA analysis becomes more sensitive (entailing the analysis of mixed traces, DNA transfer, etc.), and more open to interpretation, it becomes yet more challenging to report [52], with scientists deciding whether to apply a probative value to a result, or provide an inconclusive result that neither includes or excludes the suspect [53].

A rational assessment of what advanced DNA profiling techniques can bring to a police investigation is essential but rare. For example, in the UK, 'familial searching' has been available since 2012 and utilised in 120 cases, with just nine reported 'successes' [54], perhaps challenging proportionality equations over its use. A realistic appraisal must particularly consider how 'results', once outside of a laboratory or research setting such as in forensic validation, are applied during real-life criminal investigations. A host of confounding variables and diverse considerations come into play during the criminal process, which may distort or nullify laboratory results: "while a forensic technique may have a valid scientific basis and prove reliable in a laboratory setting, how can it be ascertained that it remains reliable when applied in the present case? What are the error rates associated with the technique, and are there any pertinent factors that could jeopardise the reliability of the testing in this particular instance?" (p. 240, [**13**]). A forensic scientist engaged in ethics as the lived practice may be required to inform these more 'realistic' appreciations of what forensic genetics can and cannot achieve within the context of police investigations, particularly when asked to deploy advanced techniques. Both law enforcement and the public should not be 'oversold' benefits that may never be realised outside of a research laboratory [**33,55**]. Our three case studies below address intelligence-related uses of forensic genetics in investigations, when arguably, the dangers of over-selling capacities by forensic scientists and service providers as well as the ready acceptance of such leads by investigators in order to progress a complex case, are high.

Assuming that DNA science moves from investigation onto a later life stage: forensic intelligence from genetic genealogy, but perhaps also from other techniques, may lead to requests by police for arrest warrants. This potential outcome would require such forensic intelligence to be carefully articulated and to take into account the context in which it may be used, including institutional policing cultures and wider political debates, e.g., about migration and criminality. In inquisitorial systems (but highly unlikely ever in adversarial systems), forensic intelligence may even make it into the report submitted to the investigating judge. At some point, if rarely, forensic experts may be called as witnesses to explain the intelligence.

We know for forensic evidence that the judge may be required to take on a "gatekeeper" role to ensure that "junk science" remains inadmissible at trial. This role and its fulfilment has been highly criticized already for adversarial systems where mechanisms are in place for contestation [**56,57**], with conclusions drawn that "scientific illiteracy on the part of the legal profession, when coupled with the flaws in forensic science, forms a 'toxic combination'" (p. 82, [**56**]). The trial process rarely affords even the scientifically literate much opportunity to interpret and apply 'correctly' complex tests (assuming there is a 'correct' application) for whether scientific evidence is 'sufficiently reliable' [**58**]. In addition, such admissibility testing may come too late in the criminal process. Consider, too, that in the inquisitorial system forensic expertise is very rarely challenged. Even where challenges to scientific information exist, decisions to admit such information at trial can appear arbitrary, or inconsistent, with decisions based upon non-scientific criteria [**59**]. Scientists, in providing intelligence as well as in giving testimony, are called upon to make ethical decisions about which information to present and how it should be best communicated. Much work has gone into, e.g., developing verbal scales as equivalents of reporting numerical values and effectively interpreting probabilities [**60,61,62**]. Forensic geneticists' adherence to ethical principles and practices in giving testimony, is a key aspect of their responsibility as practitioners (see, e.g., [**63**]), yet can be complicated by different scientific, investigative, and juridical pressures and demands on the scientific expert. A key responsibility is avoiding 'over-egging the pudding'. As such, scientific experts have personal and disciplinary responsibility for their communication, and thus at least in part, a wider responsibility for the role of intelligence in investigations and for scientific evidence in court.

In response to the exposure of mistakes, misrepresentations, and misunderstandings of DNA evidence that have led to failed investigations or miscarriages of justice, there have been efforts to regulate forensic science [**64**]. Political crises induced by a wrongful conviction or scandal have most often proven necessary to provoke regulatory intervention, but this may come too late for an individual who cannot be adequately compensated for their loss of liberty or diverse publics whose trust in the administration of justice is irreparably damaged. In addition, the implementation of such regulation, including the writing of standards and operating procedures not underpinned by rigorous research, has been criticised [**65**]. It also remains the case that error rates and limitations of techniques often remain sufficiently unarticulated, or even undetermined. Yet even assuming extra resources are provided to produce robust error rates, it is questionable whether risks can be meaningfully quantified, given the inherently contextual nature of forensic DNA evidence. As we argue, many questions are not purely scientific but include ethical considerations from the very outset of the social life of DNA. Acceptability of DNA evidence thus turns, in part, upon the criminal justice system's values and public tolerance of error, but classic regulatory risk analysis typically takes little (if any) account of sociological, economic, ethical, or even legal considerations. The countervailing difficulty of encouraging innovation and development within regulatory parameters, demands innovative solutions operating within an 'ethics as lived practice' framework. This will require embedded ethical practice, as well as regulatory flexibility and pluralism, to avoid stagnation, and setting forensic science 'in aspic' [**66**].

Such flexibility to innovate is especially critical when also facing the challenges of commercialization, leading to the confluence of competition and profit-oriented practices with those of scientific endeavour and of the criminal justice system.

This confluence can lead most obviously to conflicts of interest, which can affect the portrayal of validity/reliability and utility of forensic technologies, but also directly impact the quality of forensic analyses undertaken; the reporting of these forensic analyses, and also the scientifically robust development of existing and emerging forensic technologies [**67**,**68**,**69**]. Forensic scientists constantly make ethical decisions in this context, as discussed above for scientific and evidentiary reasons. In the commercial context, additionally, forensic practitioners make ethical decisions around accreditation for services; negotiating access to contracts; customer satisfaction; access to material and financial means in order to conduct research; contestations around other service providers, etc. [**33**]. Many scientists work together with companies in developing forensic technologies, equipment, and also at times, services. They may be delivering services for commercial providers, or they may be in direct competition. Due to the potential conflicts of interest, the commercial context makes it even more important for forensic scientists and practitioners to reflect on their decision-making in order to ensure responsible and accountable delivery of forensic genetics analyses and interpretations.

7. CASE STUDIES

Our call to adopt ethics as lived practice in decision-making processes throughout key stages in the life of forensic DNA is underscored by three case studies. These vignettes explore ethical decision-making in forensic DNA phenotyping, forensic (or: investigative) genetic genealogy, and forensic epigenetics. The three fields share a common basis: they are presented by forensic scientists and users as providing intelligence for investigations without apparent leads by generating data on a group of suspects, and they are portrayed to facilitate even cold case investigations, based on probability calculations and human (epi)genetics.

7.1. Forensic DNA Phenotyping (FDP) and Biogeographic Ancestry Testing (BGA)

FDP and BGA have provided intelligence to criminal investigations that, at times, have resulted in the detection of suspects. To date, in the USA, hundreds of investigations have drawn on intelligence from FDP and BGA analyses since 2015. In Europe, these techniques (especially BGA), have found application in cases since at least 1999, including in the Marianne Vaatstra case [**70**]; the investigation of the Madrid train bombings of 2004 [**71**]; the Phantom of Heilbronn investigation in 2008 [**72**]; and the detection of the murder of Eva Blanco Puig in Spain in 2015. In the Vaatstra case in 1999, an early form of BGA was used to the effect of taking public pressure off asylum seekers who had been considered prime suspects. In the Madrid investigation, the BGA analysis contributed some intelligence to the investigation. In the case of the Phantom of Heilbronn, the results did not contribute to the detection of perpetrators; rather, they seemed to fuel the bias of investigators. While the case of the Phantom was widely debated, unsuccessful applications of FDP or BGA tend not to be discussed publicly. Yet insight into limitations can assist the wider forensic genetics community of scientists and users, but also the public, in understanding the effective parameters of these techniques.

A more recent case is the rape and murder of Maria Ladenburger, on 16 October 2016, in the German city of Freiburg im Breisgau, Germany. At the start of the investigation, suspicion quickly fell on the large migrant community of approximately one million people who had come to Germany in 2015/16. Representatives from politics, policing and forensic genetics suggested that expanded DNA technologies would contribute to faster detection of the perpetrator if only analyses of the perpetrator's (1) appearance, using FDP, and (2) continental origin, via BGA, would have been permitted. While many German experts, including the German Spurenkommission (trace commission), initially argued that the Ladenburger murder would have been an ideal test case for application of FDP in a criminal investigation, senior German forensic geneticists working on FDP later commented that this specific case would not have benefitted from its use [**73**,**74**,**75**]. Yet, the raised expectations of audiences and investigators remained unrealistically high, leading to the legalization of FDP in Germany in 2019. Yet the Maria Ladenburger case was solved through traditional police investigative techniques, resulting in the arrest of Hussein Khavari, an Afghan migrant, previously convicted in Greece.

The discussion of FDP and BGA, in this case, has raised several questions, specifically around the communication of the utility and reliability of these emerging technologies, but also about the implications of using technology that aims to identify categories of people rather than individuals [**20**,**76**]. Overall, FDP and BGA raise a number of ethical questions for forensic scientists, questions that need to be reflected upon time and again in order to deliver responsible and accountable science. While we focus on the estimation of externally visible characteristics in this case study, we do acknowledge that the testing of biogeographic ancestry (BGA) has its own technical and ethical pitfalls, some of which it shares with FDP and some of which are specific to BGA. For example, there are significant incongruities between 'population' labels used by the forensic laboratory and those used by police forces which risk miscommunications, while opening up casework to being highly susceptible to bias (e.g., [**72**]).

First, FDP works on the basis of attribution rather than identification. However, defining visible/external attributes, e.g., hair/eye/skin colour, must be couched in ambiguity because they are subject to individual evaluation, e.g., colour perception, but also culturally informed perception of such traits (see further [**15**]). These individual attributions are then compared to non-uniform standards. Here also, as mentioned earlier, the choice of reference data has a fundamental influence, requiring forensic scientists to reflect on the biases that go into defining reference data. In many cases, DNA analysis results might be meaningless or even misleading. Articulating such crucial limitations and distinctions to users is central to ensuring the reliability and utility of FDP in an investigation, even if reporting forms seem to leave scant space for such important information. Space and time must be made for communicating these aspects since they can significantly impact the understanding investigators develop of intelligence generated via FDP analyses.

Second, the investigative value of forensic genetics arises from the capacity of technology to either deliver evidence or intelligence to detect crimes and perpetrators. In the case of FDP, intelligence is produced only in as far as the genetic analysis can either point towards (or away from) a group of shared attributes/ characteristics a perpetrator may have. FDP is about shared characteristics and as such applies suspicion to groups, collectivising suspicion [**77**]. However, since FDP deals in probabilities, the extent to which a group can be excluded can never be 100% certain. A key issue here is that the technology works best on minority groups (of an appearance distinctly different to that of a majority population) to deliver on its aim of directing the next steps in an investigation (e.g., [**78**]). It is thus vital to reflect when using technologies such as FDP, on their investigative value, which is reduced when the analysis points towards a majority group as then the expected 'reduction' in suspect populations is not achieved. Effectively, this may lead to minority groups being implicated in active investigations more often because the investigative tools work less well in the case of majority populations [**79**]. At the same time, forensic technologies such as FDP do not work in isolation from policing practices and institutional cultures [**21**]. Therefore, forensic scientists need to reflect on the discriminatory power of the technology that lies at its heart-both at policing and at public levels. This was the case in the Ladenburger investigation where non-European migrants were associated with criminal behaviour because of the way that FDP was, for a long period of time, communicated as the key technological solution to violent crime investigations.

Third, and this may be the most apparent element to forensic scientists, FDP remains an emerging technology. While pigmentation analyses may be well advanced for darker and lighter aspects of the continuum, medium levels of pigmentation remain more difficult to test for reliably (e.g., [**80**]). Furthermore, public debates between politicians, police, and scientists around the Ladenburger case confused the technology readiness level of FDP: while scientists were talking about lab-based levels of testing and aspirations for future analytical capacities, perhaps even up to "facial composites", publics, politicians and police representatives perceived of these capacities as already realised and ready for use. Key messages around the reliability and utility of FDP were mixed up. This may yet result in a significant loss of trust in the technology and forensic science in policing should its legalisation in Germany in 2019 not lead to any significant positive results in investigations in the near future.

7.2. Forensic Genetic Genealogy (FGG)

From 1974 to 1986 in California, USA, crimes including murder and rape, were committed by the same unknown individual. In 2018, investigators uploaded genetic information derived from crime scenes onto the public-access genetic ancestry database, GEDmatch. This database aggregates genetic information from citizens who upload the results of their direct-to-consumer genetic tests from genealogy companies, such as AncestryDNA, 23andMe, MyHeritage, and FamilyTreeDNA. Investigators found partial DNA matches assumed to belong to distant relatives, which enabled the creation of a 'family tree', primarily based on online genealogy database records, and including information from sources such as social media and other records. The data of these records sit outside of the forensic domain, and their use in policing raises a number of legal and ethical questions, including on the level of appropriate consent; dual/unintended use of data; and whether the capacity exists or can be introduced for blocking or deleting data from such sources if data subjects disagree with law enforcement uses of their data. Consequently, Joseph James DeAngelo, 72 years old, was identified as a suspect, and his "abandoned" DNA collected to facilitate confirmatory DNA analysis. Although the Golden State Killer was not the first case in which this technique has been used, it has become what Prainsack and Toom [**81**] term a "founding myth" and is being enthusiastically explored by police globally, and several cases have now been solved through FGG internationally. According to Kling et al. [**82**], FGG has been used to generate investigative leads in nearly 200 cold cases and some active investigations (see also [**10**,**83**,**84**]). In Sweden, prosecutors have allowed investigators to use consumer genealogy databases to solve cold cases: missing persons and a criminal case [**85**]. In Canada, two cold cases have been solved through forensic genealogy [**8**]. In the UK, a study has assessed the likely effectiveness of long-range familial searches through GEDmatch [**86**]. The Golden State Killer also sparked interest in this procedure among citizens and the companies offering these services, with existing databases now allowing for and, in some cases, encouraging law enforcement searches, as in the case with GEDmatch (recently acquired by commercial forensic service provider Verogen which raises further ethical issues which we do not have the space to explore here) and FamilyTreeDNA. In addition, services exclusively dedicated to law enforcement searches, e.g., DNASolves, DNA.Land, Geneanet, and Geni, are emerging [**82**] and some law enforcement agencies are setting up their own FGG services.

The ethical implications of FGG are wide-ranging. First, this technique increases acutely the amount of information that can be garnered from DNA data, as it makes use of a significant number of markers and uses SNPs, instead of a relatively reduced number of markers and STRs, as used in 'traditional' forensic DNA testing and databases. Genetic genealogy also allows users to identify distant relatives [**34**,**35**] and disease predispositions.

Second, FGG challenges the notion of what constitutes a DNA database used for law enforcement. Until recently, forensic DNA databases were solely constituted of DNA profiles of individuals with some degree of involvement with the criminal justice system [**36**]. Yet suspects are now being searched for in databases consisting of voluntarily uploaded genetic data from citizens who wish to know more about their health, ancestry, and/or search for relatives: a clear example of 'function creep'. Such repurposing of DNA data requires a discussion of (informed) consent [**33**], data ownership, and the accountability of genealogy companies and law enforcement in relation to human rights.

Third, the genetic databases being used/created for FGG usually have a significantly different composition than "traditional" forensic databases, the latter over-representing groups that are most likely to interact with the criminal justice system, such as racial and ethnic minorities [**87**] while genetic genealogy databases are mainly composed of an economically privileged and a European-descent population [**51**]. This leads to the expansion of affected populations as involvement with the criminal justice system is no longer a prerequisite to inclusion in law enforcement searches and follows broader patterns of surveillance expansion that encompasses ever more citizens. FGG, therefore, brings individuals whose profiles are in a recreational DNA database, under the 'genetic surveillance' gaze, making them suspects by association [**88**]. This 'function creep' illustrates the potential for wider ramifications of decisions that are particularly risky when impacts cannot be known and their diffusion across society remains uncharted. However, it should be noted that different databases have varying policies in terms of making genetic data available for law enforcement searches [**89**] GEDmatch and FamilyTreeDNA encourage such searches. GEDmatch offers users to opt-in while FamilyTreeDNA, in 2019, instituted an opt-out policy that automatically opens up users' data to law enforcement searches. However, under the jurisdiction of the European Union's General Data Protection Regulation (GDPR), users from current and recent European Union member states who have profiles on FamilyTreeDNA have been automatically opted out. This means that European Union citizens must opt-in if they wish to have their DNA profile included for criminal investigation purposes [**8**,**54**]. Adopting a different stance, companies such as AncestryDNA and 23andMe have categorically denied access for law enforcement agents to such data, unless subpoenaed by a court order to do so [**89**].

7.3. Forensic Epigenetics/-Genomics (FEpi)

Imagine: A woman is found dead, and DNA samples collected from crime scenes do not provide any investigative leads. After making a genetic estimation of appearance traits and biogeographic ancestry (see case 1), it is decided to conduct epigenetic prediction of chronological age [**90**], as well as epigenomic prediction of lifestyle and environmental factors through the DNA samples. Results indicate that the presumed suspect is approximately twenty years old, smokes tobacco, has a high alcohol intake, and probably belongs to a specific "socioeconomic status"(p. 7, [**91**]). Based on these results, police investigative work will be prioritised towards a more specific suspect pool. While hypothetical, this is the aspiration of forensic epigenetics.

Until recently, despite the enormous interest in post-genomics research, forensic applications of epigenetics were mainly restricted to body fluid identification, differentiating between monozygotic twins, and age prediction [**90,92,93**]. The potential uses of epigenetics in forensic research have only recently expanded, with expectations that the understanding of how environmental factors impact our DNA might be useful 'intelligence', by identifying suspects' lifestyle, such as smoking habits, alcohol intake, other drug use, diet, body shape and size, physical exercise, zone of residence, and socioeconomic status [**91,94**]. As with FGG, the potential ramifications of FEpi are extensive and impossible to fully predict, when technology is still in an embryonic life stage. It is thus critical that potential consequences are contemplated to develop anticipatory capacity.

First, revealing lifestyle and environmental characteristics directly interferes with the right to privacy of individuals and populations. Despite arguments—already expressed and disputed [**15**]—that "the genetic prediction of obvious appearance traits [such as smoking] because their external visibility cannot be considered private" (p. 8, [**91**]), it is clear that inferring smoking, drinking and drug and food consumption habits indirectly correlates with health information and medical conditions, further blurring the distinction between forensic and medical uses of genetic data.

Second, we might question how useful information about smoking and drinking habits might be for investigators since the indication that a suspect smokes and/or consumes alcohol does not significantly advance a criminal investigation without any suspects. Similar to FDP (case 1) and FGG (case 2), FEpi casts suspicion across entire groups by clustering 'suspect' populations that share certain lifestyles.

Third, epigenetic markers are, to a considerable extent, established in early development, potentially transmissible to subsequent generations, and sensitive to environmental factors and lifestyles [**95,96,97**]. If we look back to the hypothetical criminal case and focus upon the information that the presumed suspect is approximately twenty years old, smokes, has a significant alcohol intake, we must take into consideration (i) that maternal smoking and alcohol consumption during pregnancy could cause epigenetic changes in the offspring; (ii) passive smoking also might impact the epigenome; (iii) alcohol-dependent epigenetic signatures are partly reversible upon abstinence [**91,94**]. This suggests that the variability of epigenetic markers does not constitute reliable or consistent predictors for identifying suspects. Although FEpi is still being developed in terms of scientific validation, an approach that is anchored in ethics as lived practice implies that issues around reliability, utility, and legitimacy [**20**] must be considered focal points to be addressed and discussed.

8. CONCISE SUMMARY

As science is afforded a particularly powerful role in knowledge production and meaning-making in the criminal justice system, could this lead to a 'technological tyranny' [**98**] in which individuals and communities have to prove their innocence against probabilities? Might forensic scientists inadvertently contribute to creating new kinds of inequalities and insecurity when it comes to policing and criminal justice? (p. 365, [**99**]). In response to such questions, particularly facing forensic geneticists, in this paper we propose a move from the dominant "thin" ethics approach, to one of ethics as lived practice. We argue that forensic scientists are ethical agents, that is, they make ethical decisions daily and not just at procedural moments such as requesting ethics approval or reporting to an institutional review board. Rather, they have the agency and power to make decisions with far-reaching impacts. Such decision-making can be more apparent at some points than others and must go some way to building 'anticipatory capacity' for the governance of reliable and responsible forensic genetics. In this paper, we have articulated some of the key stages in the life of forensically relevant (epi)genetic data at which ethical decision-making takes place, and where it is vital to take the time to reflect on ethical principles and practices. Our aim has been to raise awareness of the need for continuous reflection upon ethical decisions and to articulate some of those key instances in which such reflection is clearly vital.

9. AT A GLANCE

Greater scrutiny and demands for innovation and increased productivity place pressures on scientists. Forensic genetics is advancing at a rapid pace but can only do so responsibly, usefully, and acceptably within ethical and legal boundaries. We argue that such boundaries require that forensic scientists embrace 'ethics as lived practice'. As a starting point, we critically discuss 'thin' ethics in forensic genetics, which lead to a myopic focus on procedures, and to seeing 'privacy' as the sole ethical concern and technology as a mere tool. To overcome 'thin' ethics in forensic genetics, we instead propose understanding ethics as an intrinsic part of the lived practice of a scientist. Therefore, we explore, within the context of three case studies of emerging forensic genetics technologies, ethical aspects of decision-making in forensic genetics research and in technology use. We discuss the creation, curation, and use of databases, and the need to engage with societal and policing contexts of forensic practice. We argue that open communication is a vital ethical aspect. Adoption of 'ethics as lived practice' supports the development of anticipatory capacity—empowering scientists to understand, and act within ethical and legal boundaries, incorporating the operational and societal impacts of their daily decisions, and making visible ethical decision making in scientific practice.

REFERENCES

1. Lipphardt, V.; Rappold, G.; Surdu, M. Representing vulnerable populations in genetic studies: The case of the Roma. *Sci. Context* **2020**. (Preprint). **[Google Scholar] [CrossRef]**
2. The Law Commission. *The Use of DNA in Criminal Investigations*; The Law Commission: Wellington, New Zealand, 2020; ISBN 978-1-877569-96-8. **[Google Scholar]**
3. Presser, J.R.; Robertson, K. *AI Case Study: Probabilistic Genotyping DNA Tools in Canadian Criminal Courts*; Law Commission of Ontario: Toronto, ON, Canada, 2021. **[Google Scholar]**
4. Call for Evidence Launched on New Technologies in Law Enforcement—Committees—UK Parliament. Available online: **https://committees.parliament.uk/committee/519/justice-and-home-affairs-committee/news/156778/call-for-evidence-launched-on-new-technologies-in-law-enforcement/** (accessed on 14 October 2021).
5. Independent Advisory Group on Emerging Technologies in Policing: Call for Evidence. Available online: **http://www.gov.scot/publications/independent-advisory-group-on-emerging-technologies-in-policing-call-for-evidence/** (accessed on 14 October 2021).
6. Kienzle, T. Draft DNA Phenotyping Law Set to Go before Swiss Parliament. Available online: **https://www.swissinfo.ch/eng/draft-dna-phenotyping-law-set-to-go-before-swiss-parliament/46204608** (accessed on 14 October 2021).
7. Murphy, E. Law and policy oversight of familial searches in recreational genealogy databases. *Forensic Sci. Int.* **2018**, *292*, e5–e9. **[Google Scholar] [CrossRef]**

8. Granja, R. Long-range familial searches in recreational DNA databases: Expansion of affected populations, the participatory turn, and the co-production of biovalue. *New Genet. Soc.* **2021**, *40*, 331–352. **[Google Scholar] [CrossRef]**

9. Samuel, G.; Kennett, D. The impact of investigative genetic genealogy: Perceptions of UK professional and public stakeholders. *Forensic Sci. Int. Genet.* **2020**, *48*, 102366. **[Google Scholar] [CrossRef] [PubMed]**

10. Kennett, D. Using genetic genealogy databases in missing persons cases and to develop suspect leads in violent crimes. *Forensic Sci. Int.* **2019**, *301*, 107–117. **[Google Scholar] [CrossRef]**

11. Willis, S. Forensic science, ethics and criminal justice. In *Handbook of Forensic Science*; Fraser, J., Williams, R., Eds.; Willan: London, UK, 2009; ISBN 978-1-84392-732-7. **[Google Scholar]**

12. Schiermeier, Q. Forensic database challenged over ethics of DNA holdings. *Nature* **2021**, *594*, 320–322. **[Google Scholar] [CrossRef] [PubMed]**

13. McCartney, C. The forensic science paradox. In *Counter-Terrorism, Constitutionalism and Miscarriages of Justice: A Festschrift for Professor Clive Walker*; Lennon, G., King, C., McCartney, C., Eds.; Hart Publishing: Oxford, UK, 2018; pp. 227–248. ISBN 978-1-5099-1575-0. **[Google Scholar]**

14. Guston, D.H. Understanding "anticipatory governance". *Soc. Stud. Sci.* **2014**, *44*, 218–242. **[Google Scholar] [CrossRef] [Green Version]**

15. Toom, V.; Wienroth, M.; M'charek, A.; Prainsack, B.; Williams, R.; Duster, T.; Heinemann, T.; Kruse, C.; Machado, H.; Murphy, E. Approaching ethical, legal and social issues of emerging forensic DNA phenotyping (FDP) technologies comprehensively: Reply to 'Forensic DNA phenotyping: Predicting human appearance from crime scene material for investigative purposes' by Manfred Kayser. *Forensic Sci. Int. Genet.* **2016**, *22*, e1–e4. **[Google Scholar] [CrossRef] [PubMed] [Green Version]**

16. Williams, R.; Wienroth, M. Social and Ethical Aspects of Forensic Genetics: A Critical Review. *Forensic Sci. Rev.* **2017**, *29*, 145–169. **[Google Scholar]**

17. Williams, R.; Wienroth, M. Identity, mass fatality and forensic genetics. *New Genet. Soc.* **2014**, *33*, 257–276. **[Google Scholar] [CrossRef]**

18. Kruse, C. *The Social Life of Forensic Evidence*; University of California Press: Oakland, CA, USA, 2015; ISBN 978-0-520-28839-3. **[Google Scholar]**

19. Granja, R.; Machado, H. Ethical controversies of familial searching: The views of stakeholders in the United Kingdom and in Poland. *Sci. Technol. Hum. Values* **2019**, *44*, 1068–1092. **[Google Scholar] [CrossRef] [Green Version]**

20. Wienroth, M. Value beyond scientific validity: Let's RULE (Reliability, Utility, LEgitimacy). *J. Responsible Innov.* **2020**, *7*, 92–103. **[Google Scholar] [CrossRef]**
21. Skinner, D. Forensic genetics and the prediction of race: What is the problem? *BioSocieties* **2020**, *15*, 329–349. **[Google Scholar] [CrossRef]**
22. Wienroth, M. Governing anticipatory technology practices. Forensic DNA phenotyping and the forensic genetics community in Europe. *New Genet. Soc.* **2018**, *37*, 137–152. **[Google Scholar] [CrossRef]**
23. Granja, R.; Machado, H. Forensic DNA phenotyping and its politics of legitimation and contestation: Views of forensic geneticists in Europe. *Soc. Stud. Sci.* **2020**, 1–19, (ePub ahead of Print). **[Google Scholar] [CrossRef]**
24. Guillemin, M.; Gillam, L. Ethics, reflexivity, and "ethically important moments" in research. *Qual. Inq.* **2004**, *10*, 261–280. **[Google Scholar] [CrossRef]**
25. Kahn, J. Privacy as a legal principle of identity maintenance. *Seton Hall Law Rev.* **2002**, *33*, 371. **[Google Scholar]**
26. Presidential Commission for the Study of Bioethical Issues. *Privacy and Progress in Whole Genome Sequencing*; US Department of Health & Human Services: Washington, DC, USA, 2012; p. 154.
27. Evans, J. Personal Privacy vs. Public Security. Available online: **https://techcrunch.com/2018/05/06/personal-privacy-vs-public-security-fight/** (accessed on 14 October 2021).
28. Orrù, E.; Porcedda, M.G.; Weydner-Volkmann, S. *Rethinking Surveillance and Control. Beyond the "Security vs. Privacy" Debate*; Nomos: Baden-Baden, Germany, 2017. **[Google Scholar]**
29. Timmermans, S.; Berg, M. The practice of medical technology. *Sociol. Health Illn.* **2003**, *25*, 97–114. **[Google Scholar] [CrossRef]**
30. Greenhalgh, T.; Swinglehurst, D. Studying technology use as social practice: The untapped potential of ethnography. *BMC Med.* **2011**, *9*, 45. **[Google Scholar] [CrossRef] [Green Version]**
31. Suchman, L.; Blomberg, J.; Orr, J.E.; Trigg, R. Reconstructing technologies as social practice. *Am. Behav. Sci.* **1999**, *43*, 392–408. **[Google Scholar] [CrossRef]**
32. Hunt, M.R.; Godard, B. Beyond procedural ethics: Foregrounding questions of justice in global health research ethics training for students. *Glob. Public Health* **2013**, *8*, 713–724. **[Google Scholar] [CrossRef] [PubMed]**
33. Wienroth, M. Socio-technical disagreements as ethical fora: ParabonNanoLab's forensic DNA Snapshot™ service at the intersection of discourses around robust science, technology validation, and commerce. *BioSocieties* **2020**, *15*, 28–45. **[Google Scholar] [CrossRef]**

34. The World Medical Association (WMA). *WMA Declaration of Helsinki—Ethical Principles for Medical Research Involving Human Subjects*; The World Medical Association (WMA): Helsinki, Finland, 1964. **[Google Scholar]**
35. Gaille, M.; Horn, R. The ethics of genomic medicine: Redefining values and norms in the UK and France. *Eur. J. Hum. Genet.* **2021**, *29*, 780–788. **[Google Scholar] [CrossRef]**
36. Alpaslan-Roodenberg, S.; Anthony, D.; Babiker, H.; Bánffy, E.; Booth, T.; Capone, P.; Deshpande-Mukherjee, A.; Eisenmann, S.; Fehren-Schmitz, L.; Frachetti, M.; et al. Ethics of DNA research on human remains: Five globally applicable guidelines. *Nature* **2021**, *599*, 41–46. **[Google Scholar] [CrossRef] [PubMed]**
37. Claw, K.G.; Anderson, M.Z.; Begay, R.L.; Tsosie, K.S.; Fox, K.; Garrison, N.A. A framework for enhancing ethical genomic research with Indigenous communities. *Nat. Commun.* **2018**, *9*, 2957. **[Google Scholar] [CrossRef] [PubMed] [Green Version]**
38. D'Amato, M.E.; Bodner, M.; Butler, J.M.; Gusmão, L.; Linacre, A.; Parson, W.; Schneider, P.M.; Vallone, P.; Carracedo, A. Ethical publication of research on genetics and genomics of biological material: Guidelines and recommendations. *Forensic Sci. Int. Genet.* **2020**, *48*, 102299. **[Google Scholar] [CrossRef] [PubMed]**
39. Fox, K. The illusion of inclusion—The "All of Us" research program and indigenous peoples' DNA. *N. Engl. J. Med.* **2020**, *383*, 411–413. **[Google Scholar] [CrossRef] [PubMed]**
40. Carracedo, Á.; Butler, J.M.; Gusmão, L.; Parson, W.; Roewer, L.; Schneider, P.M. Publication of population data for forensic purposes. *Forensic Sci. Int. Genet.* **2010**, *4*, 145–147. **[Google Scholar] [CrossRef]**
41. Moreau, Y. Crack down on genomic surveillance. *Nature* **2019**, *576*, 36–38. **[Google Scholar] [CrossRef]**
42. Kowal, E.; Radin, J. Indigenous biospecimen collections and the cryopolitics of frozen life. *J. Sociol.* **2015**, *51*, 63–80. **[Google Scholar] [CrossRef]**
43. Zieger, M.; Utz, S. About DNA databasing and investigative genetic analysis of externally visible characteristics: A public survey. *Forensic Sci. Int. Genet.* **2015**, *17*, 163–172. **[Google Scholar] [CrossRef]**
44. Forzano, F.; Genuardi, M.; Moreau, Y. ESHG warns against misuses of genetic tests and biobanks for discrimination purposes. *Eur. J. Hum. Genet.* **2021**, *29*, 894–896. **[Google Scholar] [CrossRef]**
45. Normile, D. Genetics papers from China face ethical scrutiny. *Science* **2021**, *373*, 727–728. **[Google Scholar] [CrossRef] [PubMed]**

46. Caliebe, A.; Walsh, S.; Liu, F.; Kayser, M.; Krawczak, M. Likelihood ratio and posterior odds in forensic genetics: Two sides of the same coin. *Forensic Sci. Int. Genet.* **2017**, *28*, 203–210. **[Google Scholar] [CrossRef] [PubMed] [Green Version]**

47. Pfaffelhuber, P.; Grundner-Culemann, F.; Lipphardt, V.; Baumdicker, F. How to choose sets of ancestry informative markers: A supervised feature selection approach. *Forensic Sci. Int. Genet.* **2020**, *46*, 102259. **[Google Scholar] [CrossRef] [PubMed]**

48. Lynch, M.; Cole, S.; Mcnally, R.; Jordan, K. *Truth Machine: The Contentious History of DNA Fingerprinting*; The University of Chicago Press: Chicago, IL, USA, 2008. **[Google Scholar]**

49. Lynch, M. God's signature: DNA profiling, the new gold standard in forensic science. *Endeavour* **2003**, *27*, 93–97. **[Google Scholar] [CrossRef]**

50. Amorim, A.; Crespillo, M.; Luque, J.A.; Prieto, L.; Garcia, O.; Gusmão, L.; Aler, M.; Barrio, P.A.; Saragoni, V.G.; Pinto, N. Formulation and communication of evaluative forensic science expert opinion—A GHEP-ISFG contribution to the establishment of standards. *Forensic Sci. Int. Genet.* **2016**, *25*, 210–213. **[Google Scholar] [CrossRef] [PubMed]**

51. Amelung, N.; Granja, R.; Machado, H. Communicating forensic genetics: "Enthusiastic" publics and the management of expectations. In *Exploring Science Communication: A Science and Technology Studies Approach*; SAGE Publishing: Thousand Oaks, CA, USA, 2020. **[Google Scholar]**

52. Lawless, C.J. The low template DNA profiling controversy: Biolegality and boundary work among forensic scientists. *Soc. Stud. Sci.* **2012**, *43*, 191–214. **[Google Scholar] [CrossRef]**

53. Gill, P.; Guiness, J.; Iveson, S. *The Interpretation of DNA Evidence (Including Low-Template DNA)*; The Forensic Science Regulator: Birmingham, UK, 2012; pp. 1–27. **[Google Scholar]**

54. Jobling, M.; Syndercombe-Court, D. *Should We Be Making Use of Genetic Genealogy to Assist in Solving Crime?* A report on the feasibility of such methods in the UK; Government of the UK: London, UK, 2020; p. 16.

55. Amankwaa, A.O.; McCartney, C. The effectiveness of the current use of forensic DNA in criminal investigations in England and Wales. *WIREs Forensic Sci.* **2021**, *3*, e1414. **[Google Scholar] [CrossRef]**

56. Cassella, J.; McCartney, C. Lowering the drawbridges: Legal and forensic science education for the 21st century. *Forensic Sci. Policy Manag. Int. J.* **2011**, *2*, 81–93. **[Google Scholar] [CrossRef]**

57. Garrett, B.L.; Neufeld, P.J. Invalid forensic science testimony and wrongful convictions. *VA Law Rev.* **2009**, *95*, 1–97. **[Google Scholar]**

58. Edmond, G. Forensic science and the myth of adversarial testing. *Curr. Issues Crim. Justice* **2020**, *32*, 146–179. **[Google Scholar] [CrossRef]**

59. Edmond, G.; Hamer, D.; Cunliffe, E. A little ignorance is a dangerous thing: Engaging with exogenous knowledge not adduced by the parties. *Griffith Law Rev.* **2016**, *25*, 383–413. **[Google Scholar] [CrossRef]**

60. Kruse, C. The Bayesian approach to forensic evidence: Evaluating, communicating, and distributing responsibility. *Soc. Stud. Sci.* **2013**, *43*, 657–680. **[Google Scholar] [CrossRef]**

61. Taylor, D.; Kokshoorn, B.; Biedermann, A. Evaluation of forensic genetics findings given activity level propositions: A review. *Forensic Sci. Int. Genet.* **2018**, *36*, 34–49. **[Google Scholar] [CrossRef] [PubMed] [Green Version]**

62. Gill, P.; Hicks, T.; Butler, J.M.; Connolly, E.; Gusmão, L.; Kokshoorn, B.; Morling, N.; van Oorschot, R.A.H.; Parson, W.; Prinz, M.; et al. DNA commission of the International society for forensic genetics: Assessing the value of forensic biological evidence—Guidelines highlighting the importance of propositions. Part II: Evaluation of biological traces considering activity level propositions. *Forensic Sci. Int. Genet.* **2020**, *44*, 102186. **[Google Scholar] [CrossRef] [Green Version]**

63. Syndercombe-Court, D.; Reed, K.; Williams, R.; Wienroth, M. *A Guide to Legal and Ethical Principles and Practices in Forensic Genetics*; EUROFORGEN-NoE: Cologne, Germany, 2016; p. 73. **[Google Scholar]**

64. McCartney, C.; Amoako, E. The UK Forensic Science Regulator: A model for forensic science regulation? *GA State Univ. Law Rev.* **2018**, *34*, 945. **[Google Scholar]**

65. Nsiah Amoako, E.; McCartney, C. The UK forensic science regulator: Fit for purpose? *WIREs Forensic Sci.* **2021**, *3*, e1415. **[Google Scholar] [CrossRef]**

66. Brown, S.; Willis, S. Complexity in forensic science. *Forensic Sci. Policy Manag. Int. J.* **2010**, *1*, 192–198. **[Google Scholar] [CrossRef]**

67. Lawless, C.J.; Williams, R. Helping with inquiries or helping with profits? The trials and tribulations of a technology of forensic reasoning. *Soc. Stud. Sci.* **2010**, *40*, 731–755. **[Google Scholar] [CrossRef]**

68. Lawless, C. Policing markets: The contested shaping of neo-liberal forensic science. *Br. J. Criminol.* **2011**, *51*, 671–689. **[Google Scholar] [CrossRef]**

69. Roberts, P. What price a free market in forensic science services—The organization and regulation of science in the criminal process. *Br. J. Criminol.* **1996**, *36*, 37–60. **[Google Scholar] [CrossRef]**

70. M'charek, A. Silent witness, articulate collective: DNA evidence and the inference of visible traits. *Bioethics* **2008**, *22*, 519–528. **[Google Scholar] [CrossRef] [PubMed]**

71. Phillips, C.; Prieto, L.; Fondevila, M.; Salas, A.; Gómez-Tato, A.; Álvarez-Dios, J.; Alonso, A.; Blanco-Verea, A.; Brión, M.; Montesino, M.; et al. Ancestry analysis in the 11-M Madrid bomb attack investigation. *PLoS ONE* **2009**, *4*, e6583. **[Google Scholar] [CrossRef] [Green Version]**

72. Lipphardt, A. The invention of the Phantom of Heilbronn. *J. Eur. Ethnol. Cult. Anal.* **2020**, 49–69. **[Google Scholar]**

73. Science Media Center. *Press Briefing: DNA-Profiling und Die Wissenschaften: Wie Weit Kann Die Erweiterte DNA-Analyse Gehen?* Science Media Center: London, UK, 2017; p. 27. **[Google Scholar]**

74. Geuther, G.; Kazmierczak, L. Advanced DNA analysis: On the trail of the perpetrators. Deutschlandfunk, 20 June 2017. Available online: **https://www.deutschlandfunk.de/erweiterte-dna-analyse-den-taetern-auf-der-spur-100.html** (accessed on 23 November 2021).

75. Vogel, G. German law allows use of DNA to predict suspects' looks. *Science* **2018**, *360*, 841–842. **[Google Scholar] [CrossRef] [Green Version]**

76. Skinner, D. Race, Racism and identification in the era of technosecurity. *Sci. Cult.* **2020**, *29*, 77–99. **[Google Scholar] [CrossRef]**

77. Hopman, R.; M'charek, A. Facing the unknown suspect: Forensic DNA phenotyping and the oscillation between the individual and the collective. *BioSocieties* **2020**, *15*, 438–462. **[Google Scholar] [CrossRef]**

78. Kayser, M.; Knijff, P. de Improving human forensics through advances in genetics, genomics and molecular biology. *Nat. Rev. Genet.* **2011**, *12*, 179–192. **[Google Scholar] [CrossRef]**

79. M'charek, A.; Toom, V.; Jong, L. The trouble with race in forensic identification. *Sci. Technol. Hum. Values* **2020**, *45*, 804–828. **[Google Scholar] [CrossRef] [Green Version]**

80. Kayser, M. Forensic DNA phenotyping: Predicting human appearance from crime scene material for investigative purposes. *Forensic Sci. Int. Genet.* **2015**, *18*, 33–48. **[Google Scholar] [CrossRef] [PubMed]**

81. Prainsack, B.; Toom, V. Performing the Union: The Prüm decision and the European dream. *Stud. Hist. Philos. Sci. Part C Stud. Hist. Philos. Biol. Biomed. Sci.* **2013**, *44*, 71–79. **[Google Scholar] [CrossRef] [PubMed] [Green Version]**
82. Kling, D.; Phillips, C.; Kennett, D.; Tillmar, A. Investigative genetic genealogy: Current methods, knowledge and practice. *Forensic Sci. Int. Genet.* **2021**, *52*, 102474. **[Google Scholar] [CrossRef] [PubMed]**
83. Katsanis, S.H. Pedigrees and perpetrators: Uses of DNA and genealogy in forensic investigations. *Annu. Rev. Genomics Hum. Genet.* **2020**, *21*, 535–564. **[Google Scholar] [CrossRef] [PubMed] [Green Version]**
84. Greytak, E.M.; Moore, C.; Armentrout, S.L. Genetic genealogy for cold case and active investigations. *Forensic Sci. Int.* **2019**, *299*, 103–113. **[Google Scholar] [CrossRef] [PubMed]**
85. Tillmar, A.; Fagerholm, S.A.; Staaf, J.; Sjölund, P.; Ansell, R. Getting the conclusive lead with investigative genetic genealogy—A successful case study of a 16 year old double murder in Sweden. *Forensic Sci. Int. Genet.* **2021**, *53*, 102525. **[Google Scholar] [CrossRef]**
86. Thomson, J.; Clayton, T.; Cleary, J.; Gleeson, M.; Kennett, D.; Leonard, M.; Rutherford, D. An empirical investigation into the effectiveness of genetic genealogy to identify individuals in the UK. *Forensic Sci. Int. Genet.* **2020**, *46*, 102263. **[Google Scholar] [CrossRef] [PubMed]**
87. Skinner, D. 'The NDNAD Has No Ability in Itself to be Discriminatory': Ethnicity and the Governance of the UK National DNA Database. *Sociology* **2013**, *47*, 976–992. **[Google Scholar] [CrossRef]**
88. Granja, R.; Machado, H. Corposrelacionais, biofamília e suspeiçãoporassociação: O caso da pesquisa familiar emgenéticaforense. In *Crime e Tecnologia: DesafiosCulturais e Políticos para a Europa*; Machado, H., Ed.; EdiçõesAfrontamento: Porto, Portugal, 2021; pp. 195–216. ISBN 978-972-36-1852-5. **[Google Scholar]**
89. Skeva, S.; Larmuseau, M.H.; Shabani, M. Review of policies of companies and databases regarding access to customers' genealogy data for law enforcement purposes. *Pers. Med.* **2020**, *17*, 141–153. **[Google Scholar] [CrossRef] [PubMed]**
90. Vidaki, A.; Daniel, B.; Court, D.S. Forensic DNA methylation profiling—Potential opportunities and challenges. *Forensic Sci. Int. Genet.* **2013**, *7*, 499–507. **[Google Scholar] [CrossRef]**
91. Vidaki, A.; Kayser, M. From forensic epigenetics to forensic epigenomics: Broadening DNA investigative intelligence. *Genome Biol.* **2017**, *18*, 238. **[Google Scholar] [CrossRef] [PubMed] [Green Version]**

92. Haddrill, P.R. Developments in forensic DNA analysis. *Emerg. Top. Life Sci.* **2021**, *5*, 381–393. **[Google Scholar] [CrossRef] [PubMed]**

93. Sabeeha, H.S.E. Forensic epigenetic analysis: The path ahead. *Med. Princ. Pract.* **2019**, *28*, 301–308. **[Google Scholar] [CrossRef] [PubMed]**

94. Vidaki, A.; Kayser, M. Recent progress, methods and perspectives in forensic epigenetics. *Forensic Sci. Int. Genet.* **2018**, *37*, 180–195. **[Google Scholar] [CrossRef]**

95. Hedlund, M. Epigenetic responsibility. *Med. Stud.* **2012**, *3*, 171–183. **[Google Scholar] [CrossRef]**

96. Loi, M.; Del Savio, L.; Stupka, E. Social epigenetics and equality of opportunity. *Public Health Ethics* **2013**, *6*, 142–153. **[Google Scholar] [CrossRef] [Green Version]**

97. Richardson, S. Maternal bodies in the postgenomic order: Gender and the explanatory landscape of epigenetics. In *Postgenomics: Perspectives on Biology after the Genome.*; Duke University Press: Durham, NC, USA, 2015. **[Google Scholar]**

98. Blake, E.T. Scientific and legal issues raised by DNA analysis. In *DNA Technology and Forensic Science*; Cold Spring Harbor Laboratory Press: New York, NY, USA, 1989. **[Google Scholar]**

99. Bowling, B. Transnational Criminology and the Globalization of Harm Production. In *What is Criminology?* Oxford University Press: Oxford, UK, 2011; ISBN 6-957182-19-0-978. **[Google Scholar]**

13 PUSHING THE BOUNDARIES: FORENSIC DNA PHENOTYPING CHALLENGED BY SINGLE-CELL SEQUENCING

1. INTRODUCTION: AN OVERVIEW

A solitary human cell harbors approximately six picograms (pg) of DNA, which corresponds to roughly 2 multiplied by 3.3 billion base pairs. The ability to sequence such a large amount of genetic material was made feasible by the rapid advancement of next-generation sequencing (NGS) technology[**1**]. The progress in the field has recently been called, by Nature, one of the technologies to watch in 2021 [**2**]. The efficacy of single-cell sequencing in medicine cannot be overestimated, and it demonstrates the possibility for implementation in forensics, where it is not commonly used. The main motivation behind introducing this technology into forensics is the potential to improve mixture deconvolution workflows and interpretation methods. If a trace is a mixture of different biological materials (for example, sperm and epithelium), there are methods to separate cell pools to investigate them separately [**3,4,5,6,7,8,9,10**]. Problems arise when the mixed trace is too complex or if it consists of the same cell type, such as blood-blood mixtures [**11**]. Implementing single-cell analyses can improve the efficacy of mixture deconvolution and, therefore, has already gained interest in the forensic community. A cell separation solution is the DEPArray™ N×T System from Menarini Silicon Biosystems, which is widely used in clinical research [**12,13,14,15**] and is suitable for forensic purposes. Current forensic studies focus on single cell separation followed by STR amplification and detection using capillary electrophoresis [**11,16,17,18,19**]. Mixtures are commonly analyzed with STR typing, but when using different, still new forensic applications, it requires a special approach. Since forensic DNA phenotyping (physical appearance and ancestry predictions) is becoming popular, as observed with legislation expansion in more countries [**20**], it is assumed that in some cases, a mixture might be the only trace submitted for phenotyping. Phenotyping workflows and prediction methods still need comprehensive research performed before they can be a fully recommended method for mixed trace analysis [**20,21,22,23**]. Here, we present an alternative approach to dealing with mixed traces submitted for forensic DNA phenotyping that introduces single-cell separation in order to deconvolute the mixture prior to genotyping and phenotyping.

2. MATERIALS AND METHODS

For the study, a fresh blood sample (female) was collected during an autopsy performed in the Institute of Legal Medicine Munich. The sampling was approved by the Bioethical Commission from the Ludwig Maximilian University of Munich. The blood was used as a reference sample (1 ng DNA input, prepared as described in **Section 2.3**), for which the white blood cells were separated and used.

2.1. Staining and Counting the Cells

In order to conduct cell separation, the biological material was first stained with the DEPArray™ Forensic SamplePrep Kit (Menarini Silicon Biosystems, Bologna, Italy), which enables the staining of epithelial cells, leucocytes, and sperm cells. Following the manufacture's protocol, 5 μL of blood was used, from which the white blood cells were marked with PE (Phycoerythrin) conjugated CD45 antibody as well as with DAPI (4′,6-diamidino-2-phenylindole; to stain the nuclei). The maximum cell input for the DEPArray™ N×T System (Menarini Silicon Biosystems) is 6000 cells; therefore, the sample was analyzed with the Countess™ II FL (Thermo Fisher Scientific, Waltham, MA, USA) automated cell counter and diluted to the recommended concentration.

2.2. Sorting the Cells

The DEPArray™ N×T System's (Menarini Silicon Biosystems) forensic protocol was used to separate the white blood cells. Only highly evaluated (of typical size, morphology, and staining) cells were chosen for the experiment. The selected cells were moved automatically by the DEPArray™ platform and placed in dedicated tubes as follows: 5 groups of 20 cells, 10 groups of 10 cells, 10 groups of 5 cells, and 20 single cells (groups and single cells were collected in two separate experiments).

Preventing Contamination

Due to working with a special type of material (single cells), ensuring that there is no contamination is extremely important for the interpretation of the data. Since the DEPArray™ N×T System is a closed system, contamination within is not possible. In order to exclude its potential occurrence in the next stages of laboratory work, all steps were carried out with a negative control included.

2.3. Library Preparation and Sequencing

For all samples used in the study, genomic DNA (gDNA) was extracted with the DEPArray™ LysePrep Kit (Menarini Silicon Biosystems) directly from the cell separation tubes to make sure the input corresponds with the number of collected cells. All sample extracts were subjected to manual library preparation using the Precision ID Library Kit and IonCode™ barcode adapters using the validated Ion AmpliSeq™ PhenoTrivium Panel [**24**]. The amplification of the targets was carried out in the same tubes as the extraction to avoid changing the DNA input. The target amplification cycle number and annealing/extension time were 23 cycles and 4 min, respectively. Libraries prepared from cell groups and half of the collected single cells were quantified using the Ion Library TaqMan Quantitation Kit (Thermo Fisher Scientific), diluted (except for libraries of <30 pM concentration), and pooled equimolarly to 30 pM for template preparation on the Ion Chef using the Ion S5™ Precision ID Chef & Sequencing Kit (Thermo Fisher Scientific). The remaining half of the single-cell-based libraries were quantified and rescued by library amplification (see **Section 2.3.1**). A range of 12–24 libraries were pooled per 530 chip and sequenced on the Ion S5 [**25**].

2.3.1. Library Amplification

Half of the undiluted single-cell-based libraries were submitted for library amplification. A 25 µl sample of the previously eluted samples was combined with 72 µL of Platinum™ PCR SuperMix HiFi and 3 µL of Library Amplification Primer Mix from the Precision ID Library Kit (Thermo Fisher Scientific). The PCR products were purified in a two-round clean-up with the Agencourt™ AMPure™ XP Reagent according to the manufacturer's manual. The amplified libraries were quantified, diluted, and pooled for automated template preparation.

2.4. Data Analysis

Primary sequence analysis was performed on Torrent Suite™ Software (TSS) 5.10.1 (Thermo Fisher Scientific) with Torrent Mapping Alignment Program (TMAP) alignment of sample reads against the hg19 genome assembly. SNP genotyping and tertiary analysis, in the form of ancestry prediction, were performed using the HIDGenotyper–2.2 plugin and Converge v2.2 (Thermo Fisher Scientific). Low-quality results were double-checked by running VariantCaller v5.10.0.18 on TSS and reviewing the raw data in IGV 2.7 (Integrative Genomics Viewer) [**26**]. Tertiary analysis was separated into two parts: phenotype prediction (which cannot be performed within Converge v2.2) and ancestry prediction by the bootstrapping admixture analysis with Converge. For the cell groups, the coverage thresholds were adjusted as follows: for the SNPs corresponding with the HIrisPlex-S panel, the analytical coverage thresholds were set based on the HIrisPlex-S panel validation for

MPS platforms [22]; for the remaining autosomal ancestry markers, the minimum coverage to call an SNP was set to 100 reads. For the single cells, the coverage threshold was lowered to 50 reads for all markers. The heterozygote balance threshold was set to 65%/35% for heterozygotes and 90%/10% for homozygotes.

2.4.1. Phenotype and Ancestry Prediction

For phenotype (HIrisPlex-S: eye, hair, and skin color) prediction, SNP genotypes were exported from Converge and used to manually generate single profiles, as described in **Section 2.4.2**, which were later converted into the input file format required by the HIrisPlex-S Webtool (**https://hirisplex.erasmusmc.nl/** accessed date 29 April 2021). The HIrisPlex-S SNP set contains an indel SNP (rs796296176) in the form of an A insertion that was manually reviewed and called using IGV 2.7. Predictions were interpreted according to the PhenoTrivium validation paper [24]. For ancestry prediction, called genotypes were merged into a consensus single SNP profile using Converge, and the prediction was performed with the bootstrapping admixture analysis [27] feature of Converge using the 75% resampling size, 1000 replications, and the Precision ID Ancestry Panel Ancestry Frequency File v1.1. The PhenoTrivium Panel contains the 145 Precision ID Ancestry SNPs used for bootstrapping admixture analysis. In the bootstrapping admixture analysis feature of Converge, admixture predictions are made based on a maximum likelihood approach used to predict the most likely admixture proportions across seven root populations (herein referred to as the core admixture algorithm): Africa (AFR), East Asia (EA), South Asia (SA), Southwest Asia (SWA), Europe (EU), America (AME), and Oceania (OCE). The predictions are bootstrapped across a random subset of sequenced SNPs, specified by the user in %, with each bootstrapping replication ran through the core admixture algorithm N times using a different subset of SNPs for each replication to capture uncertainty in the predictions. The results are displayed as an average of the bootstrapping replications for each population group and a 95% confidence interval reflecting the probable range of variability of the estimated ethnicity percentages.

2.4.2. Interpretation Models

The consensus genotypes from cell groups and single cells were used to generate a single SNP profile for tertiary analysis. Two different interpretation models were used to build the profiles: "basic", which compromised genotypes detected at least twice, and "conservative", for which genotypes were included only if detected at least four times. The cell groups were interpreted using only the "basic" approach (following the interpretation pipeline from PhenoTrivium validation), and the profiles from the single cells were generated using both models. Additionally, the tertiary analysis was performed based on single profiles obtained from single cells with the highest coverage.

3. RESULTS

3.1. Cell Groups

The cell groups were collected to evaluate the performance of the workflow combining single-cell separation using the DEPArray™ N×T System with Next Generation Sequencing using the Ion S5. The results obtained in this study were compared to the data from a previously performed validation study [**24**] of the PhenoTrivium assay. The collected cell groups approximately corresponded to 100, 50, and 25 pg DNA, which were similar to the DNA input used in the sensitivity study (125, 62, and 31 pg).

3.1.1. Coverage, Allele Frequency, and Genotype Calling

The study consisted of 5 groups of 20 cells, 9 groups of 10 cells (one sample was excluded due to a library prep error), and 10 groups of 5 cells, for a pool of 24 libraries total (one group of cells corresponds to one library), which was sequenced on a 530 Chip. Marker coverage across the 200 autosomal markers included in the panel ranged between 309,357 and 736,887 total reads for 20 cells, 216,401–549,980 reads for 10 cells, and 106,865–367,782 reads for 5 cells. The obtained values were comparable with the ones observed in the sensitivity study, where the total number of reads across the triplicates was 340,461–698,760, 374,692–637,041, and 94,120–395,664 for 125, 62, and 31 pg respectively [**24**].

Allele frequencies were calculated for 43 out of 45 heterozygote loci. Markers rs1470608 and rs10756819 (from HIrisPlex-S) were not called among all the samples due to low coverage. Frequencies were calculated by dividing the number of reads obtained for the reference allele by the total number of reads per locus obtained for both reference and alternative allele. For 20 cells, the frequency varied between 45% and 57% (average 50%), for 10 cells between 47% and 55% (avg. 51%), and for 5 cells between 43% and 58% (avg. 50%). The allele frequency for loci classified as homozygote was between 92–100% for 20 cells (avg. 99.8%), between 90% and 100% for 10 cells (avg. 99.6%) and between 87% and 100% for 5 cells (avg. 99.5%).

Previously described coverage and heterozygote balance thresholds were used to call genotypes in Converge using the HID Genotyper-2.2 plugin. For 20 cells, between 94 and 99.5% of genotypes across all 200 autosomal loci met all the thresholds. For ten cells, the values were 84–98.5%, and for five cells, 64.5–98.5%. The percentage of called genotypes in the sensitivity study was 95–98.5% for 125 pg, 88.5–90% for 62 pg, and 72–81.5% for 31 pg. All the called genotypes were in concordance with the reference sample, no incorrect genotypes were observed. In comparison, the data from the previous study showed that allelic dropouts and drop-ins occurred among the replicates with a 31 pg DNA input, which caused incorrect genotype calling.

3.1.2. Phenotype and Ancestry Prediction

Phenotype and ancestry predictions were performed for the reference sample and for 'basic' consensus profiles generated for cell groups. From 41 markers associated with eye, hair, and skin color prediction, for all cell groups, 39 SNPs were recovered, and two loci were not called due to low coverage, namely rs1470608 and rs10756819 (similar as in the sensitivity study). **Table 1. Phenotype and ancestry prediction summary for cell groups and consensus profiles from single cells. *p*-values and admixture proportions were compared with the results obtained for reference (ref.) sample (*italic* indicates identical values).**

<table>
<tr><th></th><th></th><th>Ref.</th><th colspan="4">Cells-Consensus</th></tr>
<tr><th></th><th></th><th>20</th><th>10</th><th>5</th><th>1
"basic"</th><th>1
"conservative"</th></tr>
<tr><td rowspan="3">Eye color</td><td>blue</td><td>0.000</td><td colspan="4">0.000</td></tr>
<tr><td>inter</td><td>0.002</td><td colspan="4">0.002</td></tr>
<tr><td>brown</td><td>0.998</td><td colspan="4">0.998</td></tr>
<tr><td rowspan="6">Hair color and shade</td><td>blond</td><td>0.003</td><td colspan="4">0.003</td></tr>
<tr><td>brown</td><td>0.337</td><td colspan="4">0.337</td></tr>
<tr><td>red</td><td>0.000</td><td colspan="4">0.000</td></tr>
<tr><td>black</td><td>0.661</td><td colspan="4">0.661</td></tr>
<tr><td>light</td><td>0.005</td><td colspan="4">0.005</td></tr>
<tr><td>dark</td><td>0.995</td><td colspan="4">0.995</td></tr>
<tr><td rowspan="5">Skin color</td><td>very pale</td><td>0.001</td><td colspan="4">0.001</td></tr>
<tr><td>pale</td><td>0.008</td><td colspan="4">0.008</td></tr>
<tr><td>inter</td><td>0.275</td><td colspan="3">0.275</td><td>0.284</td></tr>
<tr><td>dark</td><td>0.696</td><td colspan="3">0.687</td><td>0.676</td></tr>
<tr><td>dark to black</td><td>0.021</td><td colspan="3">0.029</td><td>0.030</td></tr>
<tr><td colspan="7"></td></tr>
<tr><td>Admixture</td><td colspan="6">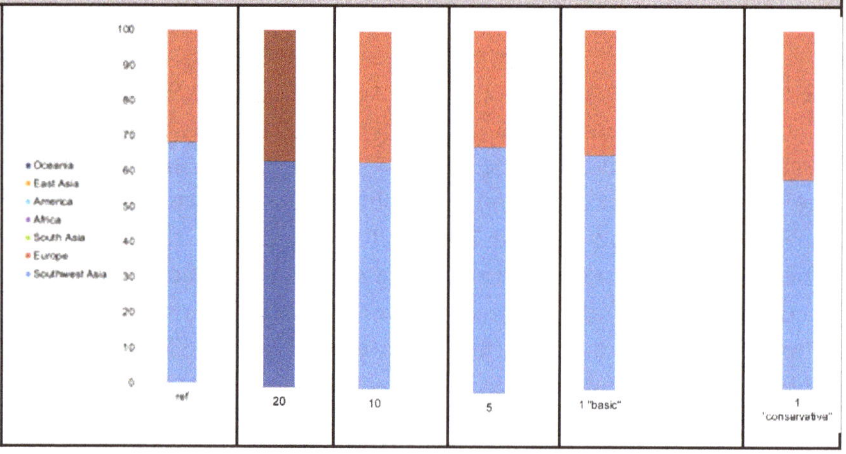</td></tr>
</table>

observed area under curve (AUC) accuracy loss (0.001 for very pale, 0.004 for pale, and 0.001 for intermediate skin) did not affect the final phenotype prediction, which was predicted to be the same as the reference, with the following *p*-values: brown eyes (0.998), black hair (0.661 for black and 0.995 for dark), and dark skin (0.696). From 145 markers used by Converge for ancestry analysis, the maximum number of SNPs was obtained from 5 cells while the profiles for 20 and 10 cells included one locus not called due to low coverage (rs2196051). The bootstrapping admixture analysis for all cell groups revealed an admixture of Southwest Asia and Europe (ratio ca. 70/30%), and the reported individual's place of birth was Iraq (no further details available). The detailed prediction results for all the samples are presented in **Table 1**.

3.2. Single Cells

3.2.1. Coverage, allele frequency, base misincorporation rates, and genotype calling

For the unamplified libraries (NA), the number of mapped reads varied between 11,979 and 58,536, of which approximately 60% were on target (**Figure 1**a). Total coverage across all the markers analyzed in Converge ranged between 1891–26,400 (avg. 8889) reads, and the mean marker coverage ranged between 9 and 132 reads (avg. 44). For the amplified libraries (AMP), the number of mapped reads increased to a range of 44,445 to 186,402 reads, of which ca. 60% were on target (**Figure 1**b). This resulted in total coverage between 13,735 and 113,398 reads (avg. 49,642) and mean marker coverage between 69 and 567 reads (avg. 248). To visualize the coverage difference between the unamplified and amplified libraries, the percentage of amplicons with a very low and very high number of reads were compared (**Figure 2**).

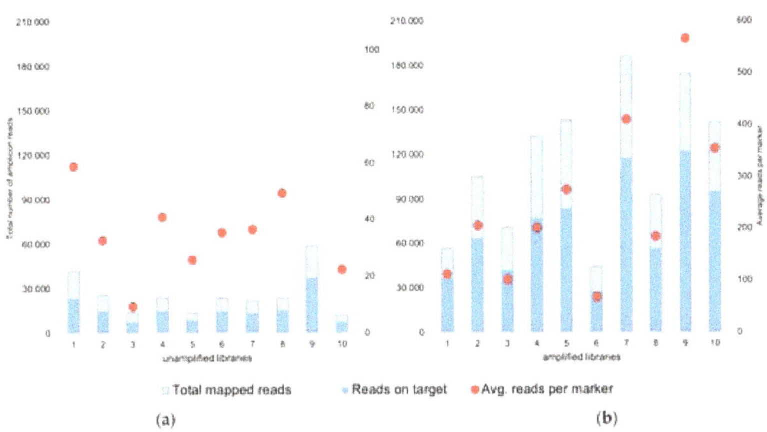

Figure 1. Comparison of the total mapped reads obtained for the unamplified (a) and amplified libraries (b), together with the number of the reads on target for both groups. The latter was comparable and reached around 60% for both

unamplified and amplified libraries but the total number of mapped reads for the amplified libraries was much higher and therefore resulted in higher total coverage. Red dots represent the average coverage per marker for each sample (coverage threshold was set as 50 reads).

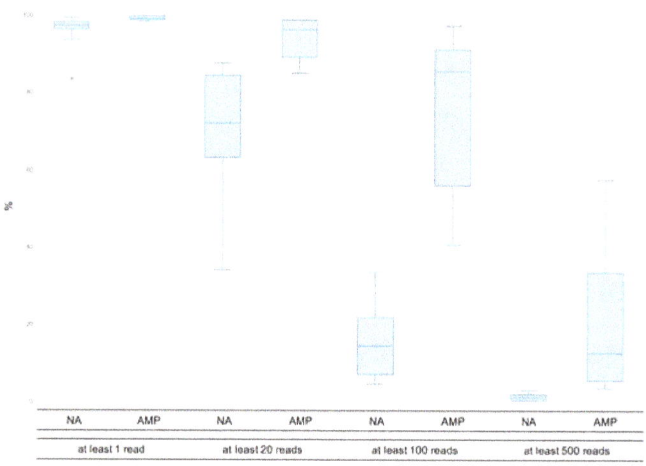

Figure 2. The minimal coverage of at least one read was observed for over 99% of all amplicons across the amplified libraries (AMP), whereas for unamplified libraries (NA), the value was 96%. At least 20 reads were reached by little less, namely 94% of the AMP libraries, while across the NA libraries, it was noticeably less, namely 70% of the amplicons. Over 75% of the amplicons across the AMP libraries had coverage of at least 100 reads, but less than 20% of the NA libraries reached the coverage. The 500X coverage was observed for almost 20% of the amplified libraries and 1% of the NA libraries.

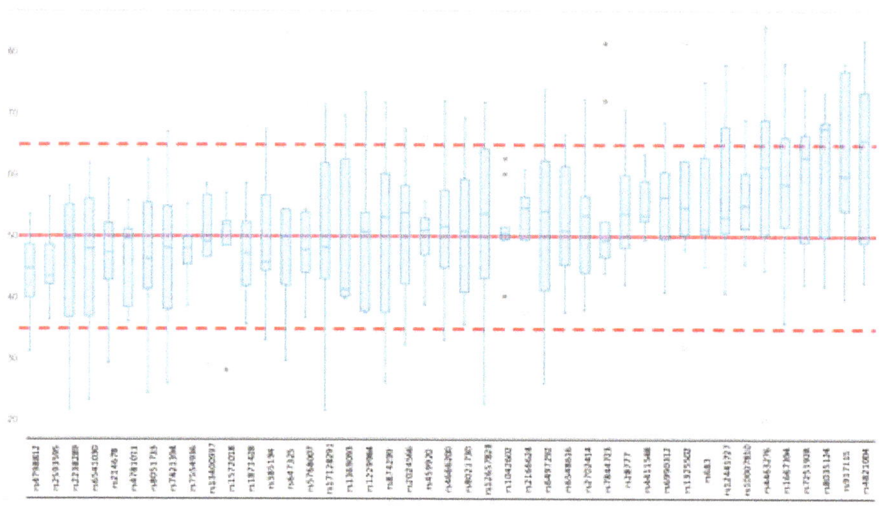

Figure 3. Allele frequency distribution across 43 heterozygote loci (rs1470608 and rs10756819 not called due to low coverage) among ten single-cell based libraries

(AMP); red lines correspond to expected (%50), upper (%65), and lower (%35) thresholds.

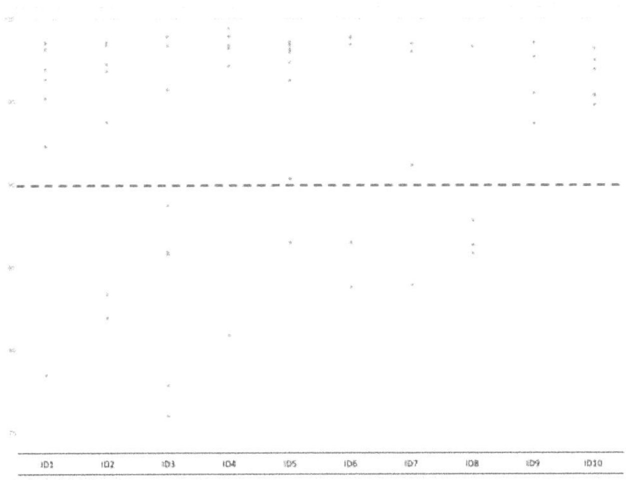

Figure 4. Allele frequency distribution across 155 homozygote loci among ten single-cell-based libraries (AMP). Grey dots represent the homozygotic loci with less than the expected %100 frequency, which is marked with blue lines. Red line corresponds to lowest (%90) accepted threshold. The minimum number of loci per sample below threshold was one; the maximum was five.

Due to a significant number of missing loci observed among the NA libraries, the allele frequency was estimated for the heterozygote loci (except for rs1470608 and rs10756819, which were not called due to low coverage) only among the AMP samples. The mean value per locus varied between 43% and 63%, and the detailed allele distribution across the loci is presented in **Figure 3**. For most of the markers, the mean heterozygote balance stayed between the accepted threshold of 35% and 65%, and only three markers (rs8035124, rs917115, rs4821004) showed imbalance. The average allele frequency for the remaining expected homozygotic loci was 100%, but for each sample, loci with frequencies below that were observed (**Figure 4**). The few loci per sample that did not meet the thresholds were marked as imbalanced, which did not result in an incorrect genotype being called.

Due to the special materials used, which were single cells, sequencing errors could be expected, like non-specific base misincorporation. When reviewing the genotypes in Converge, unexpected nucleotides were detected for two loci, namely rs7722456 (reference: C, variant: T, observed: A with maximum 20% frequency, resulting in incorrect genotyping) and rs8113143 (reference: C, variant: A, observed: G with maximum 3% frequency, discarded). The incorrect nucleotides were not observed after running the variantCaller plugin on Torrent Suite™ Software (TSS) 5.10.1 and when reviewing the raw data with IGV 2.7 (Integrative Genomics Viewer). Incorrect genotypes called in rs7722456 were discussed before [**28**] and can be explained by the homopolymer region flanking the SNP (as well as for the other SNP discussed, rs8113143). The genotypes for rs7722456 in Converge were called manually.

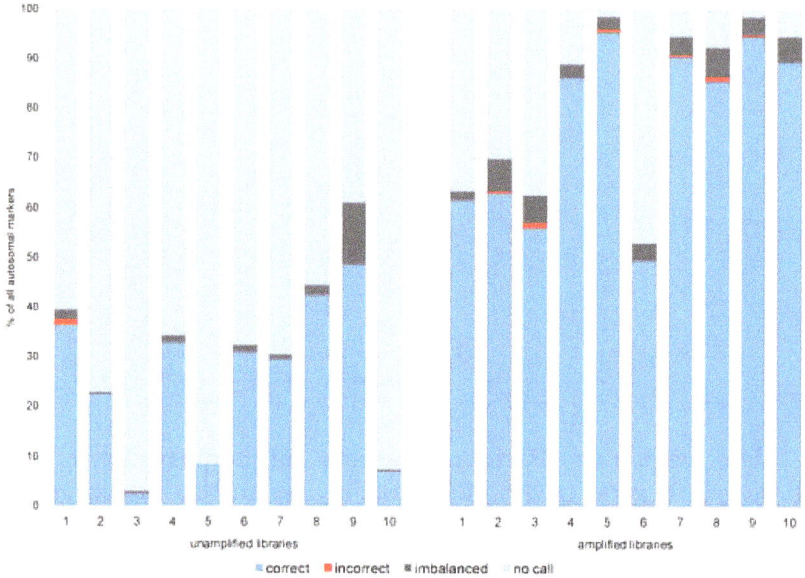

Figure 5. Genotyping rate per each sample across all 200 markers for unamplified and amplified single-cell-based libraries.

Previously described heterozygote balance and lowered coverage thresholds (50×) were used to call genotypes in Converge using the HIDGenotyper-2.2 plugin. For NA libraries, only between 2,5% and 48,5% (avg. 26%) genotypes across all 200 autosomal loci met all the thresholds, and for the AMP samples, the range was between 49,5% and 96% (avg. 78%) (**Figure 5**). Among the dropouts, most were loci with low coverage (78-185 loci for NA and 3-94 loci for AMP), and the remaining were imbalanced (max. 25 per sample for NA and 13 per sample for AMP). Across all samples, no genotype was called for one marker (rs10512572) despite passing the coverage threshold. After running the variantCaller plugin, the genotyping for this locus also failed. Upon reviewing the sequencing data in IGV 2.7, it was observed that leftover primer reads were misaligned, resulting in no call. Genotypes for this locus were called manually. Single incorrect genotypes were observed for both groups, all due to allelic dropouts.

3.2.2. Data Interpretation–"Basic" and "Conservative" Models

Due to a significant number of missing loci observed among NA libraries (the Fisher exact test statistic value is < 0.00001), only the profiles from the AMP libraries were submitted for further interpretation. For the "basic" and "conservative" models, consensus profiles from a maximum of 41 phenotypes and 145 ancestry SNPs were generated as described in **Section 2.4.1** and **Section 2.4.2** All single cells had the same genotype called for 13 out of 41 phenotype-associated markers (concordant with the reference), while for ancestry, there were 50 out of 145 markers (**Figure 6**). Only three incorrect genotypes were observed across all the SNPs, all once per marker, which did not affect the prediction.

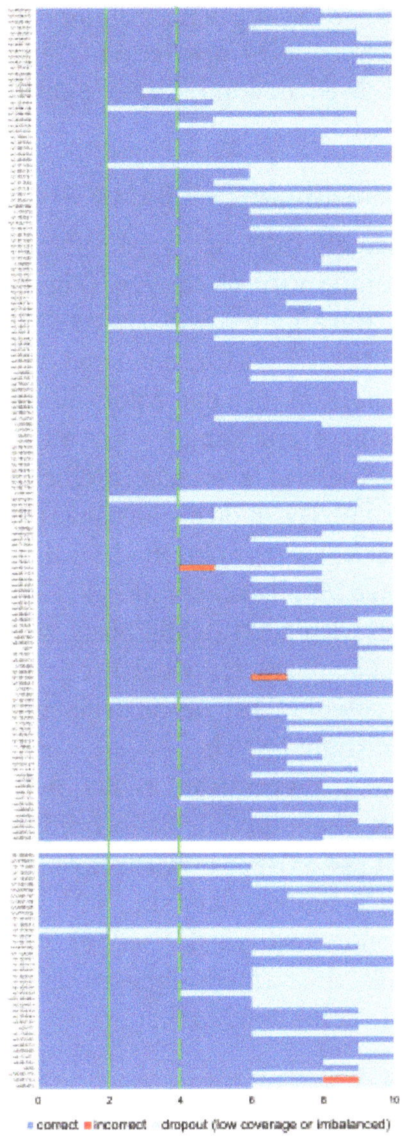

Figure 6. Genotyping rate per each locus used for ancestry prediction (max. 145 SNPs, upper part) and phenotype prediction (max. 41 SNPs, lower part). The x-axis shows the number of single cells with correct, incorrect or no genotype. Solid line: "basic" profile; dotted line: "conservative" profile.

The final "basic" profile consisted of 39 phenotype SNPs (rs1470608 and rs10756819 not called due to low coverage) and 145 ancestry SNPs, which resulted in the same *p*-values for the phenotype prediction and the same admixture (calculated by Converge) as the cell groups (**Table 1**). The "conservative" model required a genotype to be called at least four times across all the single cells, which left 38 phenotypes and 139 ancestry markers used in the final profile. The one additional missing loci among the phenotype markers, namely rs1545397 (low coverage), caused an insignificant change in the *p*-values obtained for the skin color but overall did not affect the final phenotype prediction (**Table 1**). The ancestry analysis performed by Converge showed an admixture of Southwest Asia and Europe with a proportion of ca. 60/40%.

3.2.3. Data Interpretation-Single-Cell Based Predictions

From the ten sequenced single cells, six were selected for tertiary analysis due to their high genotype calling rate of ca. 90% (**Figure 7**). All profiles used for the phenotype prediction were partial with 33–38 SNPs; therefore, the obtained predictions were reported together with calculated AUC loss as recommended by prediction models [**29**] (**Table 2**). For eye color, all cells resulted in the correct brown color prediction, where one sample showed a slightly different *p*-value than expected. The correctness of eye color predictions is strongly dependent on the rs12913832 marker [**30,31**], and this locus was correctly genotyped among all the single cells. For hair color, the obtained *p*-values were identical or almost identical as for the reference for all except one cell. For the sample named ID4, the highest *p*-value was brown with 0.521, while black was 0.465. The profile missed the rs16891982 marker (imbalanced), for which the correct CC genotype is strongly associated with black hair color [**32,33**]. Despite the difference, the final prediction was black hair color for all samples. The biggest variance was observed for skin color due to several markers relevant for prediction missing. Furthermore, skin color is the most complex phenotypic trait to predict, and incomplete profiles make it even more difficult [**34,35**]. For only three cells, the final prediction would be the same as the reference, namely dark skin with 70% probability. For samples ID4, ID7, and ID10, the *p*-value for dark skin was only around 0.5, and for the last two samples (ID7 and ID10), the second-highest was dark to black, which would incorrectly suggest the skin color to be darker.

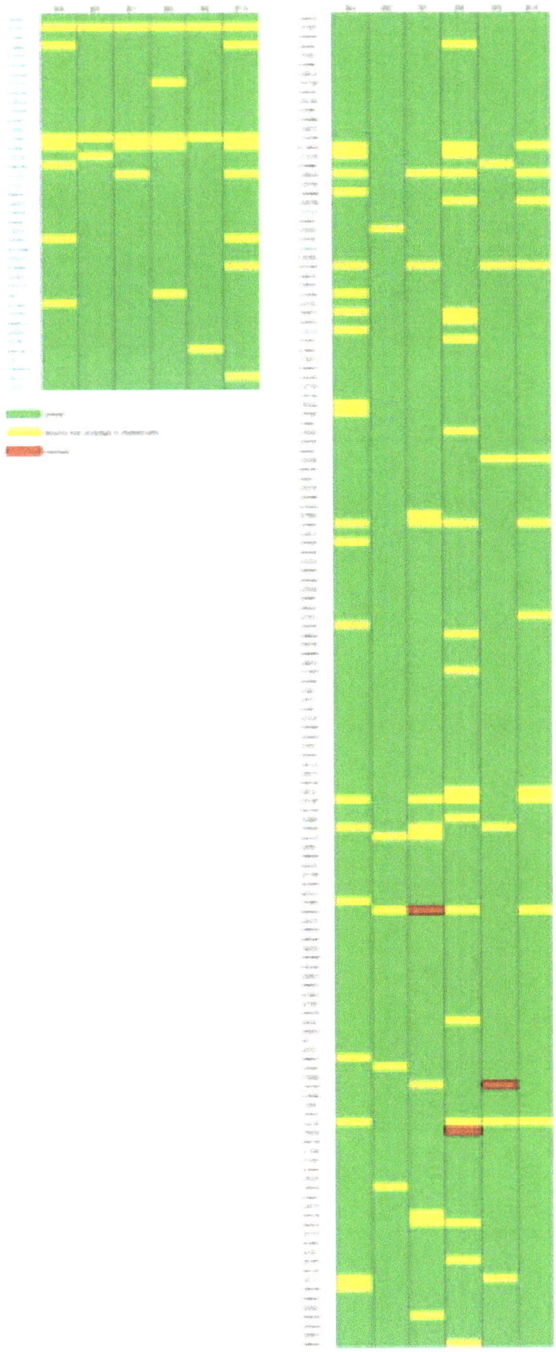

Figure 7. Genotyping summary for the selected single cells (with calling rate of at least %90); left–41 SNPs used for the phenotype prediction, right–145 SNPs used for ancestry prediction.

Table 2. Phenotype and ancestry prediction summary for selected single cells (with calling rate of at least %90). *p*-values and admixture proportions were compared with the results obtained from reference (ref.) sample (*italic* indicates identical values; * indicates different values that affected the interpretation of the results). AUC-Area Under Curve.

		Ref.	Single Cells					AUC Loss					
		ID4	ID5	ID7	ID8	ID9	ID10	ID4	ID5	ID7	ID8	ID9	ID10
Eye color	blue	**0.000**	*0.000*						0.003	0.000			
	inter	**0.002**	0.012	*0.002*					0.012	0.000			
	brown	**0.998**	0.988	*0.998*					0.003	0.000			
Hair color and shade	blond	**0.003**	0.014	*0.003*					0.006	0.000			0.004
	brown	**0.337**	0.521 *	*0.337*		0.378	*0.337*	0.343	0.007	0.000			0.003
	red	**0.000**	*0.000*						0.028	0.000			0.027
	black	**0.661**	0.465 *	*0.661*		0.619	*0.661*	0.654	0.002	0.000	0.001	0.000	0.001
	light	**0.005**	0.018	*0.005*		0.003	*0.005*		0.001	0.000			
	dark	**0.995**	0.982	*0.995*		0.997	*0.995*		0.001	0.000			
Skin color	very pale	**0.001**	0.003	0.000	0.000	*0.001*			0.012	0.004	0.002	0.003	0.002	0.002
	pale	**0.008**	0.030	0.004	0.005	0.007	0.009	0.005	0.007	0.015	0.004	0.005	0.004	0.011
	inter	**0.275**	0.418 *	0.153	0.172	0.252	0.278	0.187	0.003	0.012	0.001	0.002	0.001	0.008
	dark	**0.696**	0.520 *	0.700	0.509 *	0.710	0.689	0.430 *	0.002	0.000	0.001	0.005	0.001	0.000
	dark to black	**0.021**	*0.029*	0.143	0.315 *	0.030	0.024	0.377*	0.001	0.001	0.002	0.001	0.000	0.002

Admixture

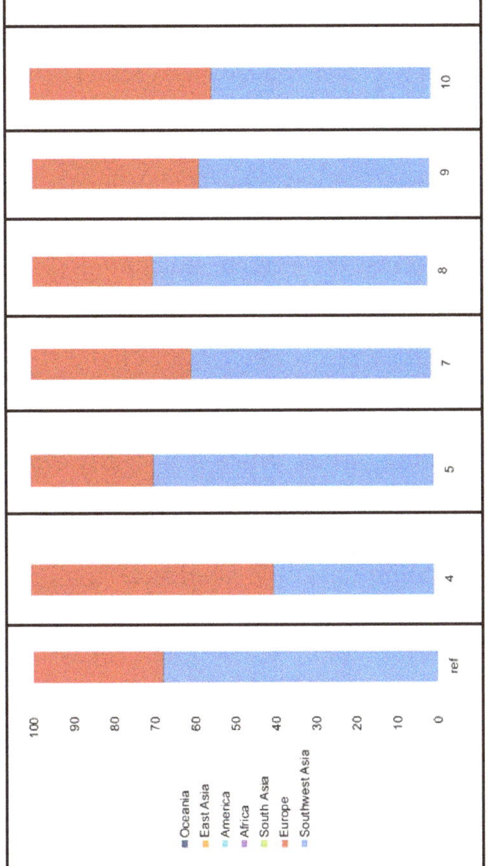
Oceania
East Asia
America
Africa
South Asia
Europe
Southwest Asia
100
90
80
70
60
50
40
30
20
10
0
ref
4
5
7
8
9
10

The profiles used for ancestry prediction were also incomplete, which compromised 120 and 140 SNPs. For all the samples, the analysis performed by Converge revealed an admixture of Southwest Asia and Europe, and for all except one sample, the proportion was close to what was detected for the reference. The result obtained for sample ID4 suggested more European than Southwest Asian ancestry and was caused by missing genotypes for rs16891982 and rs2196051, which are strongly associated with South Asian, not European, populations.

4. DISCUSSION

Mixture deconvolution is one of the most engaging topics in forensics. Over previous years different methods have been developed to assist the process, both prior to [**3,4,5,6,7,8,9,10**] and after, DNA typing of a mixed trace [**36,37,38**]. The DEPArray System, which uses a dielectrophoresis grid to isolate single cells, is a promising cell separation technology for forensic analysis. It was demonstrated that the DNA from the collected cells can successfully be processed with STR amplification kits and detected using capillary electrophoresis [**11,17,18,39**]. Single cells are considered ltDNA (low-template DNA), and handling such material is not uncommon in forensics, where specialists are often faced with low-quantity and low-quality DNA. Since such samples are treated with special lab protocols (e.g., increased number of PCR cycles), it makes them prone to stochastic effects (drop-ins and dropouts) and increased stutter ratios, the interpretation of ltDNA profiles is a common topic of discussion [**40,41,42,43,44,45**]. An alternative to traditional fragment analysis is to sequence the bi-allelic SNPs, which are less prone to artifacts and are characterized by smaller amplicons. The analysis of SNP markers is becoming more and more popular in forensics, especially thanks to the rapid development of massively parallel sequencing (MPS), which is not only more sensitive, but also features the use of high multiplex panels consisting of hundreds of markers. Successful single-cell-based STR experiments suggest additional applications in other forensic investigations. A combination of single-cell analysis and SNP sequencing could be especially useful for forensic DNA phenotyping and serve as an alternative solution for the analysis of mixed samples, for which the current interpretation of the phenotype and ancestry predictions can be complicated or impossible.

Panels containing SNPs associated with phenotype [**22,46,47,48**] and ancestry [**21,49,50,51**] have been introduced, and recently assays combining those have been published [**23,24,52**]. Forensic DNA phenotyping is now legislated in different European countries [**20**], and it is expected that police investigators will seek expert opinions on samples of interest. The desirable interpretation of phenotypic features should be based on traces derived from single individuals because mixed samples might not provide reliable results. The collection of single-source traces may not always be possible; therefore, published studies on phenotypic/ancestry SNPs should also discuss the analysis of mixtures. The first and basic indication of a mixed SNP profile is an overall increase in the number of heterozygotic loci [**21,53**]. This statement cannot easily be applied for individuals originating from admixed populations (or having biparental co-ancestry) where the number of heterozygotes is naturally higher. However, as presented in the study by Eduardoff et al. [**21**], the allele read frequencies (ARF) differ between unmixed and mixed samples. Mixtures exhibit a higher number of heterozygotes not meeting the balance threshold. Although the authors were able to successfully detect the major and minor contributors for all tested mixture ratios (from 1:1 to 1:9), they recommend careful data analysis, especially for extreme mixtures. The authors of the HirisPlex-S MPS validation paper [**22**] designed a calculator to aid an interpretation of 2-person mixtures, which is based on known major/minor contributor ratios obtained from STR typing done prior to SNP analysis. The ratios are used to separate read counts obtained during sequencing. The approach was tested on mixtures with different ratios (from 1:1 to 1:9) and using contributors with distinguishable genotypes and phenotypes. The exercise shows that the variants for 28 of the 41 SNPs included in the HirisPlex-S were successfully separated into individual profiles. The validation study of the comprehensive assay combining markers for appearance and ancestry predictions from the VISAGE consortium [**23**] also mentions mixture deconvolution, similarly of two contributors with different phenotypes and biogeographical ancestries, where the analysis was based on allele frequencies. A publication from Ralf and Kayser [**54**] presents the first actual crime scene trace of a mixed source submitted for phenotype prediction. The 2-person mixture sample of interest had almost 500 pg DNA. The phenotyping was done using the MPS-based HirisPlex-S Panel, and the sequencing included both the trace and reference material obtained from the victim. The authors were able to extract the probable suspect's genotype for all the SNP markers (one with an alternative genotype considered) and obtained distinguishable phenotypes with high probability values for all predicted traits. They also performed mixture deconvolution by using the aforementioned tool by Breslin et al. [**22**] and revealed concordance for 36/41 tested SNPs. The remaining genotypes were interpreted differently using both approaches. Overall, discussed studies show that the interpretation of the mixed samples must be made with caution, and the approach should be further evaluated, especially for low input samples or mixtures of individuals with indistinguishable phenotype/ancestry.

The advantage of implementing mixture deconvolution methods prior to the sequencing of the phenotypic/ancestry SNPs is being able to perform the predictions based on single-source profiles. In the case of single cells, the reliability of those predictions will depend on the quality of the obtained sequencing results. As already discussed before, the ltDNA requires careful and cautious analysis, and even though SNPs are less prone to stochastic effects, the following issues are expected when sequencing low template material: allelic drop-ins and dropouts, locus dropouts, imbalanced genotypes, and increased base misincorporation rates. The sensitivity studies of the large SNP assay suggest that the threshold for two of the most popular MPS platforms used in forensics, namely MiSeqFGx System and Ion S5, lies around 100 pg DNA [**22**,**23**,**24**,**52**]. In all studies, inputs as low as 5–7 pg were assessed, which roughly corresponds to one human cell. As expected, with such a low input, the obtained results consisted largely of no calls, drops in/out, and incorrect calls. Different forensically relevant SNP panels were tested for their efficacy in typing ltDNA, both using the golden forensic detection standard, namely capillary electrophoresis [**55**] and MPS [**56**,**57**]. Assuming that the DNA input is low, additional changes might be introduced to enhance the standard protocols to boost the coverage. The most common adjustment is to significantly increase the number of PCR cycles used to amplify the targets. The expected positive effect is a higher coverage, but some issues can also occur, like allelic drops in/out causing imbalances or incorrect calls [**55**,**56**]. When working with MPS and the AmpliSeq pipeline library, amplification might be applied in order to improve the read depth. The libraries which yielded low quants might be amplified with a high-fidelity PCR supermix. This step in the library preparation was previously implemented when using the AmpliSeq workflow in other molecular fields [**58**,**59**,**60**,**61**], but it does not seem to be a common practice in forensics. A study by Meiklejohn and Robertson mentions the amplification of libraries, but only in order to quantify them after using Qubit and Bioanalyzer [**62**]. The potential effect of the library amplification on the low-template samples is presented by Turchi et al. [**63**]. The authors performed a comprehensive validation of the Precision ID Identity Panel by Thermo Fisher Scientific by sequencing challenging forensic samples on the Ion PGM. The libraries from their study with less than 30 pM were amplified and again purified, similarly to our single cell-based libraries discussed in this paper. It was observed that the library amplification step helped to obtain a higher level of repeatability and improved the values of sequencing parameters. Our study shows that introducing this step resulted in over five times higher total coverage of the single cell-based libraries. The number of no-calls per sample decreased from an average of 70% for the unamplified libraries to 18% for the amplified ones. The highest number of imbalanced genotypes per sample was also half as low, and a maximum of two incorrect genotypes per sample were observed after the libraries were amplified.

The goal of sequencing a trace submitted for forensic DNA phenotyping is to obtain reliable genotypes which will be used for predictions. Incorrect phenotype and ancestry predictions might be caused by incorrect genotypes [**64**] and incomplete profiles [**29**,**65**]. The authors of the HirisPlex-S Panel explain that different genotypes missed from the profile will have different influences on the prediction model and that the incomplete data has to be reported as a probability of each trait predicted together with AUC (area under curve) accuracy loss [**22,29,35**]. We tested an approach of using incomplete single profiles generated by single cells to observe how they will affect the predictions. The calculated AUC loss would not be considered significant, but the missing genotypes resulted in *p*-values notably different than for the reference samples, causing even incorrect predictions for the skin color. In their paper, Cheung et al. tested how different percentages of genotypes missing from a profile will affect the ancestry prediction when analyzed with different classifiers [**65**]. Their study shows that an increasing number of missing markers can affect the predictions, especially in the case of admixed samples. Our method to perform the ancestry analysis was not mentioned in the paper, but our results show similar observations. We used bootstrapping admixture analysis based on a maximum likelihood approach to predict the most likely admixture proportions across seven root populations. The SNP profiles were bootstrapped using a varying number of replications, where each replication selected a random subset of SNPs to capture uncertainty in the predictions [**27**]. For the analysis of the single profiles obtained from the single cells, the number of SNPs available for bootstrapping differed between 123 and 139 out of 145. However, the number of markers missing had less impact than which markers were missing. For samples ID4 and ID8, almost the same number of SNPs were used for the analysis (123 and 125 respectively), but the prediction outcome was an admixture of Southwest Asia and Europe with a difference of 30%. For both phenotype and ancestry, the most accurate predictions in comparison to the reference sample were obtained for the "basic" consensus profile built from single cell-based libraries with the highest genotyping rate.

5. CONCISE SUMMARY

This study is a proof of concept demonstrating that single-cell sequencing can obtain the correct phenotype (HIrisPlex-S) and ancestry prediction, and therefore, can be considered a viable mixture deconvolution method for challenging samples submitted for forensic DNA phenotyping. The presented workflow combined single-cell separation with the DEPArray™ NxT System and sequencing of phenotype and ancestry-associated SNPs with the Ion S5 platform. The DNA typing was done using a previously validated custom Ion AmpliSeq™ PhenoTrivium Panel. The study was based on testing groups of 20, 10, and 5 cells, for which the results were comparable to DNA sensitivity tests performed in a previous validation study, showing the potential of the combined workflow. The number of tested single cells was 20, where half of the libraries were 'rescued' with library amplification prior to sequencing. Introducing this step helped to significantly increase the total coverage obtained for the 10 amplified libraries, which resulted in recovering more genotypes. More than half of the rescued single cells had a genotyping rate close to, or greater than, 90%. Additionally, different interpretation approaches were evaluated. The predictions for single cells were based on a "basic" and a "conservative" consensus profile (genotype called two or four times, respectively) and on single profiles themselves. The most reliable predictions were obtained when using the "basic" consensus profile, for which almost no dropouts were observed. The results suggest that collecting single cells from a mixed sample prior to forensic DNA phenotyping can be used as an alternative way of performing phenotype and ancestry predictions of the mixture's contributors. This approach will be further evaluated by sequencing mock mixtures with different ratios using contributors not only with distinguishable phenotype/ancestry but also similar ones. Additional changes to the workflow will also be evaluated, like increasing the number of PCR cycles prior to amplification, adjusting library inputs, and improving the genotyping.

6. AT A GLANCE

The discipline of single-cell sequencing is rapidly advancing and holds great promise, although its application in forensics is not yet widespread. The primary rationale for implementing this technique in forensics is to enhance the process of separating mixed samples, particularly when the trail contains cells of the same type. Effective research showcases the capacity to examine a combination by isolating individual cells and acquiring CE-based STR profiles. This suggests that the procedure could be applied to other forensic investigations, such as forensic DNA phenotyping, but it may not be entirely advisable to utilize it with mixed evidence. In this investigation, we obtained autopsy blood samples from a single source. The white blood cells in the samples were initially dyed and then separated using the DEPArray™ N×T System. Assemblages of 20, 10, and 5 cellular units, together with 20 individual cellular units, were gathered and subjected to DNA extraction. The libraries were created using the Ion AmpliSeq™ PhenoTrivium Panel, which consists of SNP-markers related with both phenotype (HIrisPlex-S: eye, hair, and skin color) and ancestry. Before sequencing, half of the libraries based on single cells underwent further amplification and purification to enhance the concentrations of the libraries. The examination of ancestry and phenotype yielded highly consistent profiles, accurately predicting the cell groupings as well as the 10 re-amplified single-cell libraries. Our findings indicate that the use of single cell sequencing has potential as an effective method for analyzing mixed traces in forensic DNA phenotyping.

REFERENCES

1. Macaulay, I.C.; Voet, T. Single Cell Genomics: Advances and Future Perspectives. *PLoS Genet.* **2014**, *10*, e1004126. **[Google Scholar] [CrossRef] [Green Version]**
2. Seven Technologies to Watch in 2021. Available online: **https://www.nature.com/articles/d41586-021-00191-z** (accessed on 26 July 2021).
3. Elliott, K.; Hill, D.S.; Lambert, D.; Burroughes, T.R.; Gill, P. Use of laser microdissection greatly improves the recovery of DNA from sperm on microscope slides. *Forensic Sci. Int.* **2003**, *137*, 28–36. **[Google Scholar] [CrossRef]**
4. Verdon, T.J.; Mitchell, R.J.; Chen, W.; Xiao, K.; van Oorschot, R.A.H. FACS separation of non-compromised forensically relevant biological mixtures. *Forensic Sci. Int. Genet.* **2015**, *14*, 194–200. **[Google Scholar] [CrossRef]**
5. Stokes, N.A.; Stanciu, C.E.; Brocato, E.R.; Ehrhardt, C.J.; Greenspoon, S.A. Simplification of complex DNA profiles using front end cell separation and probabilistic modeling. *Forensic Sci. Int. Genet.* **2018**, *36*, 205–212. **[Google Scholar] [CrossRef] [PubMed]**
6. Dean, L.; Kwon, Y.J.; Philpott, M.K.; Stanciu, C.E.; Seashols-Williams, S.J.; Dawson Cruz, T.; Sturgill, J.; Ehrhardt, C.J. Separation of uncompromised whole blood mixtures for single source STR profiling using fluorescently-labeled human leukocyte antigen (HLA) probes and fluorescence activated cell sorting (FACS). *Forensic Sci. Int. Genet.* **2015**, *17*, 8–16. **[Google Scholar] [CrossRef]**
7. Auka, N.; Valle, M.; Cox, B.D.; Wilkerson, P.D.; Dawson Cruz, T.; Reiner, J.E.; Seashols-Williams, S.J. Optical tweezers as an effective tool for spermatozoa isolation from mixed forensic samples. *PLoS ONE* **2019**, *14*, e0211810. **[Google Scholar] [CrossRef]**
8. Horsman, K.M.; Barker, S.L.R.; Ferrance, J.P.; Forrest, K.A.; Koen, K.A.; Landers, J.P. Separation of sperm and epithelial cells in a microfabricated device: Potential application to forensic analysis of sexual assault evidence. *Anal. Chem.* **2005**, *77*, 742–749. **[Google Scholar] [CrossRef]**
9. Zhao, X.-C.; Wang, L.; Sun, J.; Jiang, B.-W.; Zhang, E.-L.; Ye, J. Isolating Sperm from Cell Mixtures Using Magnetic Beads Coupled with an Anti-PH-20 Antibody for Forensic DNA Analysis. *PLoS ONE* **2016**, *11*, e0159401. **[Google Scholar] [CrossRef] [Green Version]**
10. Huffman, K.; Hanson, E.; Ballantyne, J. Recovery of single source DNA profiles from mixtures by direct single cell subsampling and simplified micromanipulation. *Sci. Justice* **2021**, *61*, 13–25. **[Google Scholar] [CrossRef]**
11. Anslinger, K.; Bayer, B.; Graw, M. Deconvolution of blood-blood mixtures using DEPArrayTM separated single cell STR profiling. *Rechtsmedizin* **2019**, *29*, 30–40. **[Google Scholar] [CrossRef] [Green Version]**

12. Philippron, A.; Depypere, L.; Oeyen, S.; Laere, B.D.; Vandeputte, C.; Nafteux, P.; de Preter, K.; Pattyn, P. Evaluation of a marker independent isolation method for circulating tumor cells in esophageal adenocarcinoma. *PLoS ONE* **2021**, *16*, e0251052. **[Google Scholar] [CrossRef]**

13. Salvianti, F.; Gelmini, S.; Mancini, I.; Pazzagli, M.; Pillozzi, S.; Giommoni, E.; Brugia, M.; di Costanzo, F.; Galardi, F.; de Luca, F.; et al. Circulating tumour cells and cell-free DNA as a prognostic factor in metastatic colorectal cancer: The OMITERC prospective study. *Br. J. Cancer* **2021**, *125*, 94–100. **[Google Scholar] [CrossRef]**

14. Rossi, T.; Gallerani, G.; Angeli, D.; Cocchi, C.; Bandini, E.; Fici, P.; Gaudio, M.; Martinelli, G.; Rocca, A.; Maltoni, R.; et al. Single-Cell NGS-Based Analysis of Copy Number Alterations Reveals New Insights in Circulating Tumor Cells Persistence in Early-Stage Breast Cancer. *Cancers* **2020**, *12*, 2490. **[Google Scholar] [CrossRef]**

15. Boyer, M.; Cayrefourcq, L.; Garima, F.; Foulongne, V.; Dereure, O.; Alix-Panabières, C. Circulating Tumor Cell Detection and Polyomavirus Status in Merkel Cell Carcinoma. *Sci. Rep.* **2020**, *10*, 1612. **[Google Scholar] [CrossRef] [Green Version]**

16. Meloni, V.; Lombardi, L.; Aversa, R.; Barni, F.; Berti, A. Optimization of STR amplification down to single cell after DEPArrayTM isolation. *Forensic Sci. Int. Genet. Suppl. Ser.* **2019**, *7*, 711–713. **[Google Scholar] [CrossRef]**

17. Williamson, V.R.; Laris, T.M.; Romano, R.; Marciano, M.A. Enhanced DNA mixture deconvolution of sexual offense samples using the DEPArray™ system. *Forensic Sci. Int. Genet.* **2018**, *34*, 265–276. **[Google Scholar] [CrossRef] [PubMed] [Green Version]**

18. Watkins, D.R.L.; Myers, D.; Xavier, H.E.; Marciano, M.A. Revisiting single cell analysis in forensic science. *Sci. Rep.* **2021**, *11*, 7054. **[Google Scholar] [CrossRef]**

19. Fontana, F.; Rapone, C.; Bregola, G.; Aversa, R.; de Meo, A.; Signorini, G.; Sergio, M.; Ferrarini, A.; Lanzellotto, R.; Medoro, G.; et al. Isolation and genetic analysis of pure cells from forensic biological mixtures: The precision of a digital approach. *Forensic Sci. Int. Genet.* **2017**, *29*, 225–241. **[Google Scholar] [CrossRef] [Green Version]**

20. Schneider, P.M.; Prainsack, B.; Kayser, M. The use of forensic DNA phenotyping in predicting appearance and biogeographic ancestry. *Dtsch. Arztebl. Int.* **2019**, *116*, 873–880. **[Google Scholar] [CrossRef]**

21. Eduardoff, M.; Gross, T.E.; Santos, C.; de la Puente, M.; Ballard, D.; Strobl, C.; Børsting, C.; Morling, N.; Fusco, L.; Hussing, C.; et al. Inter-laboratory evaluation of the EUROFORGEN Global ancestry-informative SNP panel by massively parallel sequencing using the Ion PGM™. *Forensic Sci. Int. Genet.* **2016**, *23*, 178–189. **[Google Scholar] [CrossRef] [Green Version]**

22. Breslin, K.; Wills, B.; Ralf, A.; Garcia, M.V.; Kukla-Bartoszek, M.; Pospiech, E.; Freire-Aradas, A.; Xavier, C.; Ingold, S.; De La Puente, M.; et al. HIrisPlex-S system for eye, hair, and skin color prediction from DNA: Massively parallel sequencing solutions for two common forensically used platforms. *Forensic Sci. Int. Genet.* **2019**, *43*, 102152. **[Google Scholar] [CrossRef]**

23. Xavier, C.; De La Puente, M.; Mosquera-Miguel, A.; Freire-Aradas, A.; Kalamara, V.; Vidaki, A.; Gross, T.E.; Revoir, A.; Pośpiech, E.; Kartasińska, E.; et al. Development and validation of the VISAGE AmpliSeq basic tool to predict appearance and ancestry from DNA. *Forensic Sci. Int. Genet.* **2020**, *48*, 102336. **[Google Scholar] [CrossRef]**

24. Diepenbroek, M.; Bayer, B.; Schwender, K.; Schiller, R.; Lim, J.; Lagacé, R.; Anslinger, K. Evaluation of the Ion AmpliSeq™ PhenoTrivium Panel: MPS-Based Assay for Ancestry and Phenotype Predictions Challenged by Casework Samples. *Genes* **2020**, *11*, 1398. **[Google Scholar] [CrossRef] [PubMed]**

25. Precision ID SNP Panels with the HID Ion S5™/HID Ion GeneStudio™ S5 System Application Guide. Pub. No. MAN0017767. Available online: **https://assets.thermofisher.com/TFS-Assets/LSG/manuals/MAN0017767_PrecisionID_SNP_Panels_S5_UG.pdf** (accessed on 1 July 2021).

26. Thorvaldsdóttir, H.; Robinson, J.T.; Mesirov, J.P. Integrative Genomics Viewer (IGV): High-performance genomics data visualization and exploration. *Brief. Bioinform.* **2013**, *14*, 178–192. **[Google Scholar] [CrossRef] [Green Version]**

27. Wootton, S.; Vijaychander, S.; Hasegawa, R.; Deng, J.; Lackey, A.; Gabriel, M.; Lagacé, R.; Lim, J. Analytical Improvements in Biogeographic Ancestry Inference. Presented at the 28th Congress of the International Society for Forensic Genetics; Available online: **https://www.thermofisher.com/de/de/home/products-and-services/promotions/isfg.html** (accessed on 26 July 2021).

28. Pereira, V.; Mogensen, H.S.; Børsting, C.; Morling, N. Evaluation of the Precision ID Ancestry Panel for crime case work: A SNP typing assay developed for typing of 165 ancestral informative markers. *Forensic Sci. Int. Genet.* **2017**, *28*, 138–145. **[Google Scholar] [CrossRef]**

29. Liu, F.; van Duijn, K.; Vingerling, J.R.; Hofman, A.; Uitterlinden, A.G.; Cecile, A.; Janssens, J.W.; Kayser, M. Eye color and the prediction of complex phenotypes from genotypes. *Curr. Biol.* **2009**, *19*, 192–193. **[Google Scholar] [CrossRef] [PubMed] [Green Version]**

30. Meyer, O.S.; Lunn, M.M.B.; Garcia, S.L.; Kjærbye, A.B.; Morling, N.; Børsting, C.; Andersen, J.D. Association between brown eye colour in rs12913832:GG individuals and SNPs in TYR, TYRP1, and SLC24A4. *PLoS ONE* **2020**, *15*, e0239131. **[Google Scholar] [CrossRef]**

31. Meyer, O.S.; Salvo, N.M.; Kjærbye, A.; Kjersem, M.; Andersen, M.M.; Sørensen, E.; Ullum, H.; Janssen, K.; Morling, N.; Børstin, C.; et al. Prediction of Eye Colour in Scandinavians Using the EyeColour11 (EC11) SNP Set. *Genes* **2021**, *12*, 821. **[Google Scholar] [CrossRef]**

32. Branicki, W.; Brudnik, U.; Draus-Barini, J.; Kupiec, T.; Wojas-Pelc, A. Association of the SLC45A2 gene with physiological human hair colour variation. *J. Hum. Genet.* **2008**, *53*, 966–971. **[Google Scholar] [CrossRef] [Green Version]**

33. Lin, B.D.; Mbarek, H.; Willemsen, G.; Dolan, C.V.; Fedko, I.O.; Abdellaoui, A. Heritability and Genome-Wide Association Studies for Hair Color in a Dutch Twin Family Based Sample. *Genes* **2015**, *13*, 559–576. **[Google Scholar] [CrossRef] [PubMed] [Green Version]**

34. Walsh, S.; Chaitanya, L.; Breslin, K.; Muralidharan, C.; Bronikowska, A.; Pospiech, E.; Koller, J.; Kovatsi, L.; Wollstein, A.; Branicki, W.; et al. Global skin colour prediction from DNA. *Hum. Genet.* **2017**, *136*, 847–863. **[Google Scholar] [CrossRef] [Green Version]**

35. Chaitanya, L.; Breslin, K.; Zuñiga, S.; Wirken, L.; Pośpiech, E.; Kukla-Bartoszek, M.; Sijen, T.; De Knijff, P.; Liu, F.; Branicki, W.; et al. The HIrisPlex-S system for eye, hair and skin colour prediction from DNA: Introduction and forensic developmental validation. *Forensic Sci. Int. Genet.* **2018**, *35*, 123–135. **[Google Scholar] [CrossRef] [Green Version]**

36. Benschop, C.C.; Hoogenboom, J.; Hovers, P.; Slagter, M.; Kruise, D.; Parag, R.; Steensma, K.; Slooten, K.; Nagel, J.H.; Dieltjes, P.; et al. DNAxs/DNAStatistX: Development and validation of a software suite for the data management and probabilistic interpretation of DNA profiles. *Forensic Sci. Int. Genet.* **2019**, *42*, 81–89. **[Google Scholar] [CrossRef]**

37. Bleka, O.; Storvik, G.; Gill, P. EuroForMix: An open source software based on a continuous model to evaluate STR DNA profiles from a mixture of contributors with artefacts. *Forensic Sci. Int. Genet.* **2016**, *21*, 35–44. **[Google Scholar] [CrossRef] [Green Version]**

38. Bright, J.-A.; Cheng, K.; Kerr, Z.; McGovern, C.; Kelly, H.; Moretti, T.R.; Smith, M.A.; Bieber, F.R.; Budowle, B.; Coble, M.D.; et al. STRmix™ collaborative exercise on DNA mixture interpretation. *Forensic Sci. Int. Genet.* **2019**, *40*, 1–8. **[Google Scholar] [CrossRef]**

39. Anslinger, K.; Bayer, B.; von Mariassy, D. Application of DEPArrayTM technology for the isolation of white blood cells from cell mixtures in chimerism analysis. *Rechsmedizin* **2018**, *28*, 134–137. **[Google Scholar] [CrossRef]**

40. Balding, D.J. Evaluation of mixed-source, low-template DNA profiles in forensic science. *PNAS* **2013**, *110*, 12241–12246. **[Google Scholar] [CrossRef] [Green Version]**

41. Balding, D.J.; Buckleton, J. Interpreting low template DNA profiles. *Forensic Sci. Int. Genet.* **2009**, *4*, 1–10. **[Google Scholar] [CrossRef]**

42. Kanokwongnuwut, P.; Martin, B.; Taylor, D.; Kirkbride, K.P.; Linacre, A. How many cells are required for successful DNA profiling? *Forensic Sci. Int. Genet.* **2021**, *51*, 102453. **[Google Scholar] [CrossRef]**

43. Haned, H.; Slooten, K.; Gill, P. Exploratory data analysis for the interpretation of low template DNA mixtures. *Forensic Sci. Int. Genet.* **2012**, *6*, 762–774. **[Google Scholar] [CrossRef]**

44. Pfeifer, C.M.; Klein-Unseld, R.; Klintschar, M.; Wiegand, P. Comparison of different interpretation strategies for low template DNA mixtures. *Forensic Sci. Int. Genet.* **2012**, *6*, 716–722. **[Google Scholar] [CrossRef]**

45. Benschop, C.C.G.; Haned, H.; de Blaeij, T.J.P.; Meulenbroek, A.J.; Sijen, T. Assessment of mock cases involving complex low template DNA mixtures: A descriptive study. *Forensic Sci. Int. Genet.* **2012**, *6*, 697–707. **[Google Scholar] [CrossRef]**

46. Pośpiech, E.; Kukla-Bartoszek, M.; Karłowska-Pik, J.; Zieliński, P.; Woźniak, A.; Boroń, M.; Dąbrowski, M.; Zubańska, M.; Jarosz, A.; Grzybowski, T.; et al. Exploring the possibility of predicting human head hair greying from DNA using whole-exome and targeted NGS data. *BMC Genom.* **2020**, *21*, 538. **[Google Scholar] [CrossRef] [PubMed]**

47. Kukla-Bartoszek, M.; Pośpiech, E.; Woźniak, A.; Boroń, M.; Karłowska-Pik, J.; Teisseyre, P.; Zubańska, M.; Bronikowska, A.; Grzybowski, T.; Płoski, R.; et al. DNA-based predictive models for the presence of freckles. *Forensic Sci. Int. Genet.* **2019**, *42*, 252–259. **[Google Scholar] [CrossRef] [Green Version]**

48. Meyer, O.S.; Andersen, J.D.; Børsting, C. Presentation of the Human Pigmentation (HuPi) AmpliSeq™ custom panel. *Forensic Sci. Int. Genet. Suppl. Ser.* **2019**, *7*, 478–479. **[Google Scholar] [CrossRef]**

49. Xavier, C.; de la Puente, M.; Phillips, C.; Eduardoff, M.; Heidegger, A.; Mosquera-Miguel, A.; Freire-Aradas, A.; Lagace, R.; Wootton, S.; Power, D.; et al. Forensic evaluation of the Asia Pacific ancestry-informative MAPlex assay. *Forensic Sci. Int. Genet.* **2020**, *48*, 102344. **[Google Scholar] [CrossRef]**

50. Jia, J.; Wei, Y.-L.; Qin, C.-J.; Hu, L.; Wan, L.-H.; Li, C.-X. Developing a novel panel of genome-wide ancestry informative markers for bio-geographical ancestry estimates. *Forensic Sci. Int. Genet.* **2014**, *8*, 187–194. **[Google Scholar] [CrossRef]**

51. Pereira, R.; Phillips, C.; Pinto, N.; Santos, C.; Dos Santos, S.E.B.; Amorim, A.; Carracedo, A.; Gusmão, L. Straightforward Inference of Ancestry and Admixture Proportions through Ancestry-Informative Insertion Deletion Multiplexing. *PLoS ONE* **2012**, *7*, e29684. **[Google Scholar] [CrossRef] [Green Version]**

52. Palencia-Madrid, L.; Xavier, C.; de la Puente, M.; Hohoff, C.; Phillips, C.; Kayser, M.; Parson, W. Evaluation of the VISAGE Basic Tool for Appearance and Ancestry Prediction Using PowerSeq Chemistry on the MiSeqFGx System. *Genes* **2020**, *11*, 708. **[Google Scholar] [CrossRef]**

53. Eduardoff, M.; Santos, C.; de la Puente, M.; Gross, T.; Fondevila, M.; Strobl, C.; Sobrino, B.; Ballard, D.; Schneider, P.; Carracedo, A.; et al. Inter-laboratory evaluation of SNP-based forensic identification by massively parallel sequencing using the Ion PGM™. *Forensic Sci. Int. Genet.* **2015**, *17*, 110–121. **[Google Scholar] [CrossRef] [PubMed] [Green Version]**

54. Ralf, A.; Kayser, M. Investigative DNA analysis of two-person mixed crime scene trace in a murder case. *Forensic Sci. Int. Genet.* **2021**, *54*, 102557. **[Google Scholar] [CrossRef]**

55. Børsting, C.; Mogensen, H.S.; Morling, N. Forensic genetic SNP typing of low-template DNA and highly degraded DNA from crime case samples. *Forensic Sci. Int. Genet.* **2013**, *7*, 345–352. **[Google Scholar] [CrossRef] [Green Version]**

56. Elena, S.; Alessandro, A.; Ignazio, C.; Sharon, W.; Luigi, R.; Andrea, B. Revealing the challenges of low template DNA analysis with the prototype Ion AmpliSeq™ Identity panel v2.3 on the PGM™ Sequencer. *Forensic Sci. Int. Genet.* **2016**, *22*, 25–36. **[Google Scholar] [CrossRef] [PubMed]**

57. Butler Gettings, K.; Kiesler, K.M.; Vallone, P.M. Performance of a next generation sequencing SNP assay on degraded DNA. *Forensic Sci. Int. Genet.* **2015**, *19*, 1–9. **[Google Scholar] [CrossRef]**

58. Alessandrini, F.; Caucci, S.; Onofri, V.; Melchionda, F.; Tagliabracci, A.; Bagnarelli, P.; di Sante, L.; Turchi, C.; Menzo, S. Evaluation of the Ion AmpliSeq SARS-CoV-2 Research Panel by Massive Parallel Sequencing. *Genes* **2020**, *11*, 929. **[Google Scholar] [CrossRef] [PubMed]**

59. Siji, A.; Karthik, K.N.; Chhotusing Pardeshi, V.; Hari, P.S.; Vasudevan, A. Targeted gene panel for genetic testing of south Indian children with steroid resistant nephrotic syndrome. *BMC Med. Genet.* **2018**, *19*, 200. **[Google Scholar] [CrossRef]**

60. Kluska, A.; Balabas, A.; Paziewska, A.; Kulecka, M.; Nowakowska, D.; Mikula, M.; Ostrowski, J. New recurrent BRCA1/2 mutations in Polish patients with familial breast/ovarian cancer detected by next generation sequencing. *BMC Med. Genom.* **2015**, *8*, 19. **[Google Scholar] [CrossRef] [Green Version]**

61. Tavira, B.; Gómez, J.; Santos, F.; Gil, H.; Alvarez, V.; Coto, E. A labor- and cost-effective non-optical semiconductor (Ion Torrent) next-generation sequencing of the SLC12A3 and CLCNKA/B genes in Gitelman's syndrome patients. *J. Hum. Genet.* **2014**, *59*, 376–380. **[Google Scholar] [CrossRef] [PubMed]**

62. Meiklejohn, K.A.; Robertson, J.M. Evaluation of the Precision ID Identity Panel for the Ion Torrent ™ PGM ™ sequencer. *Forensic Sci. Int. Genet.* **2017**, *31*, 48–56. **[Google Scholar] [CrossRef]**

63. Turchi, C.; Previderè, C.; Bini, C.; Carnevali, E.; Grignani, P.; Manfredi, A.; Melchionda, F.; Onofri, V.; Pelotti, S.; Robino, C.; et al. Assessment of the Precision ID Identity Panel kit on challenging forensic samples. *Forensic Sci. Int. Genet.* **2020**, *49*, 102400. **[Google Scholar] [CrossRef]**

64. Santos, C.; Fondevila, M.; Ballard, D.; Banemann, R.; Bento, A.M.; Børsting, C.; Branicki, W.; Brisighelli, F.; Burrington, M.; Capal, T.; et al. Forensic ancestry analysis with two capillary electrophoresis ancestry informative marker (AIM) panels: Results of a collaborative EDNAP exercise. *Forensic Sci. Int. Genet.* **2015**, *19*, 56–67. **[Google Scholar] [CrossRef] [PubMed] [Green Version]**

65. Cheung, E.Y.Y.; Gahan, M.E.; McNevin, D. Prediction of biogeographical ancestry from genotype: A comparison of classifiers. *Int. J. Leg. Med.* **2017**, *131*, 901–912. **[Google Scholar] [CrossRef] [PubMed]**

14 WHO WAS RESPONSIBLE FOR PACKING THE DRUGS? USING BAYESIAN NETWORKS TO EXAMINE THE TRANSFER, PERSISTENCE, AND RETRIEVAL OF DNA FROM PLASTIC BAGS AND TAPE

1. INTRODUCTION: AN OVERVIEW

When DNA from a suspect is found in samples taken from drug wrappings in criminal cases, there may be other possible reasons for how the DNA got there, how it stayed, and how it was found. Frequently, these explanations entail the transfer of evidence from an item that the individual has handled, such as a bag or clothing. Alternatively, the suspect may have come into contact with the item in a context unrelated to the crime. The hierarchy of propositions delineates the several tiers of assessment of the evidence provided for propositions at the sub-source, source, activity, and offence levels.

[**1**,**2**,**3**]. The guidelines from the International Society for Forensic Genetics (ISFG) DNA commission provide advice on evaluation and the formulation of propositions and recommend the evaluation of results at activity level in the light of the alternative propositions of the case and calculating a likelihood ratio (LR) [**2**]. A recommended tool to aid in the evaluation of evidence is Bayesian networks (BNs) [**4**] and examples of how they can be constructed and used are widely available [**5**,**6**]. To construct a BN, the relevant variables and their respective probabilities are required. Extensive research on transfer persistence, prevalence, and recovery (TPPR) of DNA has been conducted over the past 10 years [**7**], although knowledge bases that can be utilised for routine casework are currently lacking.

The indirect transfer of DNA is dependent upon several variables including: the amount of DNA present, time since initial contact, number of contacts and the type of surface of the objects involved [**8**,**9**,**10**]. DNA from a person who previously handled an object can be detected for some time after the object has been handed to a new user [**10**]. However, it is more likely that DNA from the most recent user of the object will become the major contributor over time [**11,12**]. Van Oorschot et al. [**12**] observed an average 50% drop in the DNA from the previous user immediately after handling by a new user when the surface was hard and non-porous. Mayuoni-Kirshenbaum et al. [**13**] studied the persistence of DNA from a person cutting aluminium foil when the foil was subsequently used for lock picking by another individual and only detected DNA from the cutter in 1/25 cases while the lock-picker was detected in 9/25 samples.

It has previously been demonstrated that individuals have a different propensity to transfer DNA to handled items (shedder status); shedder status is consistent over years, and it can influence transfer probabilities [**14,15,16**]. However, the classification method has been debated and a need to introduce an intermediate shedder category (between high and low) has been demonstrated [**15,16,17**].

The aim of this chapter is to provide a dataset along with a probabilistic framework to interpret DNA evidence at activity level. We present two specific scenarios related to DNA evidence in drug related cases. The first considers direct and indirect transfer to a zip-lock drugs bag that has been stored in a personal bag, whilst the second considers the persistence of DNA from a previous user of tape which was used to pack drugs by a second individual (the cardboard drug wrap experiment). The results from both experiments were examined in the light of the participant's shedder status. A probabilistic framework was developed which was based on a previously described Bayesian network [**5**] with case-specific modifications. Previous work has usually applied discrete models to measure simple presence/absence of DNA that may be attributed to a person of interest (POI). A continuous model was demonstrated by Gill et al. [**6**], based upon sub-source likelihood ratios, and this was recently complemented with a method using mean $\overline{RFU}$ (peak height) instead of sub-source likelihood ratio [**18**]. Here we provide a demonstration of continuous modelling of indirect and direct transfer with two case examples. Continuous models are the preferred method with mixture analysis [**19**] because they take more information into account. It is natural that activity level assignments should take the same route.

2. MATERIALS AND METHODS

2.1. Direct and Indirect Transfer to Zip-Lock Drugs Bags

Each participant was provided with a kit for the experiments containing two DNA-free (exposed to UV light for 10 min on each side) zip-lock bags (16.7 × 10 cm). One of the zip-lock bags contained a 15 mL centrifuge tube (VWR) filled with water to function as a weight.

2.1.1. Direct Transfer

Participants were instructed to wait at least one hour after washing hands/using antibacterial hand wash, after which they touched the upper surface of the zip-lock bag for about 30 s by opening and closing it. They were then asked to repeat this process with the same bag two more times (total of three handlings). When the procedure was completed, participants were instructed to place the zip-lock bag into a new labelled envelope.

2.1.2. Indirect Transfer

Participants were instructed to use their own personal bag/purse/backpack for this experiment. They were instructed to put on a pair of new disposable gloves before placing the zip-lock drugs bag containing the weight into their personal bag (which did not contain anything else during the experiment) and leave it there for 24 h; they were also instructed to move the personal bag during this period, e.g., carry it with them to work and home. After the 24 h had passed, the zip-lock drugs bag was removed, whilst wearing clean gloves, and placed into a new labelled envelope.

2.2. Persistence and Detection of DNA from a Previous User of a Tape Roll

In this experiment participants were divided into pairs. Participant B was given a DNA-free (treated with UV light for 20 min on each side) roll of tape. The participants were instructed to wait at least 1 h after washing hands/using antibacterial liquid; then the tape was handled for 1 min by passing it from one hand to another, stroking the sides of the tape roll towards the palm of the hands; use of the tape was mimicked by tearing off a small piece. The roll of tape was placed in a labelled envelope and passed to participant A after a minimum 10 days in storage. Participant A was provided with a piece of clean cardboard (approx. 6 × 10 cm) (UV treated on each side for 20 min), to mimic a pack of drugs. He/she was instructed to wrap the piece of cardboard with the tape (previously handled by B). The "drug wrap" was placed in a new labelled envelope. Each of the 20 participants contributed twice to the experiments swapping the roles of A and B, giving a total of 20 wrapped cardboards.

2.3. Shedder Status

Participants were asked to hold a plastic tube (15 mL centrifuge tube high performance tube, VWR) for 10 s in their dominant hand, at least one hour after washing hands/using antibacterial liquid. The experiments were repeated 3 times, with a new clean tube; there was a gap of at least three hours between each experiment.

2.4. Sample Processing

Samples were collected with a moistened (one drop of water) swab (Tubed Sterile Dryswab, Medical Wire). The sampling was performed on: (a) zip-lock bags: the top 3 cm of the outside and inside (including lock) of the zip-lock bags; (b) wrapped cardboard: the full outside and edges of the tape around the cardboard; (c): tubes: the full body of the 15 mL centrifuge tubes (not lids). One negative control sample was collected from all items (tape, cardboard, zip-lock bag and tube). For the samples collected from zip-lock bags and wrapped cardboard, the tips of the swabs were cut, placed in 1.5 mL Eppendorf tube and stored at −20 °C until further processing to avoid degradation. Samples collected from tubes were stored in ventilated evidence bags to dry before the tips of the swabs were cut directly into single PCR-tubes for a direct PCR analysis.

Samples collected from zip-lock bags and wrapped cardboard were extracted by the 5% Chelex procedure [**20**] (no prior incubation with water) where 200 μL Chelex (Bio-Rad, Hercules, CA, USA) was added to swabs. All samples were quantified with PowerQuant (Promega, Madison, WI, USA) on the 7500 Real-Time PCR system (ThermoFisher, Waltham, MA, USA) using the manufacturers' recommendations. PCR amplification was carried out using the PowerPlex Fusion 6C System kit (Promega, Madison, WI, USA) as recommended by the manufacturer (1 ng template, 25 μL reaction volume and 29 amplification cycles). Samples that had lower concentrations than the recommended template amount were amplified with the maximum template volume of 15 μL. For the shedder experiment, the PCR reaction mix was added directly to the tube containing the swab tips. Amplification was carried out using a Veriti 96-Well Thermal Cycler (ThermoFisher, Waltham, MA, USA). Samples were injected on the Applied Biosystems 3500 xL Genetic Analyzer at 1.2 kV for 24 s. The results were analysed using the GeneMapper ID-X Software version 1.6 (ThermoFisher, Waltham, MA, USA). The analytical threshold (AT) for alleles was set to 100 RFU.

2.5. Data Analysis

Results were analysed using R version 4.0.3 with the package tidyverse version 1.3.0. Likelihood ratios and mixture proportions were calculated using EuroForMix version 3.2.0 using propositions shown in in Equation (1), where U corresponds to unknown contributors, necessary to explain all alleles present. The average RFU ($\overline{RFU}$) for a contributor was calculated by dividing the total RFU (RFU_{tot}) by the number (n) loci (23, except amelogenin), Equation (2). If the sample was a mixture, the contribution from the POI was found by multiplying $\overline{RFU}$ for the sample by the mixture proportion (M_x). If a sample was diluted then the result was multiplied by a dilution factor (d_l).

$$LR = \frac{\Pr(E|H_p : POI + U_1)}{\Pr(E|H_d : U_1 + U_2)} \quad (1)$$

$$\overline{RFU}_{POI} = M_x \times \frac{RFU_{tot}}{n} \times d_l \quad (2)$$

2.5.1. List of Variables

We follow the notation used by Gill et al. [6] (**Section 4**), summarised here:

- *t* is the probability of direct transfer, persistence and recovery of DNA from the POI (under H_p only).
- *t'* is the probability of direct transfer, persistence and recovery of DNA from an unknown contributor (under H_d only).
- b is the probability of background DNA, based on observations, and is applied under both H_p and H_d. Background DNA is present from unknown sources and unknown activities. It can be described as "foreign" (non-self). For further details we refer to **Section 3.2** in [**6**].
- s is the probability of transfer, persistence and recovery; in experiment 1, this is indirect transfer under H_p and H_d, and in experiment 2, it is direct transfer under H_d only.
- Suffixes are applied to described probabilities of an event given a particular contributor, e.g., t_A refers to the probability of direct transfer, persistence and recovery of DNA from contributor A.

2.5.2. Notation Relating to the Experiments

Experiment 1: zip-lock drugs bag

$E1_{dbag}$: Dataset from the zip-lock drugs bag; direct transfer

$E2_{pbag}$: Dataset from the personal bag experiment; indirect transfer

RFU: Lognormal distribution of mean RFU from $E1_{dbag}$data; direct transfer

$\overline{RFU}_{E1dbag}$: Lognormal distribution of mean RFU from $E2_{pbag}$ data; indirect transfer

Experiment 2: Tape/cardboard drug wraps

$C1_{pack}$: Dataset from drugs wrapper; direct transfer

$C2_{tape:}$ Dataser from the tape handler; direct transfer

$\overline{RFU}_{C1pack}$: Lognormal distribution of mean RFU from $C1_{pack}$ data; direct transfer

RFU_{C2tape}: Lognormal distribution of mean RFU from $C2_{tape}$ data; direct transfer

General

- $_A$: the observed $\overline{RFU}$ value from contributor A

- $_B$: the observed $\overline{RFU}$ value from contributor B
- $_{unknown}$: the observed $\overline{RFU}$ value from an unknown contributor

2.5.3. Distribution Fitting

Log-normal distributions were fitted to the data using the R package *fitdistrplus* using function *lnorm*. The probability of background for experiment 1 was modelled using the same lognormal parameters calculated for indirect transfer data $\overline{RFU}$ $E2pbag$. To carry out sensitivity analysis, 1000× bootstraps (with replacement) were taken of datasets. For each bootstrap, a new set of log-normal parameters (mean log and SD log) were calculated using the *fitdistrplus* R package using the *lnorm* function. The probability distributions were used to substantiate the Bayesian networks described in **Section 3.5.1** and **Section 3.5.2.** The programming of these networks was carried with R-code using the formulae described in **Section 3.5.1** and **Section 3.5.2.**

3. RESULTS

3.1. Direct and Secondary Transfer to Zip-Lock Drugs Bags

The DNA from the POI could be detected in the samples collected after direct handling of a zip-lock drugs bag and after indirect transfer to the drugs bag from storage in a previously used personal bag. More alleles matching the POI along with higher DNA quantities were generally observed in the direct transfer experiment ($E1_{dbag}$) compared with indirect transfer from inside a personal bag ($E2_{pbag}$). Generally, the *RFM* as higher in $E1_{dbag}$, but a few occurrences of higher RFU_{POI} were observed in $E2_{pbag}$ (**Figure 1**).

In the direct transfer experiment four full and seven partial profiles were detected, while six samples only displayed six or less alleles and four samples had no results. Only one sample contained (two) unknown alleles. From the indirect transfer experiment three full and two partial profiles corresponding to the POI were detected. Four samples only displayed five or less alleles and ten samples had no profiling results. One sample was a mixture with 21 unknown alleles, and one sample was a partial profile with alleles corresponding to a child of the participant. The most frequently used type of personal bag used was a backpack but a large proportion of these samples (7/11) had no results. The highest $\overline{RFU}$ values on the zip-lock drugs bags were observed after storage in a personal purse/handbag, (**Figure 2**). A few participants (1 and 2, respectively) used a personal PC and shopping bag to store the zip-lock drugs bags. Most participants reported using the personal bag frequently or every day, while two participants reported rarely using their personal bags. No alleles were detected on zip lock bags after storage in rarely used personal bags.

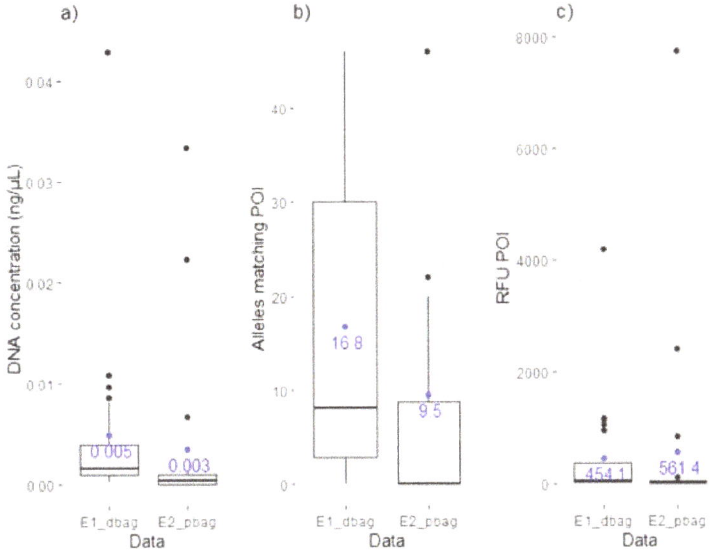

Figure 1. Boxplots displaying (a) the measured DNA concentrations (ng/µL in 200 µL), (b) the number of detected allele's matching the POI and (c) the $\overline{RFU}_{POI}$ for the samples in the $E1_{dbag}$ (direct) and $E2_{pbag}$ (indirect) data.

3.2. Persistence and Detection of DNA from a Previous User of a Tape Roll Used to Wrap Drugs

In samples collected from wrapped cardboard, DNA from both the person who previously handled the tape ($C2_{tape}$) and the person who packed the tape around the cardboard ($C1_{pack}$) could be detected. However, the quality of the profile, peak heights of detected alleles and mixture proportions varied between the cardboard wrapper and the tape handler (**Figure 3**).

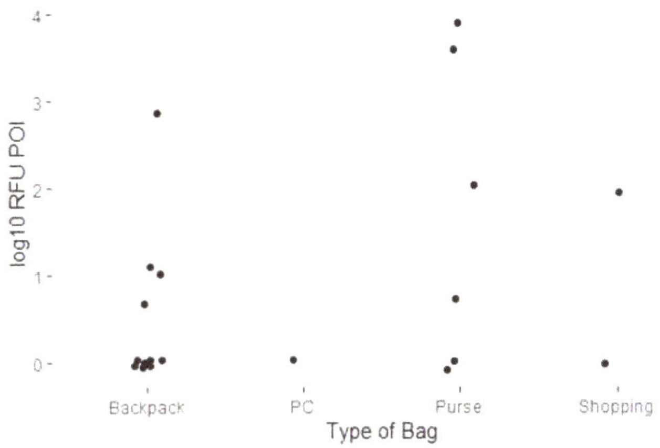

Figure 2. Dot plot displaying the log RFU_{10} values detected in the samples collected from zip-lock drugs bags after storage in a previously used personal bag (backpack, PC bag, purse/handbag or shopping bag).

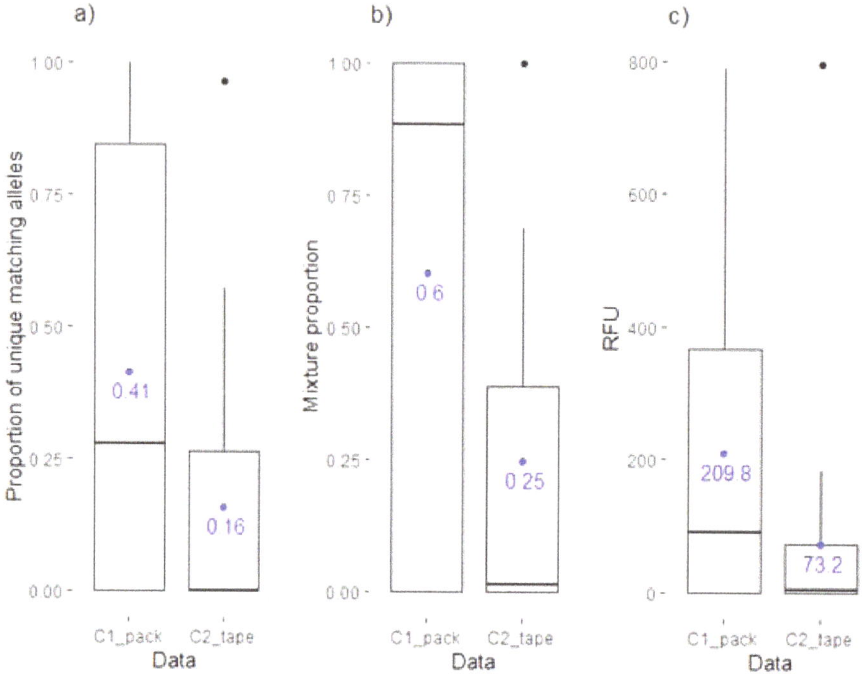

Figure 3. Boxplot displaying (a) proportion of unique alleles, (b) the mixture proportion and (c) $\overline{RFU}$ values corresponding to the wrapper ($C1_{pack}$) and the handler ($C2_{tape}$) of the tape.

Three samples had no profiling results and for an additional three samples only four alleles were detected. Seven samples were full or partial profiles corresponding to the drugs wrapper (C1), five samples were mixtures of C1 and C2. For three of these mixtures, C1 was the major contributor, whilst in the last two, the contributions were of equal proportions. Two samples were partial profiles with alleles corresponding to C2.

3.3. Shedder Status

The total *RFU* values (RFU_{tot}) from the direct PCR analysis varied from 0 to 458,705, variation was greater between individuals than within the samples from one individual (ANOVA, p = 0.04), **Figure 4**. However, some individuals, especially those that deposited low amounts of DNA, displayed lower variation. Individuals were classified as low, medium or high shedders based on the criteria defined by Johannessen et al. [**15**] with the following adjustments according to the observed values of the current dataset: high shedder class was assigned to participants that had two out of three samples above the average RFU_{tot} (62,520) and two out of three samples with 20 out of 24 full loci. Low shedder class was assigned to participants where all samples were below 8000 RFU_{tot} and two out of three with negative or partial profiles. The rest were classified as medium shedders. This resulted in 4 high, 11 medium and 5 low shedders (**Figure 3**).

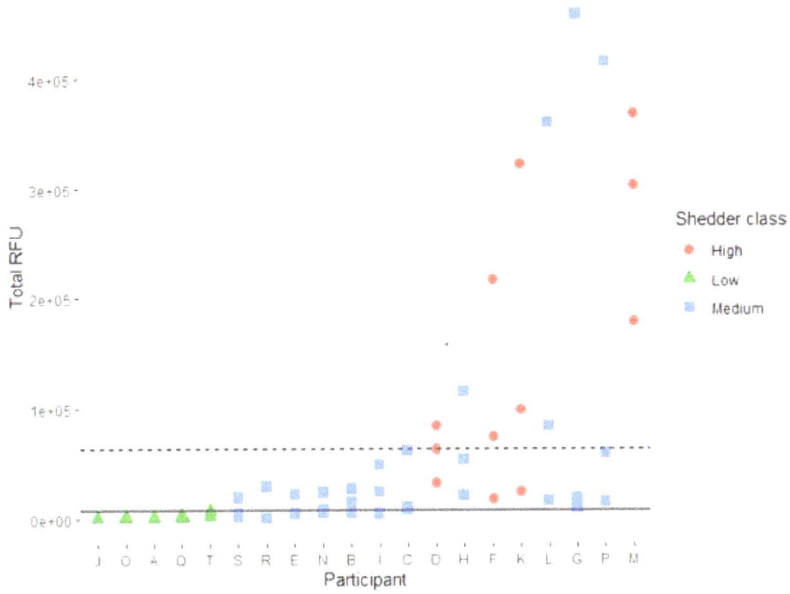

Figure 4. Scatter plot showing the total RFU_{tot} values (multiplied by M_x if sample was a mixture) for each sample collected from each participant and according to their shedder status, high (red circle), medium (blue square) low (green triangle). The average RFU_{tot} (6.25E04+) and low shedder limit of 8.00E03+ RFU_{tot} are represented with the dotted and straight lines, respectively.

DNA from unknown contributors was detected in 30/90 samples, the average M_x proportion from unknown contributors was highest for low shedders, **Table 1**.

Table 1. Detection of unknown alleles in the shedder samples and the proportion of unknown DNA detected for the three different shedder classes.

	Number of Samples	Average M_x Unknown Contrib.
High	4/12 (33%)	7.1
Medium	19/33 (58%)	8.5
Low	7/15 (47%)	52.0

3.4. Transfer in Relation to Shedder Status

The results from experiment 1, where the transfer of substances was directly done into a zip-lock bag, indicated that the highest rates of transfer were observed among those who shed a large amount of the substance (Figure 5). The middle shedder group exhibited greater variation compared to the low and high shedders. The tendency is less apparent in experiment 2, which involves the indirect transfer of substances from a personal bag to a zip-lock bag containing medications. The sample from a subject with reduced shedding had the highest quality profile. Due to the inadequate sample sizes in each group, it was not possible to assess the impact using Bayesian networks as outlined in Section 3.5.1.

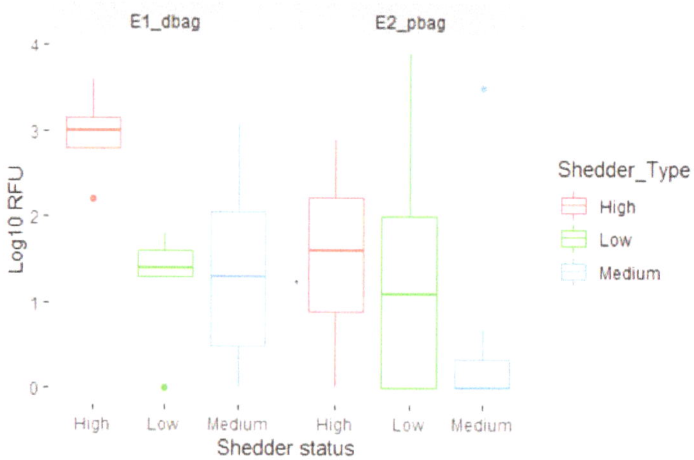

Figure 5. Boxplot displaying the log10 $\overline{RFU}$ detected in the direct transfer experiment ($E1_{dbag}$) and indirect transfer from personal bag ($E2_{pbag}$) according to the participants' shedder classification.

Data from persistence and detection of DNA from a previous user of a tape roll experiment showed that low shedders only provided low quality profiles and that higher quality profiles were provided by high shedders. Medium shedders provided both high and low quality results. (**Figure 6**).

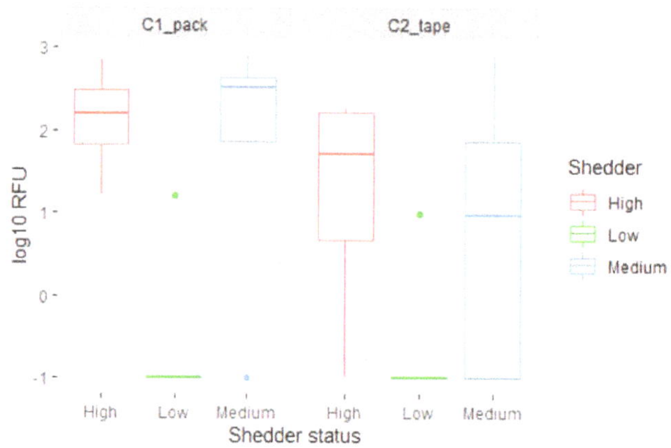

Figure 6. Boxplot displaying the log10 $\overline{RFU}$ for the drugs wrapper ($C1_{pack}$) and the previous handler ($C2_{tape}$) of the tape according to the shedder classification of the participants.

3.5. Case Examples

To demonstrate the potential use of the datasets obtained, we will use two fictive case examples that are based on actual experiences from case work, using a Bayesian network to evaluate the evidence.

3.5.1. Potential DNA Transfer from Storage in a Personal Bag

Case Circumstances

1. A large depot of drugs was found at a hideout;
2. The drugs were packed in zip-lock bags and placed in a black gym bag;
3. Upon questioning, person A claims to have no knowledge of the drugs. However, he recognizes a gym bag that the drugs had been stored in and claims that it used to belong to him but it was lost or stolen two weeks previously.

DNA Analysis

DNA samples were collected from the outside and opening of the zip-lock bag

1. Example 1: Result is a full DNA profile of a single individual. There is a candidate in the national DNA database who is identified as person A;
2. Example 2: Result is a mixture of two individuals. There is a candidate in the national DNA database who is identified as person A but there is no candidate for the second individual.

Propositions

Sub-source likelihood ratios were calculated for both contributors using EuroForMix [**21**], although this information was not used in any further analysis. To proceed with activity level analysis, it is assumed that the sub-source LR is sufficient for a court to agree the identity of a POI [**2**].

The alternative activity level propositions (for both examples) are as follows:

H_p: The suspect packed the drugs;

H_d: The suspect has no relation to the drugs but was the owner of the gym bag which was stolen two weeks previously. An unknown individual packed the drugs.

The BN used to instantiate probabilities is identical to that described by [**6**], except that the title of the nodes are changed to represent the case circumstances (**Figure 7**). Log-normal distributions were used to calculate drug bag (direct) and personal bag (indirect) TPR probabilities, conditioned upon values of $\overline{RFU}_A$ modelled with $\overline{RFU}_{E1dbag}$ RFU_{E2pbag} distributions respectively. $RFU_{unknown}$ was modelled with RFU_{E2pbag} distribution (**Table 2**), the assigned probabilities (**Figure 8**) were subsequently used in the BN (**Figure 7**) to calculate the results. To carry out sensitivity analysis, 1000× bootstraps (with replacement) were taken of the RFU_{E2_pbag} data to provide 1000 new sets of data. For each bootstrap, a new set of log-normal parameters (mean log and SD log) were calculated using the *fitdistrplus* R package and the variation was represented by percentiles in **Table 3**.

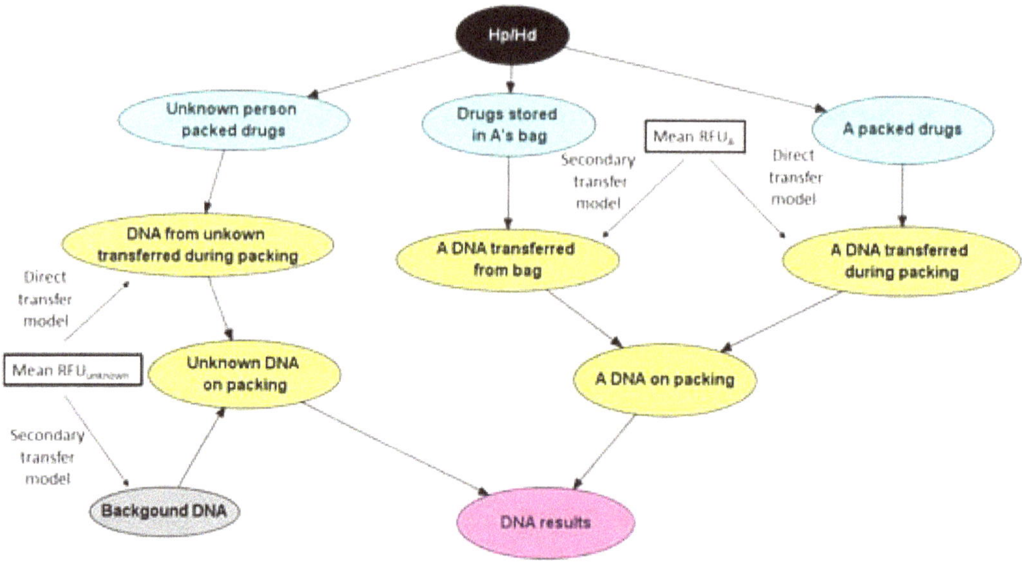

Figure 7. Bayesian network for the "drugs stored in gym bag" example showing data sets and models used to instantiate the nodes.

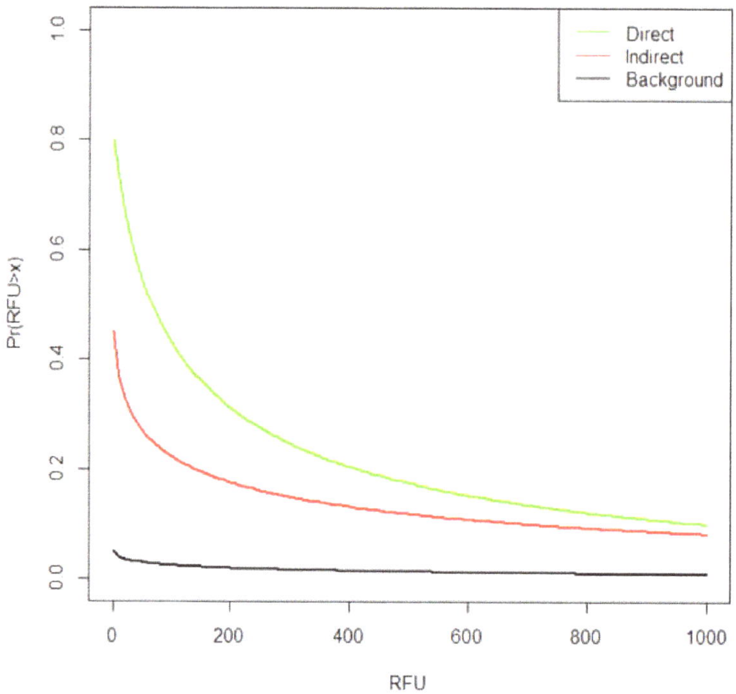

Figure 8. Comparison of probabilities of direct, indirect and background TPPR for the zip-lock drugs bag example. Direct transfer was modelled from $\overline{RFU}E1_{dbag}$; personal bag (indirect transfer) was modelled from $\overline{RFU}E_{2pbag}$ data and background was also modelled from $RFUE2_{pbag}$ data, with adjusted k = 0.95 (the proportion of observations with no background) in order to scale results as described in Supplementary Material S1.

Table 2. Relationship of probabilities of DNA transfer from individuals A (POI) and U (unknown), showing the datasets that were used to calculate probability distributions, along with the BN nodes instantiated. See nomenclature Section 2.5.1 for definitions of probabilities.

Probability	Description	Contributor	Data used to Inform Probability Distribution	BN Nodes
t_A	Packing transfer	A	RFU_{E1dbag}	A DNA transferred during packing
t'	Packing transfer	U	RFU_{E1dbag}	DNA from unknown transferred during packing
s	Bag transfer	A	RFU_{E2pbag}	A DNA transferred from bag
b	Background	U	RFU_{E2pbag}	Background DNA

Table 3. Activity level likelihood ratios where only contributor A is recovered, with sensitivity analysis, showing 97.5–2.5 percentiles. The median (50 percentile) values are those that are reported. Results for the discrete model are shown in the top row; continuous models conditioned upon $\overline{RFU}_{A>x}$ are shown in remaining rows.

$RFU_A > x$	2.50%	5%	10%	25%	50%	75%	90%	95%	97.50%
0	5	5	6	8	11	17	27	39	6×10^1
200	5	5	7	9	14	23	47	82	2×10^3
400	4	5	6	9	13	24	52	122	1×10^4
600	4	5	6	8	12	24	52	153	4×10^4
800	4	5	6	8	12	24	53	196	9×10^4
1000	4	4	5	8	12	24	54	218	2×10^5
2000	4	4	5	7	11	24	63	448	3×10^6
4000	3	4	4	6	10	23	77	1016	6×10^7
6000	3	4	4	6	9	21	87	1752	4×10^8

Derivation of the formulae used in the calculations Gill et al. [6]. Likelihood ratios were calculated as shown in Formulae 3–5 (nomenclature in **Table 2**):

a. Only the POI (A) is observed

$$LR = \frac{(s(1 - t_A) + t_A) \times (1 - b)}{s \times (1 - \Pr(U_d))} \quad (3)$$

b. POI (A) and unknown is observed

$$LR_a = \frac{(s(1 - t_A) + t_A) \times b}{s \times \Pr(U_d)} \quad (4)$$

c. Probability of recovery of DNA from an unknown contributor

d. $\Pr(U_d) = t'b + t'(1 - b) + b(1 - t') = t' + b(1 - t')$ (5)

There were only three observations of background (datasets $E1_{dbag}$ and $E2_{pbag}$). Background is from indeterminate (unknown) sources and can comprise both direct and indirect transfer. Here, it was modelled using $\overline{RFU}_{E1pbag>x}$ data (where x is a threshold value; first column in **Table 3**), except that the model was scaled relative to $k = 1 - \Pr(b)$; where k is the proportion of observations where no background was observed. From experimental observation $\Pr(b) = 0.05$ and $k = 0.95$.

Two possible outcomes of the DNA results were analysed in detail. Either a profile from the POI individual A is recovered, or A is recovered in combination with an unknown contributor which forms the basis of the proposition under H_d, else the unknown contributor is background under H_p.

Contributor A Recovered Alone

When contributor A is recovered alone, **Table 3** shows that the median (50 percentile) LRs always favour the proposition that he/she packed the drugs; the evidence provides moderate support. If a discrete model is used where allele peak height is not considered, then LR = 11 (top row of **Table 3**). Taking peak height into account has little effect in this example. The sensitivity analysis shows the evidence always favours H_p at the 2.5 percentile.

Unknown and Contributor A Recovered

A different result is obtained when a mixture of unknown and contributor A is recovered. For the discrete model (top row of **Table 4**) the LR ≈ 0.1, favouring the proposition that an unknown contributor wrapped the drugs. When allele peak height is considered, similar results are obtained (**Table 4**).

Table 4. Activity level likelihood ratios where contributor A and an unknown are recovered, with sensitivity analysis, showing 97.5–2.5 percentiles. The median (50 percentile) values are those that are reported. Results for the discrete model are shown in the top row; continuous models conditioned upon RFU_A and $RFU_{unknown}$ are shown in remaining rows.

Table 4. Activity level likelihood ratios where contributor A and an unknown are recovered, with sensitivity analysis, showing 97.5–2.5 percentiles. The median (50 percentile) values are those that are reported. Results for the discrete model are shown in the top row; continuous models conditioned upon RFU_A and $RF_{Uunknown}$ are shown in remaining rows.

RFU_A > x	$RFU_{Unknown}$ > x	2.50%	5%	10%	25%	50%	75%	90%	95%	97.50%
0	0	0.1	0.1	0.1	0.1	0.1	0.1	0.2	0.2	0.2
100	2000	0.1	0.1	0.1	0.2	0.3	0.6	1.1	1.5	2
100	7000	0.1	0.2	0.2	0.3	0.6	1.1	1.8	2.4	4
2000	100	0.1	0.1	0.1	0.1	0.1	0.2	0.5	1.4	24
7000	100	0.1	0.1	0.1	0.1	0.1	0.1	0.6	3.6	307
1000	2000	0.1	0.1	0.2	0.2	0.3	0.5	0.9	1.5	20
1000	7000	0.2	0.2	0.2	0.3	0.5	0.9	1.3	2.7	35
2000	1000	0.1	0.1	0.1	0.1	0.2	0.3	0.7	1.8	45
7000	1000	0.1	0.1	0.1	0.1	0.2	0.3	1.0	4.2	717

3.5.2. Cardboard Drug Wrap Experiment

Case Circumstances

A large depot of drugs was detected by the police during a house search of A's house.

The drugs were wrapped in cardboard and covered by packing tape;

Person A admits that he packed the drugs and does not implicate anyone else;

Person B claims to have no knowledge of the drugs. However, he worked for a moving agency and often handles packing tape. Remains of packing tapes are often left behind and could be picked up and used by others.

DNA Analysis

DNA samples were collected from the wrappings. The results of the DNA analysis were a mixture of person A and an unknown contributor. A candidate was identified as person B from a national DNA database search.

Propositions

Sub-source likelihood ratios were calculated for both contributors using EuroForMix [**20**], although this information was not used in any further analysis. To proceed with activity level analysis, it is assumed that the sub-source LR is sufficient for a court to agree the identity of a POI [**2**].

The prosecution and defence activity level propositions are as follows:

- H_p: A and B packed the drugs together and B did not previously handle the tape;
- H_d: A packed the drugs alone; B had previously handled the tape

From the case information, a Bayesian network incorporating the relevant nodes was prepared (**Figure 9**). Since both H_p and H_d agree that contributor A packed the drugs, his/her presence or absence of DNA has no effect upon the likelihood ratio. Only contributor B has an effect.

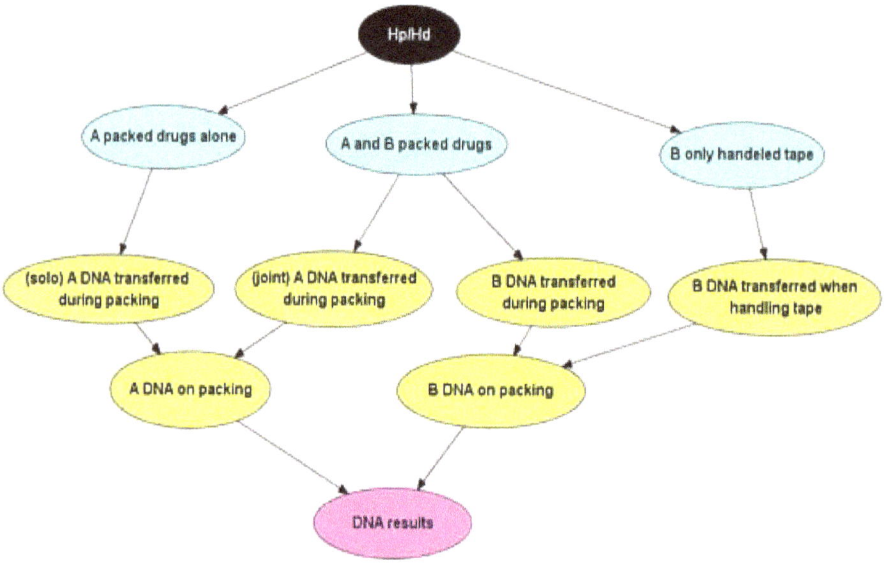

Figure 9. Bayesian network for the evaluation of evidence in the case where DNA evidence was collected from cardboard drug wrap.

Statistical Analysis

In the experiment, contributor B previously handled the tape as described in **Section 3.5.1**, and contributor A packed the drugs. The mean $\overline{RFU}$ of the observed DNA profile was split per contributor ($\overline{RFU}_A$ and $\overline{RFU}_B$), based upon their respective mixture proportions (M_x): i.e., $\overline{RFU}_{A=}$ $\overline{RFU}_{tot\times MxA}$ where M_x was calculated using EuroForMix [20]. Log-normal distributions were fitted to each set of data using the *fitdistrplus* R package (**Figure 10**). Distributions for the BN node "B DNA transferred when handling tape" (**Figure 9**) were modelled from $\overline{RFU}_{C2tape}$; distributions for BN nodes" (solo) A DNA transferred during packing" and "(joint) B DNA transferred during the packing" were both modelled from $\overline{RFU}_{C1pack}$ (**Table 5**).

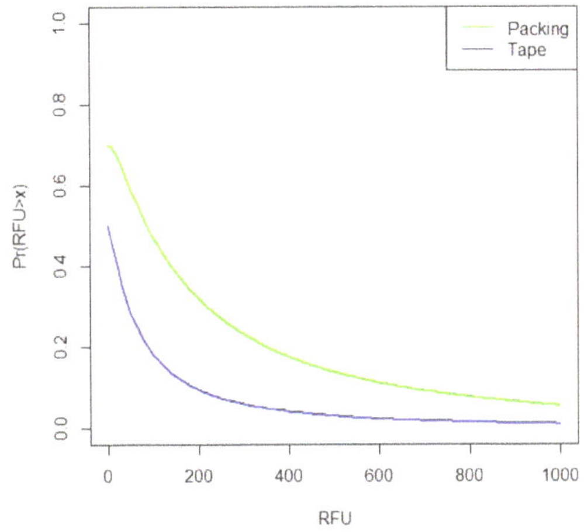

Figure 10. Lognormal distributions of $\overline{RFU}_{C1pack}$ and $\overline{RFU}_{C2tape}$.

Table 5. Relationship of probabilities of DNA transfer from individuals A and B, showing the datasets that were used to calculate probability distributions, along with the BN nodes instantiated. Nomenclature is described in Section 2.5.1.

Probability	Description	Contributor	Data Used to Inform Probability Distribution	BN Nodes
t_A	Packing transfer	A	RFU_{C1pack}	(solo) A DNA transferred during packing (joint) A DNA transferred during packing
t_B	Packing transfer	B	RFU_{C1pack}	(joint) B DNA transferred during packing
s	Tape transfer	B	RFU_{C2tape}	B DNA transferred when handling tape

Likelihood ratios were calculated as follows (see **Table 5** for context):

- A DNA is recovered: $LR=\frac{(1-t_B)}{(1-s)}$
- B DNA is recovered: $LR=t_B/s$
- A and B DNA are recovered: $LR=t_B/s$
- No DNA is recovered: $LR=\frac{(1-t_B)}{(1-s)}$

Note that likelihood ratios of (a) and (d) are the same, as are (b) and (c). This is because both H_p and H_d condition upon contributor A, hence his/her presence is cancelled out. To carry out sensitivity analysis, 1000× bootstraps (with replacement) were taken of the $\overline{RFU}_A$ and $\overline{RFU}_B$ data. For each bootstrap, a new set of log-normal parameters (mean log and SD log) were calculated using the *fitdistrplus* R package.

Results of Analysis

Log-normal distributions were used to calculate the tape handling and drug packing TPR probabilities, conditioned upon values of $\overline{RFU}_{B>x}$ from simulated profiles and subsequently used in the BN (**Figure 8**) to calculate the results (**Table 6**).

Table 6. Activity level likelihood ratios, with sensitivity analysis, showing 99–1 percentiles. The median (50 percentile) values are those that are reported. Results for the discrete model are shown in the top row; continuous model conditioned upon $\overline{RFU}_{B>x}$ shown in remaining rows.

$RFU_B > x$	1%	2.50%	5%	10%	25%	50%	75%	90%	95%	97.50%	99%
0	0.8	0.8	0.9	1.0	1.2	1.4	1.7	2.0	2.3	2.6	3.0
100	1.2	1.4	1.5	1.7	2.1	2.7	3.9	5.6	7.3	10.0	16.2
200	1.3	1.5	1.7	2.0	2.6	3.7	6.6	12.9	20.5	34.0	90.5
300	1.3	1.5	1.7	2.1	2.8	4.5	10	23	43	93	282
400	1.3	1.5	1.7	2.0	3.0	5.1	13	39	78	190	953
500	1.2	1.4	1.6	2.0	3.1	5.5	17	65	147	371	2152
600	1.1	1.3	1.5	1.9	3.0	5.9	20	97	238	603	4441
700	1.0	1.2	1.4	1.8	3.0	6.1	24	133	360	1157	8454
800	0.8	1.1	1.3	1.7	3.0	6.4	29	179	563	1872	15,118
900	0.7	1.0	1.3	1.6	3.0	6.7	35	227	833	2951	27,012
1000	0.6	1.0	1.2	1.6	2.9	6.9	40	291	1176	5202	52,019

Likelihood ratios, along with sensitivity analyses are calculated against a range of $RFU_{B>x}$ values (**Table 6**). If a profile is obtained where only contributor A is present and there is no DNA present that can be attributed to contributor B, then the LR = 0.6, i.e., supports the proposition that suspect B did not package the drugs. If DNA is present that can be attributed to B, then the LR > 1, favouring the proposition that suspect B did package the drugs. The presence/absence of individual B as a contributor to the evidence can be described either as discrete: Pr($\overline{RFU}_{B>0}$) vs. absence, or as a continuous distribution where Pr($\overline{RFU}_{B>x}$), where x is a threshold value. With a discrete model, LR=1.4 (first row of **Table 6**). Taking the value of $\overline{RFU}_{B>x}$, with the continuous model in subsequent rows of **Table 6**, a higher LR is achieved, reaching a maximum median LR=6.9 if $RFU_B > 1000$, although the evidence can only be described as "weak" following the ENFSI verbal scale [**22**]. The sensitivity analysis shows the observed range between 1–99 percentiles from 1000× bootstraps of the data. The variation increases greatly as $\overline{RFU}_B$ increases—a reflection of the small size of the datasets. In conclusion, the discrete model understates the value of the evidence, compared with the continuous model.

4. DISCUSSION

4.1. Zip-Lock Drugs Bag Experiment

The DNA of the person of interest (POI) was shown to be more commonly present on a zip-lock bag containing narcotics when the bag was immediately touched. Nevertheless, the investigation identified the highest quality profile from a zip-lock medications bag that was kept within a personal bag (purse). From the literature, indirect transfer to a sleek surface such as the zip-lock drugs bag is expected to be low [**8**]. Most of the personal bags used in the experiments had previously been frequently used by the participants, although there are few observations in each class of bags, there is an indication that larger quantities of DNA can be detected on a zip-lock drugs bag after storage in purses. More DNA could be accumulated from the user on the inside of a purse as several personal items (e.g., phone, hairbrush, wallet) containing owner DNA are frequently stored there. In addition, the surface of the inside of the bag will influence accumulation and further transfer [**8,23**]. In the case example, it is shown that personal bags that are used to store everyday items, such as clothing or personal effects that are frequently handled, accumulate amounts of DNA that may be indirectly transferred to other objects. This resulted in low likelihood ratios (moderate evidence to support H_p) when DNA from only the POI was recovered. There was little difference between the discrete model and the continuous model, the latter takes the allele peak height into account. With mixtures, both discrete and continuous models favour H_d (LR ≈ 0.1; **Table 4**). In the experiment, only low amounts of background (DNA from unknown contributors) were detected. Combined with the relatively low difference between indirect and direct transfer probabilities (**Figure 8**), especially when $\overline{RFU}$1000, this has an impact of reducing the LR if the sample is in admixture with an unknown contributor, so that it always favours the defence proposition (H_d).

4.2. Cardboard Drug Wrap Experiment

This study aimed to examine the longevity of DNA left behind by previous individuals who handled a roll of tape. Due to the tape's hard and non-porous top surface, it is expected that only little amounts of DNA would be transferred and identified upon direct contact. Additionally, the DNA from the prior user would be quickly removed [8,12,13]. The findings of this study generally correlate with these expectations: a large proportion of samples produced no results or only a few alleles. The packer (last handler) was the only contributor or the one contributing a larger amount of DNA in most of the samples that gave results. Some exceptions were observed with two profiles showing results that only corresponded to the tape handler (person B). The sides of the tape have a rougher surface where more

DNA is expected to be transferred upon initial contact [**9**,**23**]. In addition to DNA from the previous user persisting on the surface of the tape, it is possible that some of the DNA deposited on the side of tape could be transferred to the new user's hands and to the cardboard during the wrapping of the drugs. As no other items were touched in between handling the tape and performing the wrapping, there was no opportunity for the loss of person B's DNA to other surfaces. We did not monitor the wrapping procedure and recognise that the manner of contact during the procedure could influence the result. However, information regarding this will rarely be known in casework.

In the Bayesian network case example, we considered that an individual B claims that he/she handled tape which was later used to prepare drug wraps. Individual A has admitted the offence, hence, his/her presence of DNA on the drug wrap has no bearing on the value of the evidence. If contributor B's DNA is recovered, without taking account of $\overline{RFU}$ in the discrete model, the evidence is close to neutral LR = 1.4, whereas if $\overline{RFU}$ is taken into account, the value of the evidence increases with RFU, although it does not exceed LR = 7 where $\overline{RFU}$>1000 (the upper limit of experimental observations), i.e., the evidence would be described as weak using the ENFSI scale [**22**]. However, if B's DNA is absent, then this favours the defence proposition LR = 0.6.

The experimental set up is similar to that used in [**13**] where the question was if the POI previously cut an aluminium foil or used this foil in lock picking. The activity LR in the lock-picking study, calculated with a discrete model (7.4) was greater than that observed in the current study (1.4). However, the LR was in the same range as the maximum (median) level of the continuous model employed (LR = 6.9). Some of the differences could be explained by the difference in transfer to and persistence on aluminium foil vs. tape.

4.3. Shedder Status and Transfer Probabilities

A person's shedder status has previously been shown to influence the probability of transfer, persistence, and detection of DNA [**14**,**24**]. Fonneløp et al. [**14**] demonstrated that DNA was more frequently detected in samples collected from the T-shirts of a victim if the attacker was a high shedder and that the probability of detection increased further if the victim was a low shedder. While Otten et al. [**24**] observed a correspondence between shedder stratus and DNA transfer to gloves. The correspondence between shedder status and the amount of DNA transferred is further demonstrated by our findings when it comes to direct transfer to zip-lock drugs bags (**Figure 11**), direct transfer during wrapping with a tape and transfer and persistence after touching a role of tape. On the other hand, when indirect transfer from the inside of a personal bag was considered, no clear association with shedder status was observed, and the best quality profile was detected after storage in a low shedder's personal bag. We hypothesised the amount of previous use and the surface of the inside of the bag could be of higher importance for this type of transfer. The low number of samples collected in each category is also a limitation and a clearer correspondence may be seen with a larger collection of data. Shedder status was not incorporated into the Bayesian networks because dividing the results into three categories would lead to too few data in each group to analyse. Secondly, it would also be important to properly characterise the effect of shedder status on indirect transfer before applying this variable to the model.

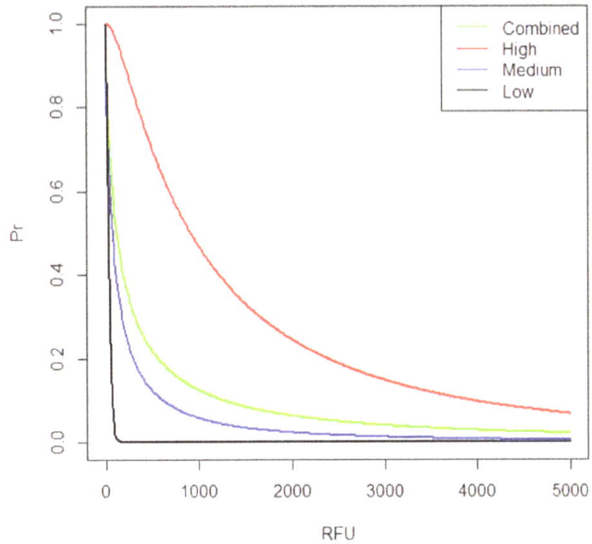

Figure 11. Effect of high/medium/low shedder status on direct transfer (E1) data.

The LRs calculated in this study are comparable to other studies where DNA transferred by hands is considered [**6**,**13**,**25**]. It is likely that including more information—especially shedder status, would, to some degree, change the LR calculations [**14**].

4.4. Detection of Unknown DNA

This experiment was performed during the COVID-19 pandemic where a general recommendation to keep a social distance of at least two meters and to wash hands frequently or use antibacterial liquid was given. It is likely that these measures could have had an impact on the detection of unknown DNA in the samples, which was low compared to previous studies [**9**,**16**,**26**].

5. CONCISE SUMMARY

When DNA from a suspect is identified in a sample obtained at a crime scene, there may be several possible explanations regarding the actions that could have caused the transfer, endurance, and retrieval of their DNA.Prior research has demonstrated that DNA can be transferred indirectly through intermediate surfaces, and that DNA can remain on an object even after it has been used by another person. Moreover, research has demonstrated that an individual's shedder status can impact the transmission, longevity, occurrence, and retrieval of DNA. This study examines the transfer, durability, and recovery of evidence on ziplock bags and tape, which are frequently seen in drugrelated situations. Additionally, the study investigates how the shedding state of the participants affects the obtained results. A probabilistic framework was formulated using a Bayesian network that has been previously defined, but with modifications tailored to specific cases.Data was continuously modeled to provide information for the Bayesian networks, and two case situations were examined. Only limited to moderate support for Hp was obtained in the specific conditions. Utilizing a continuous model that relies on the quality of the profile can alter the likelihood ratios (LRs).

6. AT A GLANCE

We have generated datasets pertaining to both direct and indirect transfer to zip-lock bags, as well as transfer and persistence on tape. Furthermore, we have demonstrated how this data may be utilized to provide information for Bayesian Networks. Given the realistic nature of the indirect and persistent scenarios examined in this study, only limited to little evidence in favor of Hp was found. We have demonstrated that utilizing a continuous model based on profile quality can modify likelihood ratios (LRs) in comparison to a discrete model and is more favorable. Insufficient datasets provide obstacles, preventing us from including shedder status into our models. Additional data is necessary to further examine the impact of shedder status in cases involving indirect transfer.

REFERENCES

1. Cook, R.; Evett, I.; Jackson, G.; Jones, P.; Lambert, J. A Hierarchy of Propositions: Deciding Which Level to Address in Casework. *Sci. Justice* **1998**, *38*, 231–239. **[Google Scholar] [CrossRef]**
2. Gill, P.; Hicks, T.; Butler, J.M.; Connolly, E.; Gusmão, L.; Kokshoorn, B.; Morling, N.; van Oorschot, R.A.; Parson, W.; Prinz, M.; et al. DNA commission of the International society for forensic genetics: Assessing the value of forensic biological evidence—Guidelines highlighting the importance of propositions. Part II: Evaluation of biological traces considering activity level propositions. *Forensic Sci. Int. Genet.* **2020**, *44*, 102186. **[Google Scholar] [CrossRef] [Green Version]**
3. Taylor, D.; Kokshoorn, B.; Biedermann, A. Evaluation of forensic genetics findings given activity level propositions: A review. *Forensic Sci. Int. Genet.* **2018**, *36*, 34–49. **[Google Scholar] [CrossRef] [PubMed] [Green Version]**

4. Biedermann, A.; Taroni, F. Bayesian networks for evaluating forensic DNA profiling evidence: A review and guide to literature. *Forensic Sci. Int. Genet.* **2012**, *6*, 147–157. **[Google Scholar] [CrossRef] [PubMed]**
5. Taylor, D.; Biedermann, A.; Hicks, T.; Champod, C. A template for constructing Bayesian networks in forensic biology cases when considering activity level propositions. *Forensic Sci. Int. Genet.* **2018**, *33*, 136–146. **[Google Scholar] [CrossRef] [Green Version]**
6. Gill, P.; Bleka, Ø.; Roseth, A.; Fonneløp, A.E. An LR framework incorporating sensitivity analysis to model multiple direct and secondary transfer events on skin surface. *Forensic Sci. Int. Genet.* **2021**, *53*, 102509. **[Google Scholar] [CrossRef] [PubMed]**
7. Van Oorschot, R.A.; Szkuta, B.; Meakin, G.E.; Kokshoorn, B.; Goray, M. DNA transfer in forensic science: A review. *Forensic Sci. Int. Genet.* **2019**, *38*, 140–166. **[Google Scholar] [CrossRef]**
8. Goray, M.; Mitchell, R.J.; van Oorschot, R.A. Investigation of secondary DNA transfer of skin cells under controlled test conditions. *Leg. Med.* **2010**, *12*, 117–120. **[Google Scholar] [CrossRef]**
9. Daly, D.J.; Murphy, C.; McDermott, S.D. The transfer of touch DNA from hands to glass, fabric and wood. *Forensic Sci. Int. Genet.* **2012**, *6*, 41–46. **[Google Scholar] [CrossRef]**
10. Fonneløp, A.; Johannessen, H.; Gill, P. Persistence and secondary transfer of DNA from previous users of equipment. *Forensic Sci. Int. Genet. Suppl. Ser.* **2015**, *5*, e191–e192. **[Google Scholar] [CrossRef]**
11. Raymond, J.; Van Oorschot, R.A.; Walsh, S.J.; Roux, C.; Gunn, P.R. Trace DNA and street robbery: A criminalistic approach to DNA evidence. *Forensic Sci. Int. Genet. Suppl. Ser.* **2009**, *2*, 544–546. **[Google Scholar] [CrossRef]**
12. Van Oorschot, R.A.; Glavich, G.; Mitchell, R.J. Persistence of DNA deposited by the original user on objects after subsequent use by a second person. *Forensic Sci. Int. Genet.* **2014**, *8*, 219–225. **[Google Scholar] [CrossRef]**
13. Mayuoni-Kirshenbaum, L.; Waiskopf, O.; Finkelstein, N.; Pasternak, Z. How did the DNA of a suspect get to the crime scene? A practical study in DNA transfer during lock-picking. *Aust. J. Forensic Sci.* **2020**, 1–11. **[Google Scholar] [CrossRef]**
14. Fonnelop, A.E.; Ramse, M.; Egeland, T.; Gill, P. The implications of shedder status and background DNA on direct and secondary transfer in an attack scenario. *Forensic Sci. Int. Genet.* **2017**, *29*, 48–60. **[Google Scholar] [CrossRef] [PubMed]**

15. Johannessen, H.; Gill, P.; Roseth, A.; Fonneløp, A.E. Determination of shedder status: A comparison of two methods involving cell counting in fingerprints and the DNA analysis of handheld tubes. *Forensic Sci. Int. Genet.* **2021**, *53*, 102541. **[Google Scholar] [CrossRef]**

16. Goray, M.; van Oorschot, R. Shedder status: Exploring means of determination. *Sci. Justice* **2021**, *61*, 391–400. **[Google Scholar] [CrossRef]**

17. Schmidt, M.; Bamberg, M.; Dierig, L.; Kunz, S.N.; Wiegand, P. The diversity of shedder tests and a novel factor that affects DNA transfer. *Int. J. Leg. Med.* **2021**, *135*, 1267–1280. **[Google Scholar] [CrossRef]**

18. Gill, P.; Bleka, Ø.; Fonneløp, A.E. RFU derived LRs for activity level assignments using Bayesian Networks. *Forensic Sci. Int. Genet.* **2021**, *56*, 102608. **[Google Scholar] [CrossRef]**

19. Gill, P.; Benschop, C.; Buckleton, J.; Bleka, Ø.; Taylor, D. A Review of Probabilistic Genotyping Systems: EuroForMix, DNAStatistX and STRmix. *Genes* **2021**, *12*, 1559. **[Google Scholar] [CrossRef]**

20. Butler, J.M. *Forensic DNA Typing: Biology, Technology, and Genetics of STR Markers*; Academic Press: Cambridge, MA, USA, 2005. **[Google Scholar] [CrossRef] [Green Version]**

21. Bleka, Ø.; Storvik, G.O.; Gill, P. EuroForMix: An open source software based on a continuous model to evaluate STR DNA profiles from a mixture of contributors with artefacts. *Forensic Sci. Int. Genet.* **2016**, *21*, 35–44. **[Google Scholar] [CrossRef] [Green Version]**

22. European Network of Forensic Science Institutions "ENFSI" Guideline for Evaluative Reporting in Forensic Science [Online]. Available online: **http://enfsi.eu/wp-content/uploads/2016/09/m1_guideline.pdf** (accessed on 1 November 2021).

23. Fonneløp, A.E.; Egeland, T.; Gill, P. Secondary and subsequent DNA transfer during criminal investigation. *Forensic Sci. Int. Genet.* **2015**, *17*, 155–162. **[Google Scholar] [CrossRef]**

24. Otten, L.; Banken, S.; Schürenkamp, M.; Schulze-Johann, K.; Sibbing, U.; Pfeiffer, H.; Vennemann, M. Secondary DNA transfer by working gloves. *Forensic Sci. Int. Genet.* **2019**, *43*, 102126. **[Google Scholar] [CrossRef] [PubMed]**

25. Szkuta, B.; Ballantyne, K.N.; Kokshoorn, B.; van Oorschot, R.A. Transfer and persistence of non-self DNA on hands over time: Using empirical data to evaluate DNA evidence given activity level propositions. *Forensic Sci. Int. Genet.* **2018**, *33*, 84–97. **[Google Scholar] [CrossRef] [PubMed]**

26. Goray, M.; Fowler, S.; Szkuta, B.; van Oorschot, R. Shedder status—An analysis of self and non-self DNA in multiple handprints deposited by the same individuals over time. *Forensic Sci. Int. Genet.* **2016**, *23*, 190–196. **[Google Scholar]** **[CrossRef]** **[PubMed]**

15 SINGLE-CELL PROFILING ALLOWS FOR ACCURATE ANALYSIS OF DNA MIXTURES

1. INTRODUCTION: AN OVERVIEW

Decoding DNA mixtures is an exceedingly challenging endeavor in the realm of forensic genetics. The current standard protocol entails the retrieval of DNA from a sample obtained at a crime scene, such as swabs, followed by the determination of the amount of DNA obtained. Subsequently, the process involves the amplification of particular sections of DNA referred to as Short Tandem Repeats (STRs). The DNA fragments are subsequently detected by Capillary Electrophoresis (CE) and alleles are identified utilizing specialist software. A skilled DNA analyst interprets the DNA profiles of the contributions using their expertise and experience. Probabilistic genotyping software systems such as STRmix, LRmix, or TrueAllele may be employed in some situations to aid in the process of interpretation. In this generalized CE-STR analysis procedure, although the DNA may still be present in individual cells when collected, the cells and the DNA are combined together during the extraction step.

A mixture is obtained when there are multiple contributors to the sample. After creating the mixture profile, an analyst(s) endeavors to interpret the information in order to ascertain various details such as the number of contributors (NOC) in the mixture, the genotypes of each individual contributor, whether a specific individual is or is not a contributor to the mixture, and other related information. Based on the information obtained from this standard process, the mixture profile is typically analyzed using indirect methods. These methods involve determining the number of observed allelic peaks or calculating the weight of the evidence using a likelihood ratio (LR) approach. This allows for the inference of the number of contributors, with one being a known person and the other being an unknown person. The likelihood ratio (LR) is used to assess the probabilities of witnessing the same evidence under different competing hypotheses. For example, it can be used to compare the hypothesis that a mixture is composed of the victim and the suspect, with the hypothesis that the mixture is composed of the victim and a random person from the population. Occasionally, the process of determining the individual genotypes can be quite difficult due to the presence of overlapping alleles, allele drop-out (ADO), allele drop-in (ADI), and ambiguity in the number of contributors (NOC). All current forensic genetic approaches, and most likely those in the foreseeable future, are affected by these events. For instance, it can be challenging using the current standard approach to ascertain the NOC (Number of Contributors) of a three-person combination consisting of two parents and their child, as the youngster shares both alleles with the parents. Furthermore, it is challenging to ascertain whether a peak in a stutter site of a primary contributor allele consists of an allele from a secondary contributor and stutter, or merely a consequence of stutter. Therefore, certain aspects of a mixture may be difficult to determine with a high level of confidence.

To reduce the limitations of current standard mixture analyses, alternate methods have been proposed, such as enzymatic digestion differential extraction [**4**,**5**], reducing the allele overlap by using sequence-based alleles instead of traditional length-based alleles [**6**], employing probabilistic genotyping software [**1**,**2**,**3**], and so forth. While these alternate methods have added some value for mixture interpretation, the recent single-cell detection and analysis technologies may also reduce the challenges of DNA mixture interpretation. In theory, no mixture is generated as each cell is independently analyzed and thus the data from a single cell are derived from a single contributor. The DNA quantity of the contributors of a mixture can be more precisely determined by counting the isolated cells and cell types; the mixture ratio can be easily determined even though ratios may not be necessary anymore; and the alleles of each single cell can be called independently. The interpretation of these results is fundamentally different from the current interpretation approaches but should be much simpler. In other words, the computational intensive probabilistic genotyping methods, which are notably superior to previous manual methods, "guess", and often quite well, the possible genotypes of the contributors [**1**,**2**,**3**]; instead, single-cell profiles can be directly compared with the reference/known profiles to support whether or not an individual is a contributor of a mixture, if a good quality profile is obtained. However, due to technology limitations, a single-cell DNA profile usually contains ADO and/or ADI; in such scenarios, the consensus profiles from clustered cells can be used in single source profile comparisons or database searches. The consensus methods have been successfully used in single cell RNA studies [**7**,**8**] and forensic applications [**9**,**10**]. Therefore, this approach can reduce uncertainty regarding the profiles contributing to a mixture and thus provide stronger support for investigative leads and the judicial processes.

In a single-cell analysis workflow, cells are individually isolated from forensic samples. There are several methods developed that are capable of isolating single cells, such as manual micromanipulation [**11**,**12**,**13**,**14**,**15**,**16**,**17**], Laser Capture Microdissection (LCM) [**18**,**19**], Magnetic Activated Cell Sorting (MACS) flow cytometry [**20**], Fluorescent Activated Cell Sorting (FACS) flow cytometry [**21**,**22**,**23**,**24**,**25**], and a dielectrophoresis system (e.g., DEPArray) [**26,27,28**]. After isolation, the single cells can be individually amplified for the targeted regions (e.g., STRs) or subjected to Whole Genome Amplification (WGA) before amplification to enrich the DNA targets [**17,29,30,31,32,33**].

Li et al. [**34**] used the Identifiler kit (Thermo Fisher Scientific, South San Francisco, CA, USA) to amplify 15 autosomal STRs from 20 single sperm cells from one donor, and Han et al. [**19**] performed a similar study on 37 single sperm cells (from three donors) typing ten autosomal STRs with amplicon sizes <300 bp. In both studies, the consensus genotypes were consistent with the known donor genotypes. In Li et al. [**34**], the ADO rate was 25% and the ADI rate was 1.3%. In Han et al. [**19**] the ADO rate was lower (15%), possibly because the overall amplicon sizes were shorter, and the ADI rate was similar (1.4%). In these studies, the ADIs did not all reside at stutter positions of the true alleles. Some ADIs may have been due to DNA fragments from the other lysed cells which were co-amplified with a single cell and/or some ADIs might be due to genotype errors or low-level contamination. Anslinger et al. [**27**] collected 11 white blood cells with the DEPArray and genotyped 16 autosomal STRs using the PowerPlexESXfast kit (Promega Corp., Madison, WI, USA). They found that 82% of alleles were obtained (i.e., 18% ADO rate), only one ADI was observed at a stutter position (i.e., 0.3% ADI rate), and the full profile was recovered by consensus. Even with ADO and ADI, full profiles of the donors were obtained by consensus of the single-cell profiles.

A recent study by Chen et al. [**33**] employed a WGA kit (Qiagen REPLI-g single cell kit, Hilden, Germany) and subsequently typed the products with both the GlobalFiler kit (Thermo Fisher Scientific, South San Francisco, CA, USA) on the CE platform and the ForenSeq DNA Signature Prep kit (Verogen, San Diego, CA, USA), on a Massive Parallel Sequencing (MPS) platform (i.e., MiSeqFGx, Verogen, San Diego, CA, USA). For the three single cells analyzed with the GlobalFiler kit only four STR alleles dropped out (i.e., 3.2% ADO rate), and no ADI was observed. For the three single cells amplified with the ForenSeq kit, all autosomal STR alleles were successfully recovered without any ADI. The better performance of the ForenSeq kit, compared with that of the GlobalFiler kit, may be due to the overall shorter amplicon sizes of the STR alleles and/or a different way noise is presented with MPS data. Deleye et al. [**31**] recently compared four commercial WGA kits, Ampli1 (Menarini Silicon Biosystems, Castel Maggiore, BO, Italy), DOPlify (PerkinElmer, Waltham, MA, USA), PicoPLEX (Takara Bio, Kusatsu, Japan), and REPLI-g (Hilden, Germany) with single B-lymphoblastoid cells and found that the REPLI-g had the lowest ADO rate for STR profiling (i.e., 8.33%), while the other kits had ADO rates varying from 35.71%–60.71%. According to Chen et al. [**35**], the major commercial WGA kits yield various genome coverages (41%–92%) and ADO rates (25%–65%). Thus, it is reasonable to assume that substantial ADO and ADI would be observed with single-cell WGA per cell if a larger number of single cells are sampled.

All these studies demonstrate that single-cell technologies can recover the STR alleles of single cells with a high success rate. However, to date, no study has provided a systematic approach to interpret the STR data from single-cells derived from mixtures. This study is the first attempt to develop and assess performance of interpretation methods for characterizing multiple donors from single-cell profiles of mixtures. The study herein evaluates these interpretation methods by a simulation approach with various mixture scenarios and reasonable ADO and ADI rates, and provides data in terms of the number of cells, NOC, and ADO and ADI rates. Through this study, the potential of the single-cell technology workflow was evaluated for the probabilities of not detecting a minor contributor during cell sampling, the accuracies of estimating the NOC with single cells and sampling, the accuracies of recovering the true alleles by single-cell profile consensus, and the capabilities of single-cell approach for deconvolving mixtures with related contributors. The results convey the high precision that a single cell workflow could achieve for mixture interpretation.

2. METHODS

2.1. Simulation Methods

The autosomal STR genotypes of the single diploid cells (e.g., epithelial cells) were simulated using the method described in [36,37] for mixtures with unrelated and related contributors. In the simulation, the genotype of each marker of an individual was randomly simulated based on the allele frequencies and the population substructure correction (i.e., Fst) following Balding et al. [38]; the markers were assumed to be independent; the genotypes of related individuals were simulated assuming the parents transmit with equal probability a single allele at each locus to their children; the Two-Phase mutation model [39] was employed in the parent–child allelic transmissions; the mutation rates in the AABB report [40] were used. First, the genotype profile of each contributor in a mixture was simulated for the 21 autosomal STR loci in the GlobalFiler PCR amplification kit [41] using Caucasian population allele frequency data [42] and a Fst of 0.01. Second, based on the simulated genotype profiles of the contributors, ADO and ADI were incorporated to simulate the profiles of every single cell with a given number of cells of the contributors (e.g., 20 cells of contributor A and 20 cells of contributor B). Third, a mixture was formed by pooling the single-cell profiles of the contributors (e.g., a mixture with 40 cells in total from contributors A and B). Multiple mixtures with various mixture ratios were simulated, including mixtures with either unrelated (UR) or related contributors, namely, parent-child (PC), full-sibling (FS), and standard family trio (Table 1). Supplementary Table S1 gives an example of a simulated mixture with two unrelated contributors (20 cells each) in the ARFF format defined by WEKA [43], in which one attribute represents one allele of a locus (i.e., two attributes for a diploid genotype).

Table 1. Accuracies of estimating the NOC with the EM algorithm and Silhouettes method for the various mixture scenarios. UR = unrelated, PC = parent-child, and FS = full-sibling. D = 0.2; e = 0.01; 10,000 simulations for each mixture scenario. The smaller allele was always assigned as the first allele for a heterozygous genotype, and alleles were listed in a defined attribute order.

No. of Cells	2-Person Mixture							3-Person Mixture			
	Mixture	Diploid			Haploid			Mixture	Diploid		Haploid
		UR	PC	FS	UR	PC	FS		UR	Trio	UR
160	80,80	100.00%	100.00%	99.90%	99.90%	98.04%	92.16%	53,53,53	99.96%	88.32%	96.55%
	106,54	100.00%	99.98%	99.74%	99.88%	96.53%	89.27%	32,32,96	99.69%	85.64%	81.79%
	128,32	100.00%	99.85%	98.87%	99.16%	91.35%	81.70%	16,48,96	98.60%	78.46%	52.55%
	144,16	99.97%	98.54%	94.86%	95.97%	87.64%	81.28%				
	152,8	99.55%	91.59%	86.12%	92.43%	87.98%	83.97%				
80	40,40	100.00%	99.93%	99.91%	99.93%	98.73%	96.32%	26,26,26	99.97%	87.94%	96.44%
	53,27	100.00%	99.94%	99.78%	99.87%	98.17%	94.40%	16,16,48	99.68%	86.51%	82.47%
	64,16	99.99%	99.96%	99.53%	99.71%	94.37%	87.68%	8,24,48	98.36%	80.54%	54.15%
	72,8	99.96%	98.99%	96.55%	95.56%	85.95%	81.09%				
	76,4	99.67%	93.47%	88.56%	88.91%	83.54%	79.10%				
40	20,20	100.00%	99.99%	99.93%	99.72%	95.25%	92.41%	13,13,13	99.96%	88.29%	91.81%
	26,14	100.00%	99.93%	99.87%	99.76%	94.56%	90.23%	8,8,24	99.79%	86.06%	76.22%
	32,8	100.00%	99.89%	99.22%	99.18%	89.52%	81.95%	4,12,24	98.44%	79.79%	46.55%
	36,4	99.90%	98.03%	95.37%	93.40%	76.58%	71.13%				
	38,2	95.49%	78.98%	74.69%	84.31%	68.47%	65.89%				

No. of Cells	2-Person Mixture							3-Person Mixture			
	Mixture	Diploid			Haploid			Mixture	Diploid		Haploid
		UR	PC	FS	UR	PC	FS		UR	Trio	UR
20	10,10	100.00%	99.54%	99.02%	97.21%	82.30%	79.50%	6,6,6	99.81%	85.59%	69.19%
	13,7	99.98%	99.48%	98.84%	97.30%	82.77%	79.09%	4,4,12	99.52%	82.22%	51.34%
	16,4	99.96%	99.62%	98.92%	95.34%	73.85%	70.30%	2,6,12	89.80%	72.16%	21.69%
	18,2	99.61%	97.00%	94.29%	81.92%	59.54%	58.65%				
	19,1	98.51%	86.74%	82.46%	60.82%	53.45%	53.40%				
10	5,5	99.95%	96.77%	94.02%	82.78%	56.59%	55.61%	3,3,3	97.16%	66.81%	33.58%
	6,4	99.84%	96.65%	93.90%	80.38%	55.79%	53.41%	2,2,6	99.05%	76.26%	38.63%
	8,2	99.68%	94.61%	91.60%	78.59%	49.60%	48.24%	1,3,6	59.31%	37.65%	12.44%
	9,1	98.91%	89.19%	85.09%	65.71%	41.74%	42.05%				

The ADO model described in Balding et al. [44] was used in the simulation, in which D represents the probability of drop-out of one allele at a heterozygous locus, and D_2 represents the probability of drop-out at a homozygous locus, with $D_2 \approx ½D^2$. The rare scenarios that a locus of a diploid cell or a haploid cell presents more than two alleles or more than one allele, respectively, were ignored in this study (i.e., only up to two alleles or one allele with strongest signal(s) for diploid and haploid cells, respectively, would be called), allele size dependent degradation was ignored, and somatic mutations (i.e., the genotypes of the diploid cells from the same contributor may be different) were also ignored. The ADI was incorporated as a random genotyping error (with an ADI rate of e); namely, one allele may be randomly called as any other allele at this locus in the population data with equal chance. For example, changes to genotypes 10,10 → 10,15 or 10,11 → 10,15 for a homozygous and heterozygous locus, respectively, for diploid cells were produced to approximate ADI. Thus, a heterozygous locus with ADI has to have one ADO involved. However, the ADO and ADI events in general were assumed to be independent for the purposes of this study. When the ADO and ADI rates are 0, the genotype profiles of single diploid cells from the same contributor should be identical.

For the profiles of the single haploid cells the interpretation is slightly different because only one allele out of two at each autosomal locus is randomly distributed in the haploid cells. Therefore, the interpretation of haploid single-cell mixture profiles may be more complicated than that of diploid cells, even without ADO or ADI. In the haploid profile simulations, the alleles were generated by randomly selecting one allele of the two simulated diploid alleles at each locus and assuming independence among the loci. The same population data, parameters, ADO model, and ADI model as in the diploid simulations were used in simulating the haploid single-cell profiles.

2.2. Clustering Single Cells

To estimate the NOC of a mixture and allele consensus of each contributor, the single cells need to be grouped into clusters (or contributors). There is a large number of clustering algorithms that may serve this purpose. This study tested two classic clustering algorithms, Expectation–Maximization (EM) [45,46] and K-Means [47]. The EM algorithm assigns a probability distribution for the latent variables to each instance in the dataset (i.e., each single cell), which indicates the probability of it belonging to each of the clusters (i.e., each contributor in a mixture), computes parameters to optimize the likelihood of the model, and then repeats these two steps until convergence [43,45]. The EM algorithm implemented in WEKA uses multivariate Gaussians with diagonal covariance matrices as mixture components [46]. The EM algorithm does not explicitly require a distance measure between the instances. The K-Means algorithm randomly chooses an initial centroid for each cluster, each instance is assigned to its nearest centroid to form clusters, the centroids are updated with the instances in the same clusters, and these two steps are repeated until convergence [**47**]. The K-Means algorithm requires a distance measure between instances, which is Euclidean distance by default. The EM and SimpleKMeans algorithm implementations with default setting in the WEKA software package [**43**] were used in this study.

The clustering algorithms usually require a predefined number of clusters. To determine the best number of clusters, the Silhouettes method [**48**] implemented in ClusterMaker [**49**] was used. For a given number of clusters, the Silhouettes coefficient measures how similar an instance is to its own cluster compared to other clusters. The maximum value of the Silhouettes coefficient indicates the most likely number of clusters for a given range of the number of clusters [**50**]. For the mixtures in **Table 1**, the Silhouettes coefficients of two to six clusters were calculated and compared, and the number of clusters with the highest Silhouettes coefficient was determined as the NOC.

2.3. Single-Cell Profile Visualization

Multidimensional Scaling (MDS) was used to visualize the clustering of single-cell profiles. The three-dimensional (3D) MDS coordinates of every single-cell profile of a mixture were calculated using MDSJ [**51**], in which the distance between two profiles was measured as the number of mismatched alleles normalized by the total number of alleles. Then, the 3D coordinates were visualized by JTableSaw [**52**].

2.4. Consensus Method

For the single-cell profiles clustered into the same cluster, the genotype profiles of these single cells were determined by consensus and then compared with the true genotypes to assess the accuracies of recovering the true alleles by the clustering approach. For both diploid and haploid cells, the consensus of each allele was determined independently (i.e., alleles at the same position defined in the order. The consensus allele at a position (e.g., the first allele of the vWA locus in a diploid cell, or the only allele of the vWA locus in a haploid cell) in a cluster was defined as the most frequent allele at this position among the single-cell profiles in this cluster. The missing data due to ADO were ignored during consensus.

3. RESULTS

3.1. The Probability of Not Detecting a Contributor during Sampling

The existing CE-STR mixture analysis procedure may lack the ability to accurately determine the exact DNA input quantities of the contributors, as well as the likelihood of a modest contributor going undetected in a mixture. The limitations arise mostly from the fact that only contributors with a significant amount of input display allele peaks that surpass a detection threshold. Additionally, the presence of allele sharing can obscure the identification of a minor contributor. These issues become more pronounced as the number of contributors grows. Therefore, this procedure may fail to detect a modest contribution. Through single-cell analysis, the amount of a sample can be accurately measured by directly counting the cells present, however there is a possibility of sampling error during the process of isolating the cells. Hence, the likelihood of failing to identify a minor contribution can be approximated using single-cell profiling.

Assuming a large number of cells are available in a forensic sample (i.e., the sampling process follows a binomial distribution) and each cell has an equal chance to be sampled, the probability that none of the cells from a contributor with a mixture proportion of p is sampled in the total sampled n cells is $(-1p)^n$. **Table 2** estimates these probabilities for various mixture proportions of a contributor(s) and different total numbers of sampled cells, either diploid or haploid cells, with 6.6 pg DNA for a diploid cell and 3.3 pg DNA for a haploid cell [**53**]. The mixtures containing both diploid and haploid cells (e.g., a mixture of epithelial cells and sperm cells) were not considered (but can be readily determined if desired); these cellular mixtures can be separated either by counting the number of alleles across the loci or by technologies (e.g., DEPArray) to select cells from different tissues.

Table 2. The probability of not detecting a minor contributor in a set of single-cell samplings.

No. of Cells	DNA Quantity (pg)		The Mixture Proportion of a Contributor				
	Diploid	Haploid	1%	5%	10%	20%	50%
1	6.6	3.3	99.00%	95.00%	90.00%	80.00%	50.00%
5	33	16.5	95.10%	77.38%	59.05%	32.77%	3.13%
10	66	33	90.44%	59.87%	34.87%	10.74%	0.10%
15	99	49.5	86.01%	46.33%	20.59%	3.52%	0.00%
20	132	66	81.79%	35.85%	12.16%	1.15%	0.00%
40	264	132	66.90%	12.85%	1.48%	0.01%	0.00%
80	528	264	44.75%	1.65%	0.02%	0.00%	0.00%
160	1056	528	20.03%	0.03%	0.00%	0.00%	0.00%
500	3300	1650	0.66%	0.00%	0.00%	0.00%	0.00%

In general, as more cells are sampled the chance of not detecting a contributor decreases. For example, if 80 diploid cells are sampled (~528 pg DNA in total, close to the recommended DNA input of some commercial CE-STR kits, that is, 500 pg), there is a 44.75% chance that none of the cells from a 1% minor contributor is sampled; however, for a 5% minor contributor there is only a 1.65% chance of not detecting the minor contributor. For a low quantity DNA sample with only 15 cells (~99 pg), there is a 46.33% chance that a 5% minor contributor would not be detected. To obtain a high confidence of detecting a minor contributor with 1% or lower proportion, ≥500 single cells may need to be sampled.

3.2. The Accuracies of NOC Estimation

3.2.1. NOC Estimation by Clustering

The two-person and three-person mixtures were simulated with various numbers of cells, various mixture ratios, unrelated or related contributors, and both diploid and haploid cells (**Table 1**). In the simulation, the ADO and ADI rates were set as 0.2 and 0.01, respectively, which were similar to the estimations of several studies [**19**,**27**,**34**] and reflected the current performance of single-cell studies. For each mixture scenario, 10,000 cases were simulated, and the NOC for each case was estimated by the EM and K-Means clustering algorithms, together with the Silhouettes method. **Table 1** shows the accuracies of estimating the NOC for the simulated mixture scenarios with the EM algorithm, which provides higher accuracies compared with the K-Means algorithm.

For diploid cells, in general, higher accuracies were achieved with more cells and/or more balanced mixture ratios. If contributors were unrelated, almost all scenarios had >98% accuracy, except for three-person mixtures with a very few number of cells sampled and severely imbalanced mixture ratios (e.g., 2,6,12 and 1,3,6). Almost 100% accuracies were obtained if the contributors were unrelated, and there were more than 20 cells in total sampled. For PC and FS pairs, most of the estimated accuracies were higher than 98%, although, as expected, lower than those of UR pairs. Estimating the NOC for a mixture composed of a family trio can be particularly challenging with CE-STR analyses as well as with MPS; however, the accuracies with single cell analyses and the EM algorithm together with Silhouettes method were mostly >80%, except for scenarios with an extremely low number of sampled cells (e.g., ten cells) or highly imbalanced mixture ratios (e.g., 1:3:6).

The NOC accuracies with haploid cells were lower than those of diploid cells, because the same contributor may have many different haploid profiles due to meiosis, and the ADO and ADI would add more variants making the clustering of haploid cells more challenging. However, even with these limitations, high accuracies were achieved when the mixture ratios were relatively balanced and the number of sampled cells was high (e.g., 98.17% accuracy with a (53,27) Parent-Child mixture). Even for highly imbalanced mixtures (for example, 72,8), NOC accuracies were 95.56% and 85.95% for UR and PC contributors, respectively, which were still reasonably high.

3.2.2. Impact of ADO and ADI on NOC Estimation by Clustering

The accuracies of estimating the NOC can be impacted by the ADO and ADI rates. To investigate the impact of ADO and ADI, the same simulations as above were conducted for three mixture examples (**Figure 1**) with the same number of contributor cells (two, six, and 12 cells) for a three-person mixture, with different relationships among the contributors, different cell types, and various ADO and ADI rates.

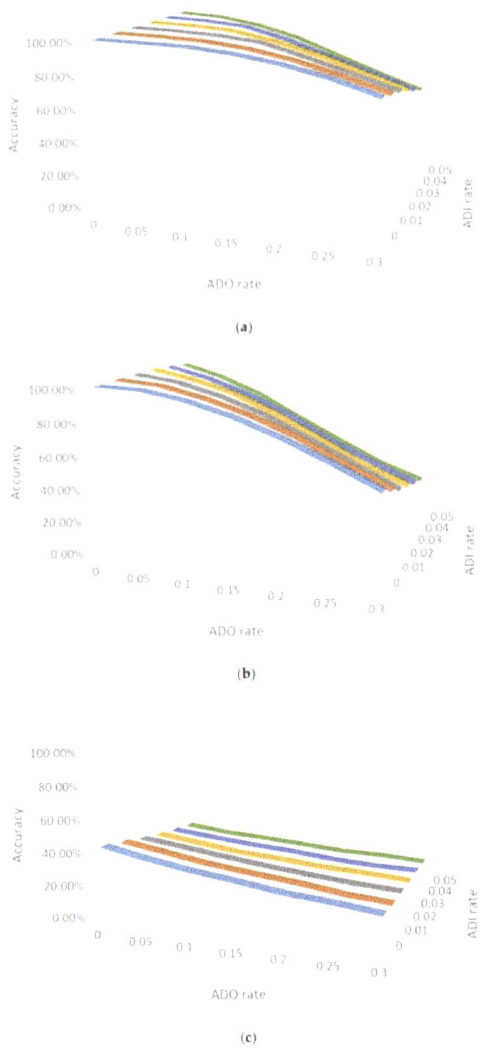

Figure 1. The accuracies of NOC estimation for 3-person mixtures with 2, 6, and 12 single cells with various ADO and ADI rates in three scenarios: (a) Diploid cells with three unrelated contributors, (b) Diploid cells with three family trio contributors (father, mother, and child), and (c) Haploid cells with three unrelated contributors. D = 0.2; e = 0.01; 10,000 simulations for each mixture scenario.

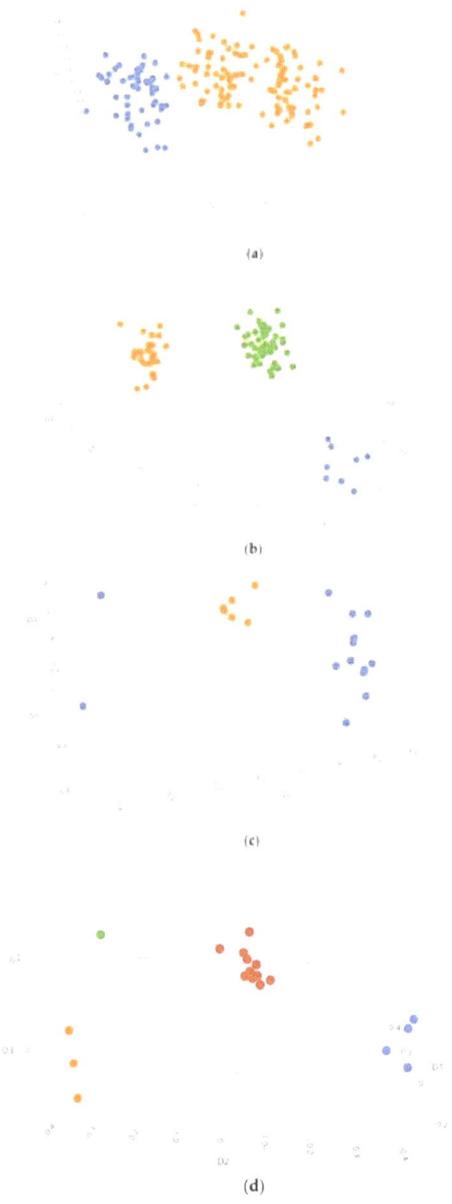

Figure 2. Mixture visualization examples with 3D MDS plots: (a) A family trio diploid cell mixture (53, 53, 53) incorrectly clustered with the EM algorithm and Silhouettes method as two contributors. (b) A family trio diploid cell mixture (8,24,48) correctly clustered with the EM algorithm and Silhouettes method as three contributors. (c) A 3-person haploid cell mixture (2,6,12) incorrectly clustered with the EM algorithm and Silhouettes method as two contributors. (d) A 3-person diploid cell mixture (4, 4, 12) incorrectly clustered with the EM algorithm and Silhouettes method as four contributors. Each color represents each cluster. Rotatable plots of these figures can be found in Supplementary Figure S1.

3.2.3. NOC Estimation by Visualization

Among these tested mixtures, the accuracies dropped substantially when the mixture ratios were extremely imbalanced (i.e., 9:1 and 19:1), which is mostly because the classic clustering algorithms do not perform well with imbalanced data (i.e., large sample size differences between or among clusters) [**54**]. The related contributors (i.e., the data points are closer to each other), haploid data (i.e., genotype uncertainty), and small numbers of cells (i.e., clustering with sparse data) can further reduce the accuracies. In such scenarios, particularly when the number of sampled cells is small, visualization of the distance between and among single-cell profiles may facilitate human judgement on the NOC. Visualization has been shown to be an effective way to estimate the NOC and differentiate cell types in single-cell applications [**55,56**].

Figure 2 displays several examples of 3D MDS plots. In **Figure 2**a, the EM algorithm and Silhouettes method together incorrectly determined a family trio mixture as a two-person mixture, because many cells of two specific contributors (i.e., one of the parents and the child) were very close to each other and the algorithm considered them as one contributor. However, with visualization, the NOC most likely could be determined as three. **Figure 2**b is an example with a correct NOC estimation, in which all contributors were clearly separated. In **Figure 2**c, two contributors (2 and 12) were incorrectly clustered into a single contributor, but with the MDS plot they appear separated. In **Figure 2**d, a three-person mixture was incorrectly determined as a four-person mixture. In one of the minor contributors, one single-cell profile was distant from the other three single-cells, which led to an incorrect NOC estimation. Even with MDS plot, it was challenging to determine the NOC, as both three and four clusters appeared possible.

3.2.4. NOC Estimation by Identity-by-State (IBS) Distance Measure

Another way to estimate the NOC is to measure the distances between single-cell profiles using predefined IBS thresholds. **Figure 3** shows the IBS distributions of unrelated and related pairs of individuals (UR, PC, and FS) and pairs of the known alleles of a contributor versus the consensus alleles with a given number of sampled cells, assuming the clustered profiles were from the same contributor. Apparently, there was no overlap between the IBS distribution of the unrelated and one cell with diploid cells, which means two unrelated diploid cells can be clearly separated simply by a reasonable IBS threshold (e.g., IBS $\leq$ 24 for the scenario of D = 0.2 and e = 0.01). This simple approach can readily determine if a group of single cells is from one contributor or multiple unrelated contributors (i.e., single source or mixture), as well as if a cell profile and a known profile from an unrelated individual may be from the same source. Thus, the NOC can be precisely determined by measuring the IBS distance between every pair of diploid cell profiles and a predefined maximum IBS distance for cells within the same cluster (e.g., IBS ≤ 24), assuming all contributors are unrelated. However, this approach may not work well for haploid cell profiles, because the consensus diploid profile relying on a single haploid cell would be homozygous across the loci. There would be a good proportion of the IBS distributions of UR and 1 haploid cell profile that would overlap. For example, %9.6 of unrelated pairs had IBS ≥ 16, and %9.7 of the consensus profiles from one haploid cell had IBS ≤ 17 (**Figure 3**b).

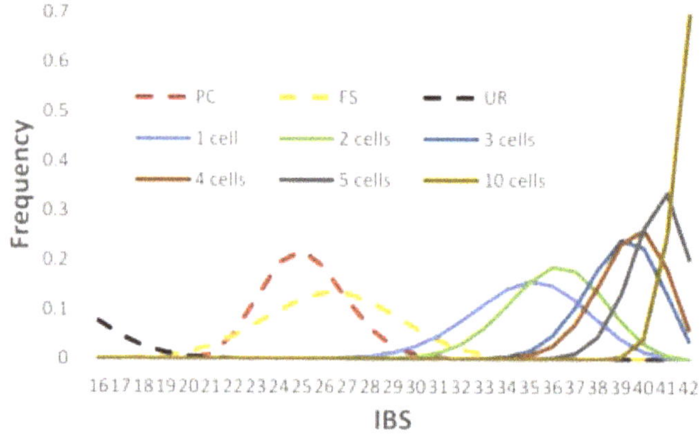

(a)

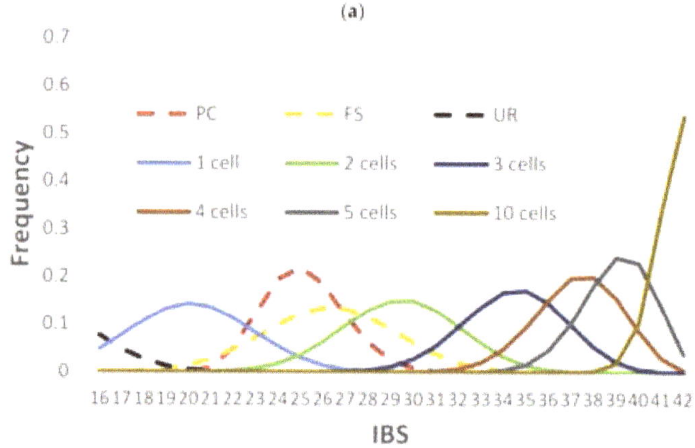

(b)

Figure 3. IBS distributions of unrelated and related pairs (UR, PC, and FS, in dotted lines) and pairs of the true profile versus the consensus profile with a given number of cells (in solid lines) based on (a) diploid cells and (b) haploid cells. $D = 0.2$; $e =$ **;0.01 1,000,000 simulations for each distribution. The IBS of "*n* cell(s)" is the IBS between true genotypes and the consensus genotypes from *n* cell(s). These distributions are based on 21 GlobalFiler markers and thus the maximum IBS is 42. The consensus profile might contain missing alleles, and these missing alleles were excluded in counting IBS.**

If the contributors are related, the IBS measure may not reach 100% accuracy to differentiate two unrelated cells from two related single cells (i.e., there were overlaps between the "one cell" distribution and the FS or PC distributions (**Figure 3**)). However, if multiple cells could be associated by either clustering, measured distances, or visualization, the consensus profile could differentiate a profile from his/her parent or full-siblings simply by IBS counting (e.g., a false rate less than 0.05% with a consensus profile from five cells for both PC and FS, and almost no error with ten cells (**Figure 3**)), which was similar to haploid cells (**Figure 3**b).

3.3. Accuracies of Consensus

3.3.1. IBS Distributions with Consensus

Once the NOC is estimated for a mixture, the consensus profile of each contributor can be generated by common alleles of the clustered single cells. As shown in **Figure 3**a, one single diploid cell as a single contributor mostly shared at least 28 alleles (with an average of 35 alleles) with the true profile, and the mismatches were mostly due to the high ADO rate (i.e., 0.2). Thus, there is sufficient precision to differentiate one single-cell profile from a random person profile (~0.03% average error rate with a threshold of IBS ≤ 24; the average error rate was the average of the false positive rate of determining a consensus profile and an unrelated profile as from the same source, and the false negative rate of determining the two profiles as being from different sources).With more cell profiles clustering in the same contributor, more shared alleles are found between the consensus profile and the true profile.

The consensus of haploid cell profiles would be more challenging, but would have more success than the current standard process. It is unlikely to determine the source of a single haploid cell profile (**Figure 3**b). However, the consensus profile of two haploid cells in a cluster, if they truly belong to the contributor, could be differentiated from a random person (<0.13% average error rate with a threshold of IBS ≤ 21), even though, on average, ~30% of the consensus alleles might not be correct (i.e., different from the true alleles or missing). For a contributor with ≥1 haploid cells, the consensus profile most likely would have ≤3 inconsistent alleles compared with the true profile. In general, these results were consistent with the simulation study by Lucy et al. [**57**], although ADO and ADI were not included in their study.

In addition, the results in **Figure 3** indicated that the mixture ratio estimation of the sperm samples with the current CE-STR analysis might not be accurate for the extremely low quantity DNA samples, even if there is no ADO, ADI, or degradation. Each sperm cell has only one allele at a locus, and a contributor with a few sperm cells may have highly imbalanced allele sampling. For example, three out of four sampled sperm cells may share one particular allele of a heterozygote and the other sperm could have a different allele, which could make this contributor appear as a 3:1 mixture with two homozygotes. The mixture ratio at different loci may randomly vary when the number of sampled haploid cells is small, which likely could lead to an incorrect interpretation or more alternate genotypes to consider with the CE-STR mixture analysis.

3.3.2 Suspects or their immediate family members as potential contributors

In certain cases, it may be necessary to determine if a suspect or one of the suspect's close relatives is a potential contributor to a mixture. Based on the IBS distributions in **Figure 3**a, with only a few diploid cell profiles, an IBS threshold of ≤34 can precisely differentiate the related individuals if a consensus profile is from a known profile or a full-sibling of this known profile. With only three diploid cell profiles, 0.68% of the consensus profiles would have an IBS ≤ 34 (i.e., false negatives), 0.18% of the FS pairs would have IBS ≥ 35 (i.e., false positives or more appropriately adventitious associations). Sampling more cells can reduce the false rates. For haploid mixtures, if the same threshold is used (i.e., IBS ≤ 34), five haploid cell profiles would be needed to achieve high accuracy (i.e., 0.64% false negative rate). Thus, it only requires a consensus accuracy as low as 83.3% (35/42) to precisely differentiate if a consensus profile is from a known individual (e.g., suspect) or this individual's first-degree relative(s).

The average accuracies of consensus for all diploid mixtures and most haploid mixtures with balanced contributors were higher than 83.3% (**Table 3**). Therefore, except for minor contributors with ≤2 diploid cells or ≤4 haploid cells sampled, the consensus profiles of each contributor in the mixtures should be able to determine the source of the profile with very high accuracies (**Table 3**). Again, more cells sampled would increase these accuracies.

Table 3. The average of the average accuracies of consensus alleles compared with the true alleles of the contributors. Only the mixtures with the correct NOC estimation were included. The NOC was determined using the EM algorithm and Silhouettes method. D = 0.2; e = 10,000 ;0.01 simulations for each mixture scenario.

No. of Cells.	2-Person Mixture							3-Person Mixture			
	Mixture	Diploid			Haploid			Mixture	Diploid		Haploid
		UR	PC	FS	UR	PC	FS		UR	Trio	UR
160	80,80	100.00%	100.00%	100.00%	100.00%	99.93%	99.28%	53,53,53	100.00%	100.00%	100.00%
	106,54	100.00%	100.00%	99.99%	100.00%	99.80%	98.83%	32,32,96	100.00%	100.00%	99.94%
	128,32	100.00%	99.98%	99.90%	99.55%	98.64%	97.17%	16,48,96	99.96%	99.95%	99.47%
	144,16	99.85%	98.91%	97.91%	94.96%	93.92%	91.86%				
	152,8	98.01%	90.16%	89.05%	81.77%	85.70%	84.88%				
80	40,40	100.00%	100.00%	100.00%	100.00%	99.96%	99.73%	26,26,26	99.99%	99.99%	99.97%
	53,27	100.00%	100.00%	99.99%	99.99%	99.90%	99.49%	16,16,48	99.91%	99.91%	99.73%
	64,16	99.94%	99.93%	99.89%	99.75%	99.09%	97.90%	8,24,48	99.47%	99.45%	98.53%
	72,8	99.19%	98.65%	98.01%	95.15%	92.18%	90.78%				
	76,4	96.40%	93.09%	91.63%	81.77%	83.45%	83.02%				
40	20,20	99.96%	99.96%	99.96%	99.93%	99.80%	99.52%	13,13,13	99.69%	99.68%	99.00%
	26,14	99.88%	99.87%	99.87%	99.67%	99.41%	98.90%	8,8,24	98.94%	98.95%	97.56%
	32,8	99.21%	99.18%	99.08%	98.37%	96.61%	95.43%	4,12,24	97.79%	97.78%	94.55%
	36,4	96.79%	96.17%	95.85%	91.16%	87.64%	86.57%				
	38,2	85.79%	81.23%	81.22%	76.48%	78.91%	78.88%				

No. of Cells.	Mixture	2-Person Mixture						3-Person Mixture			
		Diploid			Haploid			Mixture	Diploid		Haploid
		UR	PC	FS	UR	PC	FS		UR	Trio	UR
20	10,10	99.16%	99.17%	99.13%	98.20%	96.75%	95.68%	6,6,6	96.98%	96.97%	92.68%
	13,7	98.90%	98.87%	98.85%	97.53%	95.57%	94.51%	4,4,12	95.72%	95.71%	89.05%
	16,4	96.83%	96.80%	96.71%	93.02%	89.51%	88.51%	2,6,12	94.02%	94.02%	84.83%
	18,2	92.78%	92.31%	92.01%	81.99%	77.81%	78.35%				
	19,1	90.85%	88.80%	87.86%	69.30%	68.57%	69.78%				
10	5,5	96.33%	96.34%	96.30%	91.56%	87.57%	86.61%	3,3,3	92.68%	92.74%	78.80%
	6,4	95.37%	95.33%	95.33%	90.70%	86.72%	85.95%	2,2,6	89.45%	89.47%	75.58%
	8,2	92.11%	91.99%	91.97%	82.58%	78.46%	77.78%	1,3,6	90.86%	90.42%	66.12%
	9,1	90.81%	90.36%	90.17%	71.50%	67.26%	67.31%				

3.3.3. Accuracies of Consensus by Clustering

The distributions depicted in Figure 3 were based on the assumption that the cells grouped together as contributors were indeed the cells originating from that specific contributor. Cells can potentially be grouped together with an inaccurate source. Assuming there is a sufficient number of cells, the incorrectly clustered cells should be outnumbered by the cells that actually belong to a contributor. If a good clustering algorithm is used and there are enough cells clustered to this contributor, the incorrect alleles should be eliminated through consensus filtering.

To test this hypothesis, the consensus accuracies of clustering results were estimated using the same simulations conducted as in **Table 1**, and the average accuracies of consensus alleles were calculated for each simulated mixture. For example, a mixture with two contributors may have consensus accuracies of 0.8 and 0.9, respectively, and the average accuracy would be 0.85. For mixtures with an imbalanced mixture ratio (e.g., 1:3:6), the minor contributor(s) usually had lower consensus accuracies, mostly due to the small numbers of sampled cells. **Table 3** shows the average of 10,000 simulations for the average accuracies of the consensus profile. Only the cases with the correct NOC estimation were included in these calculations.

As expected, given the correct NOC estimation, mixtures with a relatively high number of cells have very high consensus accuracies. Full profiles can be precisely obtained by consensus of the cells in the same clusters, if there are enough sampled diploid cells for a mixture with relatively balanced contributor ratios. The mixtures with relatively low consensus accuracies were mainly due to the low accuracies of the minor contributors with really low numbers of sampled cells. For example, the mixtures with (152,8) had an average accuracy of 98.01%, but the average consensus accuracy of the major contributor (i.e., 152) was 100%, but the minor contributor (i.e., eight) had a lower accuracy (i.e., ~96%).

Even with imbalanced contributors, all mixtures with at least five diploid cell profiles for the minor contributors achieved >96% average consensus accuracy, mostly greater than 97.6% (i.e., one allele mismatch or missing, 41/42). Particularly, the individual profiles of the family trio mixtures could be recovered even with only 40 sampled cells and imbalanced contributors, if one allele mismatch or missing was allowed. These observations were consistent with those in **Figure 3**, which indicates that consensus was able to filter out most incorrect alleles, if not all, caused by either ADO and ADI or incorrect clustering. These consensus profiles with just a few mismatched or missing alleles should be precise enough to generate reasonable partial matches (or hits) in a forensic DNA database (e.g., CODIS) search and support the potential identities of the contributors.

The consensus accuracies of the haploid cell mixtures, in line with expectations, were lower than those of diploid cell mixtures. However, the accuracies were still high for mixtures with enough sampled cells (e.g., ≥20 cells) and relatively balanced contributors, even with related contributors (e.g., 94.51% for FS mixtures of 13:7). Similar to the NOC estimation with haploid cell profiles, the accuracy dropped substantially for extremely imbalanced contributors. This observation is most likely because the EM algorithm does not perform well for imbalanced data. When the number of cells of a cluster is few (e.g., ≤4), together with ADI and relatively high ADO, 1 or 2 incorrectly clustered cells may dramatically change the consensus alleles. Better clustering algorithms specifically designed for imbalanced data may be able to improve the consensus accuracies.

Moreover, the relationships between the contributors appear to have little impact on the consensus accuracies when the number of cell profiles of each contributor is high (e.g., ≥16) or the contributors are relatively balanced; the consensus accuracies of unrelated and related contributors of the mixtures were almost identical for diploid cells (**Table 3**). In other words, the relationships between the contributors could impact the NOC estimation, but once NOC was correctly estimated, the consensus within each cluster solely relied on the profiles in the same cluster. However, when the number of cells from a contributor is small, the consensus accuracies vary more so, since a few incorrectly clustered cells may change the consensus alleles.

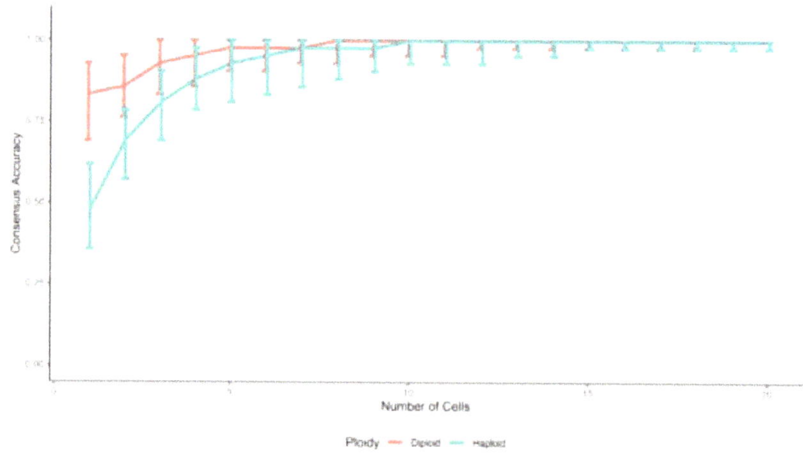

Figure 4. Consensus accuracy as a function of the number of cells in the cluster, with 1:1 ratio mixtures. D **= 0.2;** e **= 10,000 ;0.01 simulations for each mixture. The bars represent the %95 confidence intervals.**

In addition, the two-person mixtures with 1:1 ratio were used as examples to investigate the consensus accuracy as a function of the number of cells in the cluster (**Figure 4**), for both diploid and haploid cells. The accuracies increase fastly with more cells added into the same clusters when the number of cells in a cluster is relatively small. With these balanced mixtures, a cluster with ≥5 diploid cells or ≥7 haploid cells had ≥97.6% accuracy (i.e., one allele mismatched or missing), which was consistent with the results in **Table 3**. For the extremely imbalanced mixtures, the consensus accuracy of the minor contributor can fluctuate more due to the EM algorithm's poor performance with imbalanced mixture ratios.

3.3.4. Impact of ADO and ADI on Consensus Accuracies

The above analysis used the typical ADO and ADI rates as an example ($D = 0.2$ and $e = 0.01$) to investigate the performance of consensus. The consensus accuracies are expected to be higher with lower ADO and ADI rates. **Figure 5** shows the IBS distributions of the consensus profiles from one diploid cell (**Figure 5**a) and four haploid cells (**Figure 5**b), as examples, compared with the IBS distributions of UR, PC and FS. The IBS distribution of two diploid cell profiles was only slightly higher than that of one diploid cell profile, particularly when the ADO and ADI rates were low. Therefore, only one diploid cell profile scenario was used here to assess whether one diploid cell profile can provide high accuracies of determining the sources of contributors. If $D = 0$ and $e = 0$, the IBS should always be 42 for diploid cells. Thus, this scenario is not included in **Figure 5**a.

As expected, lower ADO and ADI rates provided higher IBS values on average. ADO had a larger impact on the distributions, mostly because the ADO rates were much larger than ADI rates in the simulated scenarios. Same as shown in **Figure 3**a, with $D = 0.2$ and $e = 0.01$, one diploid cell profile would be good enough to differentiate if it is from a known reference person or an unrelated person. If contributors are related, $D \le 0.05$ may be required to precisely differentiate a known individual from his/her full-siblings (i.e., ~%0.15 average error rate with a threshold of IBS ≤ 34), although $D = 0.1$ may also be acceptable (i.e., ~%0.8 average error rate with a threshold of IBS ≤ 32). With more cells sampled and clustered, the consensus accuracies are expected to be higher, which would tolerate higher ADO and ADI rates to obtain similar accuracies.

With four haploid cells, even without ADO and ADI, the full profile may not be able to be recovered in most cases. However, the consensus profile, even partial, can be differentiated from an unrelated person, even with an ADO rate of 0.2. To differentiate a known individual from his/her full-siblings, with four haploid cells in a single cluster, it would be desirable to have an ADO ≤ 0.05 (i.e., ~0.35% average error rate with a threshold of IBS ≤ 34).

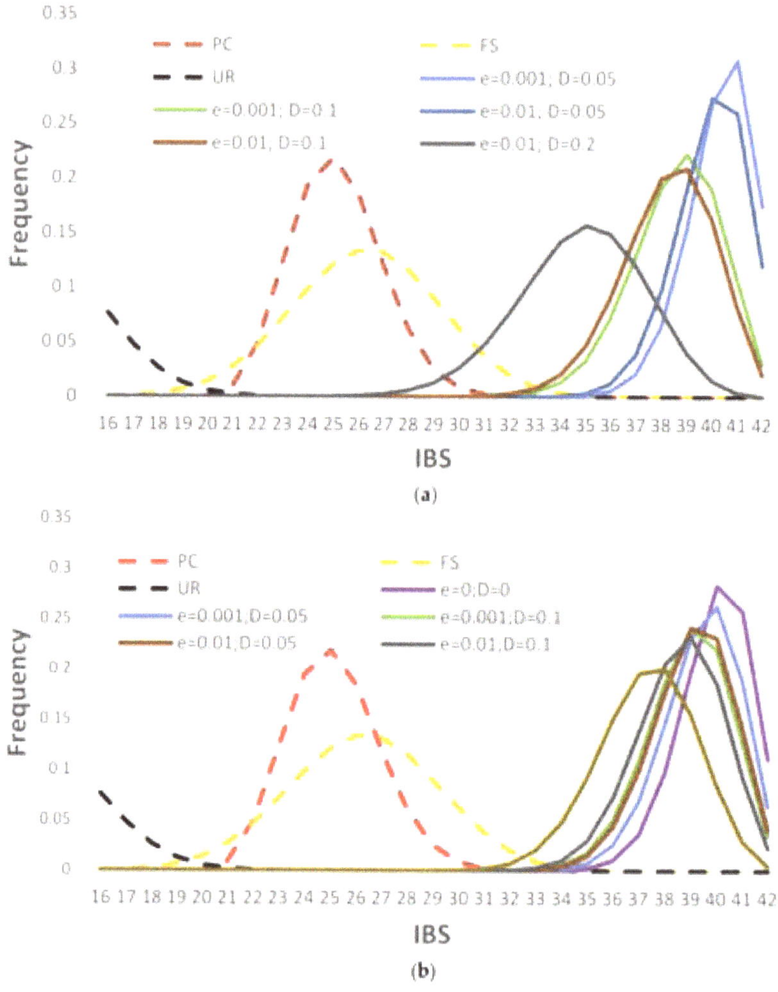

Figure 5. IBS distributions of unrelated and related pairs (UR, PC, and FS, in dotted lines) and pairs of the true profile versus the consensus profile with a given number of cells (in solid lines) for (a) 1 diploid cell and (b) 4 haploid cells. 1,000,000 simulations for each distribution. Missing alleles are excluded in counting IBS.

Additionally, the same examples as in **Figure 1** were simulated to assess the consensus accuracies with potential incorrectly clustered cells even with correct NOC estimations (**Figure 6**). The accuracies decrease with either increasing ADO or ADI rates. As expected, the accuracies of diploid cell mixtures decrease slower than those of haploid cell mixtures, because haploid cells have half the alleles of diploid cells. The ADI seems to have a bigger impact on haploid mixtures compared with diploid mixtures. Therefore, for genotyping mixture samples comprised of sperm, it would be better to have a low ADI rate, while a relatively high ADO rate may be acceptable.

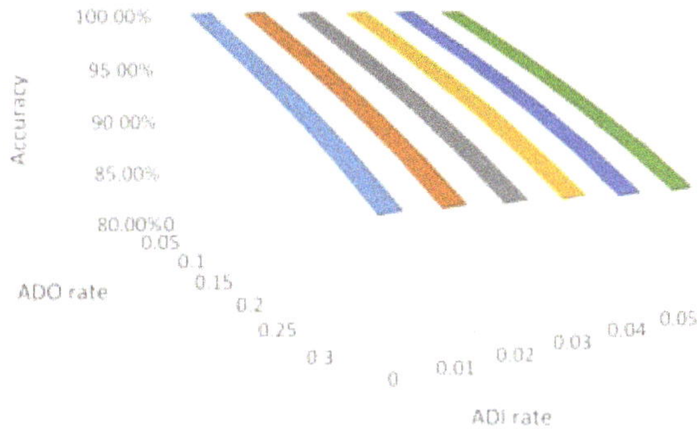

(a)

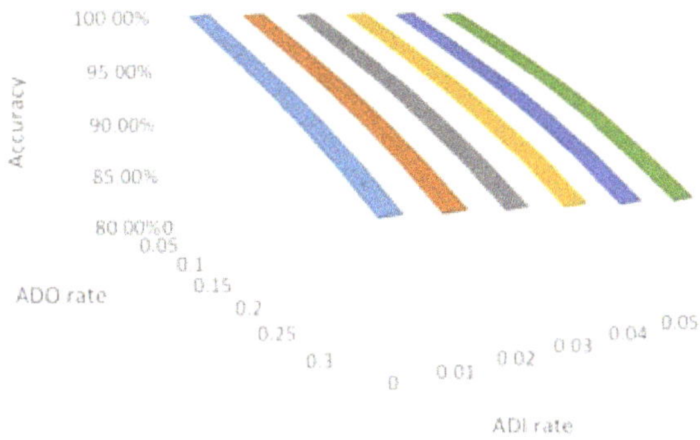

(b)

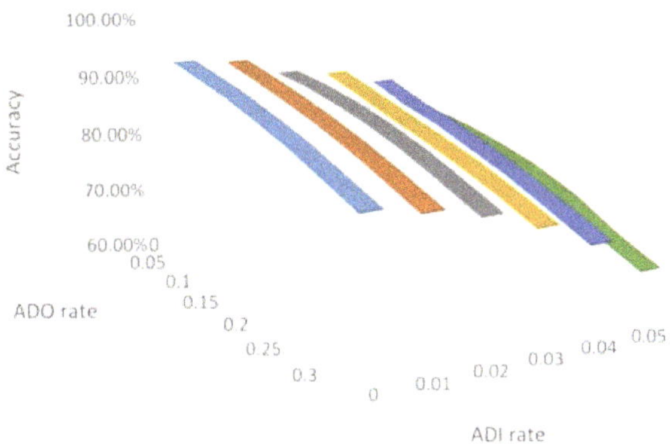

(c)

Figure 6. The accuracies of consensus for -3person mixtures with 6 ,2, and 12 single cells with various ADO and ADI rates in three scenarios: (a) diploid cells with three unrelated contributors, (b) diploid cells with three family trio contributors (father, mother, and child), and (c) haploid cells with three unrelated contributors.

Furthermore, akin to the information presented in Table 3, the consensus accuracies for diploid cells were nearly equivalent in both the unrelated and familial trio diploid mixtures. Put simply, clustering is more effective in estimating the number of contributors (NOC) for mixtures with unrelated contributors compared to family trio mixtures (89.80% vs. 72.16%, respectively). However, if the NOC is accurately predicted, clustering works equally well for both unrelated and family trio contributor mixtures.

4. DISCUSSION

4.1. A Paradigm Shift of Mixture Interpretation

Single-cell technologies have been utilized in several research for forensic applications, mostly focusing on wet-lab procedures to produce genotype profiles from individual cells. The simulations conducted in this study are the first to offer guidance on a comprehensive workflow for interpreting single-cell mixture profiles. They assess the effectiveness of single-cell profiles in interpreting mixtures and demonstrate that single-cell based interpretation can provide a level of precision that surpasses that of manual or probabilistic genotyping approaches used with DNA profiles generated by standard CE-STR analyses.

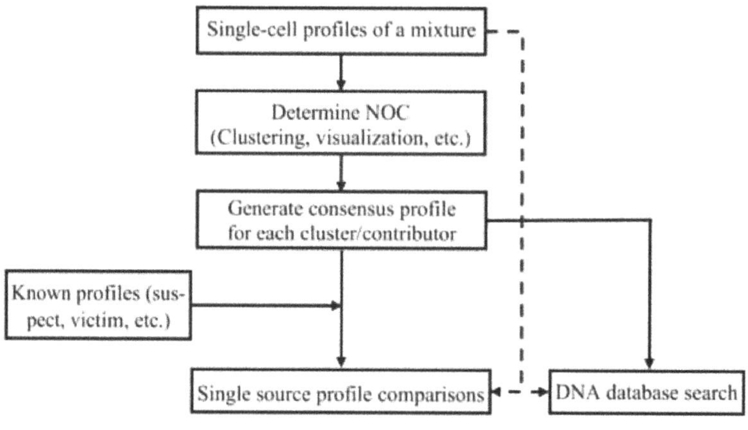

Figure 7. The interpretation workflow of single-cell profiles generated from a mixture.

Figure 7 illustrates the general mixture interpretation workflow with the single-cell mixture profiles. First, with the single-cell profiles generated from a mixture, the NOC of the single-cell profiles is estimated, either by clustering (e.g., EM algorithm and Silhouettes method), distance measure (e.g., IBS), or visualization. Second, the consensus profile of each cluster/contributor is generated from the single-cell profiles in each cluster. Then, the consensus profile(s) would be either directly searched against a DNA database(s) for partial or full matches (or hits), or compared with known profiles (e.g., suspect's profile) to determine if these profiles, or these profiles' relatives, potentially could be the contributor(s) of the mixture. In cases in which a full, or close to full profile, is obtained from a single cell(s), the profile may be directly compared with a known reference profile(s) or searched in a DNA database(s).

With single-cell technologies, a DNA mixture is not formed during the DNA extraction process. Thus, the interpretation is simplified and based on single source profile comparisons. Therefore, the need for methods, such as probabilistic genotyping, that indirectly determine the genotypes of the contributors, may not be required in some cases. Instead, the single-cell based approach can directly determine the profiles of the contributors from single-cell profile clusters or from the single source profiles of the individual cells.

For the single source comparisons to confirm or exclude individuals, IBS counting can be employed to determine the potential source or relationship between a consensus profile from single cells and a reference profile to be compared by a predefined IBS threshold as shown in **Figure 3** and **Figure 4**. A LR based approach still can apply in these comparisons [**58**]. Let A be the consensus profile and B be a reference profile (e.g., a suspect's profile). The hypotheses can be framed, for example, as A and B are from the same source or from different and unrelated sources. Then, the following equation can be used to calculate an LR, with ADO and ADI incorporated to explain the allele difference between A and B, if there are any. However, the ADO and ADI rates of the consensus profile are different from the rates for single cell profiles. These rates should be calculated by consensus similar to the methods of estimating quality scores in MPS [**59**].

$$\mathrm{LR} = \frac{\Pr(\mathrm{A}, \mathrm{B} | \mathrm{A} \ \& \ \mathrm{B} \text{ are from same source})}{\Pr(\mathrm{A}, \mathrm{B} | \mathrm{A} \ \& \ \mathrm{B} \text{ are unrelated})}. \tag{1}$$

In Equation (1), the numerator is the product of the transition probabilities in **Table 1** of [**60**] and the genotype frequency of B, and the denominator is simply the product of the genotype frequencies of A and B. When all alleles of A and B are identical and there is no ADO and ADI, this equation reduces to the inverse of the traditional Random Match Probability (RMP) of A or B.

In cases when a relative of, for example, the suspect is hypothesized as the contributor, both IBS and LR approaches can still apply. For the IBS approach, a different threshold can be used, as shown in **Figure 3** and **Figure 4**. For the LR approach, the denominator hypothesis would be different (Equation (2)). The numerator can be calculated by expanding **Table 1** in [**61**] with the incorporation of ADO and ADI, using a similar model to that described in [**60**].

$$\mathrm{LR} = \frac{\Pr(\mathrm{A}, \mathrm{B} | \mathrm{A} \ \& \ \mathrm{B} \text{ are related as a specific relationship})}{\Pr(\mathrm{A}, \mathrm{B} | \mathrm{A} \ \& \ \mathrm{B} \text{ are unrelated})}. \tag{2}$$

If there is no known profile to compare, the consensus profile (full or partial) may be searched against a forensic DNA database(s) (e.g., CODIS). If a consensus profile is generated from a cluster with a good number of sampled cells (e.g., ≥10), the partial matches (or hits) highly likely belong to family members of the sample donor or the sample donor.

4.2. The Capabilities and Limitations

This study provides evidence that, utilizing single-cell technology, DNA mixtures can be accurately disentangled using appropriate methods of interpretation. The probability of overlooking a tiny contributor in a DNA mixture can be measured using the single-cell profiles. Additionally, the proportion of each contributor in the mixture can be calculated at the individual cell or chromosomal level by analyzing the single-cell profile clusters. The proportions observed in the isolated cells may not accurately represent the proportions in the entire mixture sample prior to cell isolation. This is especially true when the proportion is small or the number of sampled cells is small. This discrepancy can be attributed to the efficiency of cell extraction and the presence of sampling errors. It is possible that only a small percentage of the cells are actually isolated for genotype profiling. Nevertheless, in certain situations, the significance or relevance of mixing proportions may be overshadowed by the mere task of recognizing the cells that are there.

The probabilities of not detecting a minor contributor estimated in **Table 2** assume a large number of cells are available for analysis. For low-quantity DNA samples, which may contain a few cells, hypergeometric distribution may be a better model to estimate these probabilities. However, the hypergeometric distribution requires knowing the total number of cells in a forensic sample, which may not be possible or precisely determined. In contrast, binomial distribution does not require the total number of cells in the estimation, and thus it may still be a good option in low-quantity DNA samples.

NOC estimation is the key component in this interpretation workflow. If the NOC is overestimated (e.g., a two-person mixture is determined as a three-person mixture), most likely the single cells of a contributor(s) are separated into two or more clusters, and the consensus profile of each cluster may still be valid. However, if the NOC is underestimated, cells from multiple contributors would be merged into one cluster, which could lead to an incorrect consensus profile. Nevertheless, by the clustering approach, the NOC of a mixture can be precisely determined for the mixtures with a high number of cells and relatively balanced mixture ratios, even for mixtures with related contributors.

In this study, for the tested classic clustering algorithms, EM performed better than K-Means in general, but K-Means performed better in some mixtures, such as the (5,5) mixture with unrelated contributors. Both of these classic algorithms do not perform well with imbalanced mixture ratios [**54**]. The EM algorithm does not explicitly require a distance measure, and the Euclidean distance was used in the K-Means algorithm. While the Euclidean distance measure may not be the best distance for this application, it was used herein as a proof of concept. Specifically designed distances for STR alleles, such as counting the number of step differences between STR alleles, only counting the heterozygous loci to mitigate the impact of ADO, or allele or genotype matching based measures (e.g., IBS), could be better choices for algorithms which require a distance measure. The Silhouettes method was used in this study to determine the number of clusters, but it does not work for the scenarios when NOC = 1 and is not necessarily the best method for single-cell STR data. Other methods, such as the Davies–Bouldin index, the Dunn index, and the Akaike information criterion (AIC), may help determine the number of clusters and may be tested in the future. Overall, better clustering algorithms (such as t-SNE clustering [**62**], semi-supervised and self-supervised learning [**63,64**]) and better methods of determining the number of clusters should be pursued for the single-cell profiles with imbalanced mixtures. In addition, similar to that of other single-cell applications, visualization can be helpful to estimate the NOC, as human judgment may do better deciding the NOC when the number of instances is small (e.g., few to hundreds), which is the case for many forensic mixture samples. However, human judgment can be subjective; different individuals may reach different conclusions. Thus, visualization is more of a tool to assist the users to decide NOC or limit the range of possible NOCs.

With a correct NOC, a high precision consensus profile can be obtained through the clustered cell profiles. The sources of these consensus profiles can be precisely determined by comparing them with the known profiles or considering the untyped relatives. The accuracy relies on the number of cell profiles in the clusters, as well as the ADO and ADI rates. But even with $D = 0.2$ and $e = 0.01$, the consensus accuracies of all contributors approach %100 for mixtures with ≥40 cells and relatively balanced mixture ratios. Particularly, the family trio mixtures can be deconvoluted to recover the genotypes of the individual contributors, even though related, which in theory is more challenging with CE-STR analysis without conditioning on one of the contributors [**65**]. This study did not investigate the consensus accuracies when the estimated NOC was incorrect. It was mainly because, if the number of the true profiles (e.g., two) and the number of consensus profiles (e.g., three) are different, it is not easy to decide which consensus profiles should compare with which true profiles. The accuracies with NOC overestimation and underestimation will be investigated in future studies.

The success of the single-cell mixture interpretations was based on the assumption that a reasonable number of intact single cells can be isolated from a DNA sample, there are no doublets or triplets of cells during cell isolation, and most of the alleles of these isolated single cells can be successfully genotyped or sequenced. Although these assumptions may not be held, low quality data per call could be filtered out using a threshold or with algorithms. For example, doublets can be detected by an outlier detection algorithms [**66**] or simply counting the number of alleles (i.e., triplet and/or quadruplet allele genotypes per locus) or heterozygotes. Regardless, the interpretation methods developed in this study allowed for imperfect data, such as small sample sizes, ADO, and ADI.

For some degraded forensic samples, many of the cells may be lysed, and thus isolating intact cells may be difficult. Additionally, if extracellular DNA is present, the DNA fragments released from the lysed cells may be isolated with an intact cell(s), which could cause extra alleles to appear for some of the single-cell profiles. Validation studies should be able to determine the impact of such events occurring. However, the impact may be small because the extracellular fragments are a small subset of the genome; forensic applications currently target only a very small percentage of the human genome (i.e., ~10k bp out of 3 billion bp). In addition, if the NOC is estimated correctly and the consensus works well, the drop-in alleles still may be able to be filtered out.

Even if a good number of cells are isolated, genotype profiling of the single copy alleles in each single cell can be challenging, depending on the ADO and ADI rates. The ADO and ADI rates can have a substantial impact on the analysis, such as NOC estimation and allele consensus. The current ADO and ADI rates observed in various studies [**19,27,34**] may still be high for each single-cell profile compared with the profiles generated from bulk cells. However, WGA could help enrich targets [**31,33**] for higher accuracies of genotype profiling, and single-cell isolation and amplification technologies are likely to become more robust.

Generating consensus profiles from multiple single cells that are clustered can mitigate the effects of high ADO and ADI rates, as described in **Section 3.3**. If there are ≥10 cell profiles in a cluster, with $D = 0.2$ and $e = 0.01$, a full consensus profile can be generated with no mismatched alleles or only one mismatched allele on average, with up to three mismatched alleles, in ~%99.96 of the diploid mixtures and ~%99.7 of the haploid mixtures. The process is also highly successful for developing the consensus profile among related profiles.

In addition, the NOC estimation and allele consensus with single-cell profiles do not need to assume independence among the tested alleles and markers, since the machine learning algorithms by nature are able to accommodate dependent variants. Therefore, in contrast to the complexity of current practices to incorporate dependence among the markers, the single-cell based mixture interpretation can directly use the genetically linked markers and markers in Linkage-Disequilibrium (LD) together, as well as the dependence among the alleles at the same locus, without any statistical corrections. This feature makes the single-cell based interpretation readily able to include as many markers as possible to increase the discriminating power and better identify individual contributors. However, current methods, including the current probabilistic genotyping approach, do not readily address dependent markers.

4.3. Future Improvements

As a proof-of-concept, this study only simulated the 21 STR loci in the GlobalFiler kit by simplified but reasonable ADO and ADI models to demonstrate the capabilities of using single-cell profiles to deconvolve mixtures. The same principles should apply for other autosomal markers as well, either more STRs, Single Nucleotide Polymorphisms (SNPs), or Insertion-Deletions (InDels). With more markers typed, the accuracy of NOC estimation should increase. ADO may have a higher impact on performance using SNPs than using STRs, since most of the SNPs have only two alleles at a locus and three possible genotypes at each locus. But the huge numbers of SNPs in the human genome should be able to overcome this limitation quite well. Indeed, whole genome sequencing (WGS) of single cells has been described for other applications [**67**].

The lineage markers on mitochondrial DNA (mtDNA) and the Y chromosome can be very useful for NOC estimation and allele consensus due to their unique features. The diploid cells from the same contributor should have the same Y chromosome markers (if a male contributor) and mtDNA haplotype (except for heteroplasmy or somatic mutations). Use of Y markers in the haploid cells is more complicated. Half of all sperm from a male contributor should have the same Y chromosome barring mutation, and the other half of the sperm should carry the same X chromosome barring mutation. Therefore, an interpretation process could be designed that first clusters profiles based on Y chromosome markers, followed by autosomal or X chromosome markers. X chromosome markers have been useful in certain cases, such as incest cases. However, mixture interpretation using X chromosome markers with CE-STR analysis has been problematic. With single-cell profiling, X chromosome markers could be used routinely for mixture deconvolution, especially when combined with Y markers.

The ADO model employed in this study did not consider degradation due to different allele and locus sizes. Accuracies could increase if degradation was included in the ADO model, such as including the pattern of the ADO rates for different sizes of alleles could improve the modeling of ADO rates for different alleles. A detection threshold still needs to be determined to differentiate alleles from noise. The ADI model used in this study assumed a random genotyping error while a large proportion of the ADIs are likely exaggerated stutters for specific STR alleles. It would be desirable to incorporate a stutter model, which should increase the accuracies of NOC estimation and allele consensus; less distinctly different alleles would be observed in each cluster with a stutter model allowing the profiles to cluster more effectively. However, not all wet-lab technologies would generate stutters. For SNPs and Indels, there is no stutter. For STRs, the stutters may be minimized by the recently developed Unique Molecular Index (UMI) technology [**68**]. There are many factors that may impact ADO and ADI, such as types of markers, sequencing platforms, library preparation kits, amplification kits, sequencing kits, protocols, software programs to call alleles, detection thresholds, and so forth. Thus, technology and marker specific ADO and ADI models may be the better options. In addition, the independence of the ADO and/or ADI events among the cells may not be assumed with real samples [**69**]. Therefore, more sophisticated ADO and ADI models that incorporate the underlying dependence between the cells may be better refracting the actual scenarios. Furthermore, this study used all single cells in analysis. In real casework, a filter should be implemented to remove the single cells with the number of missing alleles or loci beyond a certain threshold, and thus, only good quality data would be used in clustering and profile comparisons, which in general should improve interpretation accuracies.

Another improvement in accuracy could be to use clustering algorithms with a better performance on imbalanced data [**70**]. Particularly, when a minor contributor is represented with an extremely small number of sampled cells (e.g., one or two), outlier detection algorithms (e.g., Isolation Forest [**71**]) or self-supervised learning [**72**], instead of the classic clustering algorithms, may be used to select the single cell(s) of the minor contributor(s). Improving accuracies of determining the number of clusters, other NOC methods, such as Davies–Bouldin index, Dunn index, Akaike information criterion (AIC), and so forth, may also be considered. The current allele consensus method treats two alleles at the same locus independently. Performance may be improved if consensus is performed at the genotype level. With such improvements, it is reasonable to expect that this single-cell interpretation workflow should be able to interpret mixtures with a higher number of contributors (i.e., NOC ≥ 4). In addition, the measures of several different NOCs may be very close to each other, and multiple alleles may be close likely being the consensus allele in the consensus profile. Picking one among multiple similarly likely NOCs or alleles could lead to error. Thus, the interpretation methods may allow multiple possible scenarios with estimated weights, similar to the continuous probabilistic genotyping approaches. These scenarios may be evaluated, compared, and combined in a Bayesian or LR fashion.

4.4. Costs

The comprehensive process of interpreting mixtures using current standard procedures involves wet-lab experimentation, data analysis, legal processes, law enforcement investigation, and other related activities. It is important to do a cost benefit analysis in a holistic manner, taking into account not just the laboratory expenses but also other factors. The cost of the wet-lab workflow may be $50–100 per sample currently (including sample collection, DNA extraction, quantification, amplification, genotyping, but not including the labor), depending on the costs of the reagents and local infrastructure. There are more costs that tend to occur after the DNA mixture profile is generated, such as DNA interpretation (including software and hours of work interpreting mixture profiles), evidence presentation in the court, and so forth. Additional costs may also occur, including the law enforcement investigation, reanalyzing samples/aliquots, hiring external reviewers for complex mixture interpretation, and so forth.

Compared with the current routine methods, the single-cell based mixture interpretation methods will add more costs to the wet-lab workflow, depending on the number of cells to be analyzed and the reagents used in the workflow. In general, the more cells that are analyzed, the higher are the costs. However, the cost increase may not be linear to the number of cells to be tested, as the volume of reagents may be reduced in many steps (such as amplification), and some steps of the workflow will not be needed (such as extraction and quantification). More importantly, the single-cell based methods provide a much more simplified interpretation with higher precision. Thus, the costs of the subsequent work (such as data interpretation, additional wet-lab analysis, law enforcement investigation, external review, judicial proceedings, etc.) may be substantially reduced, and the overall costs of mixture cases, at a systems level, may be reduced.

Therefore, although it may be more expensive to sample several individual cells compared to evaluating a single mixture sample in the wet lab, the costs of analysis, such as MPS, are decreasing, and it is probable that more efficient single-cell technologies will be created. Nevertheless, it is crucial to consider the cost of analysis in relation to the benefits of reducing uncertainty in mixture interpretation. This includes saving time and resources for law enforcement investigations and minimizing the expenses of judicial proceedings. Ultimately, these measures would result in cost savings for the government's overall investigation.

5. Concise Summary

Emerging single-cell technologies have been utilized in wet-lab experiments to tackle the difficulties associated with assessing forensic mixed evidence. Nevertheless, there has been a lack of commitment towards creating a methodical strategy for evaluating the individual cellular profiles obtained from the combinations. This study is the initial endeavor to establish a thorough interpretation procedure that analyzes single-cell profiles from mixtures both individually and holistically. This method involves evaluating the genotypes of individual cells, estimating the number of contributors in the single-cell profiles, creating a consensus profile for each contributor, and ultimately using the consensus profile(s) for searching a DNA database or comparing them with known profiles to identify their possible origins. The efficacy of this workflow for interpreting single-cell data was evaluated through simulations involving different mixture scenarios and empirical measurements of allele drop-out and drop-in rates. The assessment included determining the accuracy of estimating the number of contributors (NOC), the accuracy of identifying the true alleles through consensus, and the ability to separate mixtures with related contributors. The results affirm that the interpretation of mixtures based on single cells can offer a level of precision that is unattainable with current routine CE-STR tests. A novel framework for interpreting mixtures is now accessible to improve the interpretation of forensic DNA casework.

5. AT A GLANCE

Our study has shown that the interpretation of mixtures using single-cell analysis can offer a level of accuracy that is not attainable with the present conventional CE-STR analyses. A novel framework for interpreting mixtures is now accessible to augment the efficacy of forensic DNA casework. Greater emphasis should be placed on enhancing the intricacies of the interpretive techniques, as well as refining the methods for isolating individual cells and characterizing their genotypes. Furthermore, similar to other recent advancements like familial searching or investigative genetic genealogy, this single-cell profiling method can be employed when previous methods have failed to yield the desired result. This assumes that there is an adequate sample for further testing or that a portion of the biological evidence has been preserved for reanalysis using an alternative approach.

REFERENCES

1. Bright, J.-A.; Taylor, D.; McGovern, C.; Cooper, S.; Russell, L.; Abarno, D.; Buckleton, J. Developmental validation of STRmix™, expert software for the interpretation of forensic DNA profiles. *Forensic Sci. Int. Genet.* **2016**, *23*, 226–239. **[Google Scholar] [CrossRef] [PubMed]**
2. Gill, P.; Haned, H.; Eduardoff, M.; Santos, C.; Phillips, C.; Parson, W. The open-source software LRmix can be used to analyse SNP mixtures. *Forensic Sci. Int. Genet. Suppl. Ser.* **2015**, *5*, e50–e51. **[Google Scholar] [CrossRef] [Green Version]**
3. Perlin, M.W.; Legler, M.M.; Spencer, C.E.; Smith, J.L.; Allan, W.P.; Belrose, J.L.; Duceman, B.W. Validating TrueAllele DNA mixture interpretation. *J. Forensic Sci.* **2011**, *56*, 1430–1447. **[Google Scholar] [CrossRef]**
4. Gill, P.; Jeffreys, A.J.; Werrett, D.J. Forensic application of DNA ‘fingerprints’. *Nature* **1985**, *318*, 577–579. **[Google Scholar] [CrossRef] [PubMed]**

5. Voorhees, J.C.; Ferrance, J.P.; Landers, J.P. Enhanced elution of sperm from cotton swabs via. enzymatic digestion for rape kit analysis. *J. Forensic Sci.* **2006**, *51*, 574–579. **[Google Scholar] [CrossRef] [PubMed]**
6. Novroski, N.M.; King, J.L.; Churchill, J.D.; Seah, L.H.; Budowle, B. Characterization of genetic sequence variation of 58 STR loci in four major population groups. *Forensic Sci. Int. Genet.* **2016**, *25*, 214–226. **[Google Scholar] [CrossRef] [PubMed]**
7. Gan, Y.; Li, N.; Zou, G.; Xin, Y.; Guan, J. Identification of cancer subtypes from single-cell RNA-seq data using a consensus clustering method. *BMC Med. Genom.* **2018**, *11*, 117. **[Google Scholar] [CrossRef]**
8. Cui, Y.; Zhang, S.; Liang, Y.; Wang, X.; Ferraro, T.N.; Chen, Y. Consensus clustering of single-cell RNA-seq data by enhancing network affinity. *Brief. Bioinform.* **2021**, bbab236. **[Google Scholar] [CrossRef]**
9. Kling, D.; Tillmar, A.O.; Egeland, T. Familias 3-extensions and new functionality. *Forensic Sci. Int. Genet.* **2014**, *13*, 121–127. **[Google Scholar] [CrossRef] [PubMed]**
10. Vullo, C.M.; Catelli, L.; Rodriguez, A.A.I.; Papaioannou, A.; Merino, J.C.Á.; Lopez-Parra, A.; Gaviria, A.; Baeza-Richer, C.; Romanini, C.; González-Moya, E. Second GHEP-ISFG exercise for DVI: "DNA-led" victims' identification in a simulated air crash. *Forensic Sci. Int. Genet.* **2021**, *53*, 102527. **[Google Scholar] [CrossRef]**
11. Budowle, B.; Eisenberg, A.J.; Van Daal, A. Validity of low copy number typing and applications to forensic science. *Croat. Med. J.* **2009**, *50*, 207–217. **[Google Scholar] [CrossRef]**
12. Li, C.; Qi, B.; Ji, A.; Xu, X.; Hu, L. The combination of single cell micromanipulation with LV-PCR system and its application in forensic science. *Forensic Sci. Int. Genet. Suppl. Ser.* **2009**, *2*, 516–517. **[Google Scholar] [CrossRef]**
13. Huffman, K.; Hanson, E.; Ballantyne, J. Recovery of single source DNA profiles from mixtures by direct single cell subsampling and simplified micromanipulation. *Sci. Justice* **2021**, *61*, 13–25. **[Google Scholar] [CrossRef] [PubMed]**
14. Xu, C.; Feng, L.; Yang, F.; Jia, J.; Ji, A.-Q.; Hu, L.; Li, C.-X. Mucosal cell isolation and analysis from cellular mixtures of three contributors. *J. Forensic Sci.* **2015**, *60*, 783–786. **[Google Scholar] [CrossRef] [PubMed]**
15. Brück, S.; Evers, H.; Heidorn, F.; Müller, U.; Kilper, R.; Verhoff, M.A. Single cells for forensic DNA analysis-from evidence material to test tube. *J. Forensic Sci.* **2010**, *56*, 176–180. **[Google Scholar] [CrossRef] [PubMed]**
16. Schneider, C.; Müller, U.; Kilper, R.; Siebertz, B. Low copy number DNA profiling from isolated sperm using the aureka®-micromanipulation system. *Forensic Sci. Int. Genet.* **2012**, *6*, 461–465. **[Google Scholar] [CrossRef] [PubMed]**

17. Theunissen, G.M.; Rolf, B.; Gibb, A.; Jäger, R. DNA profiling of sperm cells by using micromanipulation and whole genome amplification. *Forensic Sci. Int. Genet. Suppl. Ser.* **2017**, *6*, e497–e499. **[Google Scholar] [CrossRef] [Green Version]**

18. Vandewoestyne, M.; Deforce, D. Laser capture microdissection in forensic research: A review. *Int. J. Leg. Med.* **2010**, *124*, 513–521. **[Google Scholar] [CrossRef] [PubMed] [Green Version]**

19. Han, J.-P.; Yang, F.; Xu, C.; Wei, Y.-L.; Zhao, X.-C.; Hu, L.; Ye, J.; Li, C.-X. A new strategy for sperm isolation and STR typing from multi-donor sperm mixtures. *Forensic Sci. Int. Genet.* **2014**, *13*, 239–246. **[Google Scholar] [CrossRef]**

20. Xu, Y.; Xie, J.; Chen, R.; Cao, Y.; Ping, Y.; Xu, Q.; Hu, W.; Wu, D.; Gu, L.; Zhou, H.; et al. Fluorescence- and magnetic-activated cell sorting strategies to separate spermatozoa involving plural contributors from biological mixtures for human identification. *Sci. Rep.* **2016**, *6*, 36515. **[Google Scholar] [CrossRef] [PubMed] [Green Version]**

21. Schoell, W.M. Separation of sperm and vaginal cells with flow cytometry for DNA typing after sexual assault. *Obstet. Gynecol.* **1999**, *94*, 623–627. **[Google Scholar] [CrossRef]**

22. Verdon, T.J.; Mitchell, R.J.; Chen, W.; Xiao, K.; van Oorschot, R.A. FACS separation of non-compromised forensically relevant biological mixtures. *Forensic Sci. Int. Genet.* **2015**, *14*, 194–200. **[Google Scholar] [CrossRef]**

23. Miller, J.M.; Brocato, E.R.; Yadavalli, V.K.; Greenspoon, S.A.; Ehrhardt, C.J. Testing hormone-specific antibody probes for presumptive detection and separation of contributor cell populations in trace DNA mixtures. *bioRxiv* **2019**. **[Google Scholar] [CrossRef]**

24. Stokes, N.A.; Stanciu, C.E.; Brocato, E.R.; Ehrhardt, C.J.; Greenspoon, S.A. Simplification of complex DNA profiles using front end cell separation and probabilistic modeling. *Forensic Sci. Int. Genet.* **2018**, *36*, 205–212. **[Google Scholar] [CrossRef]**

25. Dean, L.; Kwon, Y.J.; Philpott, M.K.; Stanciu, C.E.; Seashols-Williams, S.J.; Cruz, T.D.; Sturgill, J.; Ehrhardt, C.J. Separation of uncompromised whole blood mixtures for single source STR profiling using fluorescently-labeled human leukocyte antigen (HLA) probes and fluorescence activated cell sorting (FACS). *Forensic Sci. Int. Genet.* **2015**, *17*, 8–16. **[Google Scholar] [CrossRef] [PubMed]**

26. Williamson, V.R.; Laris, T.M.; Romano, R.; Marciano, M.A. Enhanced DNA mixture deconvolution of sexual offense samples using the DEPArray™ system. *Forensic Sci. Int. Genet.* **2018**, *34*, 265–276. **[Google Scholar] [CrossRef] [Green Version]**

27. Anslinger, K.; Graw, M.; Bayer, B. Deconvolution of blood-blood mixtures using DEPArrayTM separated single cell STR profiling. *Rechtsmedizin* **2019**, *29*, 30–40. **[Google Scholar] [CrossRef] [Green Version]**

28. Fontana, F.; Rapone, C.; Bregola, G.; Aversa, R.; de Meo, A.; Signorini, G.; Sergio, M.; Ferrarini, A.; Lanzellotto, R.; Medoro, G.; et al. Isolation and genetic analysis of pure cells from forensic biological mixtures: The precision of a digital approach. *Forensic Sci. Int. Genet.* **2017**, *29*, 225–241. **[Google Scholar] [CrossRef] [PubMed] [Green Version]**

29. England, R.; Nancollis, G.; Stacey, J.; Sarman, A.; Min, J.; Harbison, S. Compatibility of the ForenSeq™ DNA signature prep kit with laser microdissected cells: An exploration of issues that arise with samples containing low cell numbers. *Forensic Sci. Int. Genet.* **2020**, *47*, 102278. **[Google Scholar] [CrossRef]**

30. Lim, H.H.; Srisiri, K.; Rerkamnuaychoke, B.; Bandhaya, A. A comparative study of whole genome amplification and low-template DNA profiling. *Forensic Sci. Int. Genet. Suppl. Ser.* **2019**, *7*, 509–511. **[Google Scholar] [CrossRef]**

31. Deleye, L.; Plaetsen, A.-S.V.; Weymaere, J.; Deforce, D.; Van Nieuwerburgh, F. Short tandem repeat analysis after whole genome amplification of single B-lymphoblastoid cells. *Sci. Rep.* **2018**, *8*, 1–6. **[Google Scholar] [CrossRef] [Green Version]**

32. Schneider, P.; Balogh, K.; Naveran, N.; Bogus, M.; Bender, K.; Lareu, M.; Carracedo, A. Whole genome amplification—the solution for a common problem in forensic casework? *Int. Congr. Ser.* **2004**, *1261*, 24–26. **[Google Scholar] [CrossRef]**

33. Chen, M.; Zhang, J.; Zhao, J.; Chen, T.; Liu, Z.; Cheng, F.; Fan, Q.; Yan, J. Comparison of CE- and MPS-based analyses of forensic markers in a single cell after whole genome amplification. *Forensic Sci. Int. Genet.* **2020**, *45*, 102211. **[Google Scholar] [CrossRef]**

34. Li, C.-X.; Han, J.-P.; Ren, W.-Y.; Ji, A.-Q.; Xu, X.-L.; Hu, L. DNA profiling of spermatozoa by laser capture microdissection and low volume-PCR. *PLoS ONE* **2011**, *6*, e22316. **[Google Scholar] [CrossRef] [PubMed]**

35. Chen, C.; Xing, D.; Tan, L.; Li, H.; Zhou, G.; Huang, L.; Xie, X.S. Single-cell whole-genome analyses by linear amplification via. transposon insertion (LIANTI). *Science* **2017**, *356*, 189–194. **[Google Scholar] [CrossRef] [PubMed] [Green Version]**

36. Ge, J.; Budowle, B.; Chakraborty, R. Choosing relatives for DNA identification of missing persons. *J. Forensic Sci.* **2011**, *56*, S23–S28. **[Google Scholar] [CrossRef] [PubMed]**

37. Ge, J.; Budowle, B. Kinship index variations among populations and thresholds for familial searching. *PLoS ONE* **2012**, *7*, e37474. **[Google Scholar] [CrossRef] [PubMed]**

38. Balding, D.J.; Nichols, R.A. DNA profile match probability calculation: How to allow for population stratification, relatedness, database selection and single bands. *Forensic Sci. Int.* **1994**, *64*, 125–140. **[Google Scholar] [CrossRef]**

39. Ge, J.; Budowle, B.; Chakraborty, R. DNA identification by pedigree likelihood ratio accommodating population substructure and mutations. *Investig. Genet.* **2010**, *1*, 8. **[Google Scholar] [CrossRef] [Green Version]**

40. AABB. *Relationship Testing Annual Reports*; AABB: Bethesda, MD, USA, 2008. **[Google Scholar]**

41. Wang, D.Y.; Gopinath, S.; Lagacé, R.E.; Norona, W.; Hennessy, L.K.; Short, M.L.; Mulero, J.J. Developmental validation of the GlobalFiler® express PCR amplification kit: A -6dye multiplex assay for the direct amplification of reference samples. *Forensic Sci. Int. Genet.* **2015**, *19*, 148–155. **[Google Scholar] [CrossRef]**

42. Moretti, T.R.; Moreno, L.I.; Smerick, J.B.; Pignone, M.L.; Hizon, R.; Buckleton, J.S.; Bright, J.-A.; Onorato, A.J. Population data on the expanded CODIS core STR loci for eleven populations of significance for forensic DNA analyses in the United States. *Forensic Sci. Int. Genet.* **2016**, *25*, 175–181. **[Google Scholar] [CrossRef] [Green Version]**

43. Eibe, F.; Hall, M.A.; Witten, I.H. *Online Appendix for Data Mining: Practical Machine Learning Tools and Techniques*; The WEKA Workbench; Morgan Kaufmann: Burlington, MA, USA, 2016. **[Google Scholar]**

44. Balding, D.J.; Buckleton, J. Interpreting low template DNA profiles. *Forensic Sci. Int. Genet.* **2009**, *4*, 1–10. **[Google Scholar] [CrossRef]**

45. Dempster, A.P.; Laird, N.M.; Rubin, D.B. Maximum likelihood from incomplete data via the EM algorithm. *J. R. Stat. Soc. Ser. B* **1977**, *39*, 1–22. **[Google Scholar]**

46. Witten, I.H.; Frank, E.; Hall, M.A. *Data Mining: Practical Machine Learning Tools and Techniques*, 3rd ed.; Elsevier/Morgan Kaufmann: San Francisco, CA, USA, 2011. **[Google Scholar] [CrossRef] [Green Version]**

47. Kanungo, T.; Mount, D.; Netanyahu, N.; Piatko, C.; Silverman, R.; Wu, A. An efficient k-means clustering algorithm: Analysis and implementation. *IEEE Trans. Pattern Anal. Mach. Intell.* **2002**, *24*, 881–892. **[Google Scholar] [CrossRef]**

48. Rousseeuw, P.J. Silhouettes: A graphical aid to the interpretation and validation of cluster analysis. *J. Comput. Appl. Math.* **1987**, *20*, 53–65. **[Google Scholar] [CrossRef] [Green Version]**

49. Morris, J.H.; Apeltsin, L.; Newman, A.M.; Baumbach, J.; Wittkop, T.; Su, G.; Bader, G.D.; Ferrin, T.E. ClusterMaker: A multi-algorithm clustering plugin for cytoscape. *BMC Bioinform.* **2011**, *12*, 436. **[Google Scholar] [CrossRef] [Green Version]**
50. Kaufman, L.; Rousseeuw, P.J. *Finding Groups in Data: An Introduction to Cluster Analysis*; John Wiley and Sons: Hoboken, NJ, USA, 2009. **[Google Scholar]**
51. Pich, C. MDSJ: Java Library for Multidimensional Scaling (Version 0.2). 2009. Available online: **http://www.inf.uni-konstanz.de/algo/software/mdsj** (accessed on 21 October 2020).
52. Java Data Frame and Visualization Library. Available online: **https://jtablesaw.github.io/tablesaw/** (accessed on 21 October 2020).
53. Cairns, P. Gene methylation and early detection of genitourinary cancer: The road ahead. *Nat. Rev. Cancer* **2007**, *7*, 531–543. **[Google Scholar] [CrossRef] [PubMed]**
54. Xuan, L.; Zhigang, C.; Fan, Y. Exploring of clustering algorithm on class-imbalanced data. In Proceedings of the 2013 8th International Conference on Computer Science & Education, Colombo, Sri Lanka, 26–28 April 2013; pp. 89–93. **[Google Scholar]**
55. Wang, B.; Zhu, J.; Pierson, E.; Ramazzotti, D.; Batzoglou, S. Visualization and analysis of single-cell RNA-seq data by kernel-based similarity learning. *Nat. Methods* **2017**, *14*, 414–416. **[Google Scholar] [CrossRef]**
56. Amir, E.-A.D.; Davis, K.L.; Tadmor, M.D.; Simonds, E.; Levine, J.H.; Bendall, S.C.; Shenfeld, D.K.; Krishnaswamy, S.; Nolan, G.P.; Pe'Er, D. ViSNE enables visualization of high dimensional single-cell data and reveals phenotypic heterogeneity of leukemia. *Nat. Biotechnol.* **2013**, *31*, 545–552. **[Google Scholar] [CrossRef] [PubMed] [Green Version]**
57. Lucy, D.; Curran, J.; Pirie, A.; Gill, P. The probability of achieving full allelic representation for LCN-STR profiling of haploid cells. *Sci. Justice* **2007**, *47*, 168–171. **[Google Scholar] [CrossRef] [PubMed]**
58. Council, N.R. *An Update: The Evaluation of Forensic DNA Evidence*; National Academies Press: Washington, DC, USA, 1996. **[Google Scholar]**
59. Wall, J.D.; Tang, L.F.; Zerbe, B.; Kvale, M.N.; Kwok, P.-Y.; Schaefer, C.; Risch, N. Estimating genotype error rates from high-coverage next-generation sequence data. *Genome Res.* **2014**, *24*, 1734–1739. **[Google Scholar] [CrossRef] [PubMed] [Green Version]**
60. Ge, J.; Budowle, B. Modeling one complete versus triplicate analyses in low template DNA typing. *Int. J. Leg. Med.* **2013**, *128*, 259–267. **[Google Scholar] [CrossRef] [PubMed]**

61. Ge, J.; Chakraborty, R.; Eisenberg, A.; Budowle, B. Comparisons of familial DNA database searching strategies. *J. Forensic Sci.* **2011**, *56*, 1448–1456. **[Google Scholar] [CrossRef] [PubMed]**

62. Kobak, D.; Berens, P. The art of using t-SNE for single-cell transcriptomics. *Nat. Commun.* **2019**, *10*, 1–14. **[Google Scholar] [CrossRef] [Green Version]**

63. Chen, L.; He, Q.; Zhai, Y.; Deng, M. Single-cell RNA-seq data semi-supervised clustering and annotation via. structural regularized domain adaptation. *Bioinformatics* **2021**, *37*, 775–784. **[Google Scholar] [CrossRef]**

64. Chen, L.; Zhai, Y.; He, Q.; Wang, W.; Deng, M. Integrating deep supervised, self-supervised and unsupervised learning for single-cell RNA-seq clustering and annotation. *Genes* **2020**, *11*, 792. **[Google Scholar] [CrossRef]**

65. Lin, M.-H.; Bright, J.-A.; Pugh, S.N.; Buckleton, J.S. The interpretation of mixed DNA profiles from a mother, father, and child trio. *Forensic Sci. Int. Genet.* **2020**, *44*, 102175. **[Google Scholar] [CrossRef] [Green Version]**

66. Stoeckius, M.; Zheng, S.; Houck-Loomis, B.; Hao, S.; Yeung, B.Z.; Mauck, W.M.; Smibert, P.; Satija, R. Cell hashing with barcoded antibodies enables multiplexing and doublet detection for single cell genomics. *Genome Biol.* **2018**, *19*, 1–12. **[Google Scholar] [CrossRef] [Green Version]**

67. Zhang, L.; Dong, X.; Lee, M.; Maslov, A.Y.; Wang, T.; Vijg, J. Single-cell whole-genome sequencing reveals the functional landscape of somatic mutations in B lymphocytes across the human lifespan. *Proc. Natl. Acad. Sci. USA* **2019**, *116*, 9014–9019. **[Google Scholar] [CrossRef] [Green Version]**

68. Woerner, A.E.; Mandape, S.; King, J.L.; Muenzler, M.; Crysup, B.; Budowle, B. Reducing noise and stutter in short tandem repeat loci with unique molecular identifiers. *Forensic Sci. Int. Genet.* **2021**, *51*, 102459. **[Google Scholar] [CrossRef]**

69. Sheth, N.; Swaminathan, H.; Gonzalez, A.J.; Duffy, K.R.; Grgicak, C.M. Towards developing forensically relevant single-cell pipelines by incorporating direct-to-PCR extraction: Compatibility, signal quality, and allele detection. *Int. J. Leg. Med.* **2021**, *135*, 727–738. **[Google Scholar] [CrossRef] [PubMed]**

70. Lin, W.-C.; Tsai, C.-F.; Hu, Y.-H.; Jhang, J.-S. Clustering-based undersampling in class-imbalanced data. *Inf. Sci.* **2017**, *409-410*, 17–26. **[Google Scholar] [CrossRef]**

71. Liu, F.T.; Ting, K.M.; Zhou, Z.-H. Isolation forest. In Proceedings of the 2008 Eighth IEEE International Conference on Data Mining, Washington, DC, USA, 15–19 December 2008; pp. 413–422. **[Google Scholar]**

72. Yang, Y.; Xu, Z. Rethinking the value of labels for improving class-imbalanced learning. *arXiv* **2006**, arXiv:2006.07529 2020. **[Google Scholar]**

16 CROSS-AMPLIFICATION IN STRIGIFORMES: A NEWLY DEVELOPED STR PANEL SPECIFICALLY TAILORED FOR USE IN FORENSIC APPLICATIONS

1. INTRODUCTION: AN OVERVIEW

Habitat loss and fragmentation are the main threats to the survival of wildlife species. Multiple studies have thoroughly established the impact of their actions on the decline of biodiversity [1]. Furthermore, the global wildlife trade further exacerbates the exhaustion of natural resources. This results in the over exploitation of natural populations, when individuals are extracted from their natural habitats, and is considered one of the primary factors contributing to the extinction of species. [2,3,4]. Seafood, furniture, and fashion are the primary categories sought after in international trade. Additionally, the sale of pets has a significant impact on several individuals. Approximately 5% of import/export requests pertain to parrots and birds of prey, which includes raptors and owls. This percentage is equivalent to the combined total of all other commercially traded bird species [1]. Baker and colleagues [**5**] calculated that animals' demand for pets and entertainment purposes contributed to at least one-fifth of the wildlife trade [**6**,7,**8**], and they showed that the removal of wild individuals from native populations was accounted in migrating species for the second threat that pushes species into extinction, lower only to the habitat loss [**9**].Species over-exploitation increases the negative effects produced by habitat unsuitability, causing the reduction of individuals in wild populations that, in turn, can produce a fast loss of genetic variability in a short timescale. Domínguez et al. [**10**] showed that decreasing populations and pet trade could have generated a bottleneck, reducing genetic variability in the yellow cardinal, *Gubernatrix cristata*, an endangered passerine from the southern region of South America. Harris et al. [**11**] found that 14 species routinely traded in Indonesian wildlife markets had undergone population decline differently from the other 22 untraded species, which did not suffer any number reduction. Evans and Sheldon [**12**] found a correlation between the decline of mean heterozygosity with an increasing extinction risk and identified the smaller population sizes as the main cause of this correlation.

The Washington Convention, implemented in 1976, has offered effective measures to govern and prohibit the trade of endangered species, thereby safeguarding them from extinction. All the nations comprising the European Community have been signatories to the convention that has been implemented through various Council (EU) Regulations. The species that get the highest level of protection are included in Annex A, which comprises the majority of species listed in Appendix I, as well as many species from Appendix II that are classified as severely endangered within the European Community.

One of the orders suffering a strong decline in Europe is Strigiformes, which includes two main families: Tytonidae and Strigidae. While the first family is split into two subfamilies, each of which including one genus (*Tyto* and *Phodilus*), the taxonomy of Strigidae is quite controversial [**13**]. Differently from traditional systematics, Wink et al. [**14**] recognized the existence of three subfamilies (Ninoxinae, Striginae, Surninae) with several tribes in the last two through a molecular phylogenetic approach. Concerning the substantial decline of Strigiformes in Europe, it is caused mainly by pesticides, changing agricultural techniques with the loss of the structures for nesting and lower rodent availability than in traditional farming, and caught of individuals from the wild. Several main factors threatening owl survival in the Gaza Strip were identified by Abdel Rabou [**15**], including habitat loss and destruction, poaching and trapping, myths and superstitions, secondary poisoning, road kills, and fence agricultural lands. Wan et al. [**16**] found a negative influence of fragmentation on the population size and genetic diversity. As for other species included in the CITES Appendix II, only individuals imported into the EU before the Convention of Washington entered into force can be traded. This permission is also applicable to these individuals' descendants, but only if their birth in captivity is proved.

The demand for pets among birds gradually increased in the past decades [**17**]. Panter, Atkinson and White [7] recorded a wild-caught export of 18,948 individuals belonging to 86 owl species from 1975 to 2015. In 2019, Ribeiro and colleagues [**18**] denounced a future trend in legal bird trade driven by sociocultural motivations with an increased demand that will interest wild-caught rather than captive-bred birds. This preference is expected to drive towards an increase in the illegal trade of the rarest or more popular species. Also, the reduced number of wild individuals increases their commercial value, encouraging, even more, their illegal trade. Remarkably, the illicit traffic involves the smuggling of eggs and the laundering through captive breeding facilities of wild-caught animals [**7**].

Owls trade increased in the last decades and has also been influenced by a change in habits; for example, Nijman and Nekaris [**19**] recorded an increased owl trade, from <0.06% before 2002 to >0.43% post 2008, in wildlife markets in Java and Bali and suggested a delayed "Harry Potter effect". Siriwat and colleagues [**20**] identified the same increasing trade in Thailand without finding a correlation with the "Harry Potter effect"; nevertheless, they found an association between the increasing owl request, the novel online market and more popularity of owls as a pet. Moreover, they related the higher market price of some species to lower individual availability.

Even though the main actions aim to prevent and hamper over-exploitation and illegal trade represent the first steps in preserving biodiversity, a biomolecular investigation can give valuable hints for either verifying the depletion of genetic variability or tracking illegal trade. This can be carried out through individual identification using molecular techniques and kinship analyses. One of the most used genetic markers to study genetic variability is microsatellite loci (i.e., short tandem repeats or STRs). They are multiallelic and PCR multiplexable, allowing the description of genetic variability with reduced costs. Although the comparability between different laboratories has often proved to represent a limit for their use, this hampering can be easily resolved by using an allelic ladder for each marker—that identifies the position of all the alleles—and by sharing reference samples.

Despite the availability of species-specific panels of microsatellite loci in Strigiformes [**21,22,23,24,25,26**] and the cross-amplification of some markers in species lacking their markers [**27**], a unique panel permitting the description of the genetic variability across species has not been found yet. Such a protocol would permit the recording of comparable variability indices among different species and speed the process, reducing molecular analyses costs; it could also be informative in detecting illegal trade of individuals collected from the wild.

Thus, we tested 32 highly polymorphic microsatellite markers available from literature in 11 species of Strigiformes with the following aims:

- To evaluate the reliability of a unique panel of microsatellite loci for several species of Strigiformes;
- To test its use for conservation and forensic purposes;
- To assess the use of the panel in confirming phylogenetic relationships among species.

2. MATERIALS AND METHODS

The overall experimental strategy used in this work was based on the following steps:

i. Thirty-two microsatellite markers selected from the literature were tested on two unrelated individuals of a restricted panel of three species (*Athene noctua*, *Strixaluco*, *Bubo bubo*) (**Table 1**).

Table 1. List of species and individuals used in each step: family groups from 1 to 9 were tested in the first and second screening; additional species and family groups (from 10 to 28) were tested in the third step.

Family	Subfamily	Tribe	Species	ID	Gender	Relationships	Family Group
Strigidae	Striginae	Strigini	*Strixaluco*	S_al 1	Male	Father	1
				S_al 2	Female	Mother	
				S_al 3	unknown	Offspring	
				S_al 4	Male	Father	2
				S_al 5	Female	Mother	
				S_al 6	unknown	Offspring	
				S_al 7	Male	Father	3
				S_al 8	Female	Mother	
				S_al 9	unknown	Offspring	
		Bubonini	*Bubo bubo*	B_bu 1	Male	Father	4
				B_bu 2	Female	Mother	
				B_bu 3	unknown	Offspring	
				B_bu 4	Male	Father	5
				B_bu 5	Female	Mother	
				B_bu 6	unknown	Offspring	
				B_bu 7	Male	Father	6
				B_bu 8	Female	Mother	
				B_bu 9	unknown	Offspring	
	Surninae	Athenini	*Athene noctua*	A_no 1	Male	Father	7
				A_no 2	Female	Mother	
				A_no 3	unknown	Offspring	
				A_no 4	Male	Father	8
				A_no 5	Female	Mother	
				A_no 6	unknown	Offspring	
				A_no 7	Male	Father	9
				A_no 8	Female	Mother	
				A_no 9	unknown	Offspring	

Family	Subfamily	Tribe	Species	ID	Gender	Relationships	Family Group
	Striginae	Strigini	*Strixnebulosa*	S_ne 1	Male	Father	10
				S_ne 2	Female	Mother	
				S_ne 3	unknown	Offspring	
				S_ne 4	Male	Father	11
				S_ne 5	Female	Mother	
				S_ne 6	unknown	Offspring	
				S_ne 7	Male	Father	12
				S_ne 8	Female	Mother	
				S_ne 9	unknown	Offspring	
			Strixuralensis	S_ur 1	Male	Father	13
				S_ur 2	Female	Mother	
				S_ur 3	unknown	Offspring	
				S_ur 4	Male	Father	14
				S_ur 5	Female	Mother	
				S_ur 6	unknown	Offspring	
		Bubonini	*Bubo scandiacus*	B_sc 1	Male	Father	15
				B_sc 2	Female	Mother	
				B_sc 3	unknown	Offspring	
				B_sc 4	Male	Father	16
				B_sc 5	Female	Mother	
				B_sc 6	unknown	Offspring	
				B_sc 7	Male	Father	17
				B_sc 8	Female	Mother	
				B_sc 9	unknown	Offspring	
		Otini	*Otus scops*	O_sc 1	Male	Father	18
				O_sc 2	Female	Mother	
				O_sc 3	unknown	Offspring	
				O_sc 4	Male	Father	19
				O_sc 5	Female	Mother	
				O_sc 6	unknown	Offspring	
				O_sc 7	Male	Father	20
				O_sc 8	Female	Mother	
				O_sc 9	unknown	Offspring	

Family	Subfamily	Tribe	Species	ID	Gender	Relationships	Family Group
		Asionini	*Asiootus*	A_ot 1	Male	Father	21
				A_ot 2	Female	Mother	
				A_ot 3	unknown	Offspring	
				A_ot 4	Male	Father	22
				A_ot 5	Female	Mother	
				A_ot 6	unknown	Offspring	
	Surninae	Surnini	*Surniaulula*	S_ul 1	Male	Father	23
				S_ul 2	Female	Mother	
				S_ul 3	unknown	Offspring	
				S_ul 4	Male	Father	24
				S_ul 5	Female	Mother	
				S_ul 6	unknown	Offspring	
				S_ul 7	Male	Father	25
				S_ul 8	Female	Mother	
				S_ul 9	unknown	Offspring	
			Glaucidium passerinum	G_pa 1	Male	Father	26
				G_pa 2	Female	Mother	
				G_pa 3	unknown	Offspring	
				G_pa 4	Male	Father	27
				G_pa 5	Female	Mother	
				G_pa 6	unknown	Offspring	
Tytonidae	Tytoninae		*Tyto alba*	T_al 1	Male	Father	28
				T_al 2	Female	Mother	
				T_al 3	unknown	Offspring	

ii. Microsatellite markers giving polymorphic profiles in all the species tested at step (i) were used for genotyping three family groups per species (**Table 1**).

iii. Microsatellite primers giving polymorphic amplification patterns at step (ii) were then used to evaluate the STR panel for forensic purposes on a larger panel of family groups, comprising different subfamilies, tribes and species, aiming at identifying a unique shared panel of polymorphic loci (**Table 1**). The selected species are the most commonly traded in the Italian market.

According to the manufacturer's instructions, DNA was isolated from feathers using DNeasy Blood and Tissue Kit (Qiagen, Hilden, Germany). After digestion in 180 µL ATL buffer, 20 µL proteinase K and 20 µL DTT and incubation overnight at 56 °C, the lysate was loaded in a QIAcube HT robotic station (Qiagen, Hilden, Germany) for further purification steps.

DNA was amplified in an 8 µL final volume reaction. The PCR reactions were carried out as follows: in 8 µL of the final volume, we added 1x reaction buffer, 0.02% BSA, 1.5 mM $MgCl_2$, 0.125 mM of dNTPs, Hot Start Taq Polymerase 0.025 U (Qiagen, Hilden, Germany), 0.125 mM primer (forward and reverse), 1 µL DNA template and nuclease-free water to reach the final volume.

Twenty-five out of 32PCR primers used in this work were originally isolated from five species of Strigidae *(Bubo bubo, Otus elegans, Strix occidentalis, Strixnebulosa, Glaucidium brasilianum),* one from the goshawk (*Accipiter gentilis).* Six were previously designed to and/or shown to display high cross utility throughout many genetically distant passerine and shorebirds and were found polymorphic in *Tyto alba*. (**Table 2**); each forward primer was labelled with fluorescent ABI dyes. Amplification reactions were performed in simplex to test for the right allelic range and no specific amplification. According to melting temperature and reference bibliography, we chose the following thermal profile: 94 °C for 15′, followed by 35 cycles at 94 °C for 40″, 60 °C for 40″, ending with a final extension at 72 °C for 10′.

Amplicons were separated through capillary electrophoresis in an ABI 3130xl genetic analyzer (Thermo Fisher Scientific); alleles were scored in GeneMapper 4.0 using GeneScan 500 ROX size standard (Thermo Fisher Scientific, Waltham, MA, USA). An allelic ladder was constructed in Genemapper, merging the sample electropherograms at each locus. Scoring has been reported for each peak, allowing the data comparison among laboratories.

Genotypes resulting from the screening were analyzed for the selected loci. Allele number (NA), effective allele number (NE), number of private alleles, observed and expected heterozygosities (Ho, He) were computed using GenAlEx 6.41 [**32**]. Gimlet 1.3.3 [**33**] was used to evaluate the probability of identity (PID) among individuals of each cluster species, to calculate the minimum number of markers necessary to achieve a reliable individual genotyping for unrelated (PID) and related samples ($PID_{\_sib}$, sibling).

A factorial correspondence analysis (FCA) was carried out in Genetix 4.05 [**34**] to check the genetic distances between species, tribes and subfamilies.

Table 2. Loci chosen from literature and tested for the cross-species amplification study.

Primer ID	Forward Primer Sequence	Reverse Primer Sequence	Original Species	References
54f2	GAGAGATGTTGGGCTCTTGTG	TGTTAATTGCATTGAATGTCAGC	Passerine and shorebird species	[25]
BOOW19	GGAAACTTACTAGAAAATAAATGACTGG	CTTTCTAACTTTCCCATGCAAC	Passerine and shorebird species	[25]
Calex05	TCCAGCTGAAGTCTTCCGTGAAT	TCCACACCTGTTCGACAGTTCAATA	Passerine and shorebird species	[25]
HvoB1	AAGCAAGGACTTTCCTTCCAG	TCTCAAATTGGAACAGAGAAAGG	Passerine and shorebird species	[25]
Tgu06	CGAGTAGCGTATTTGTAGCGA	AGGAGCGGTGATTGTTCAGT	Passerine and shorebird species	[25]
TG04-061	GACAATGGCTATGAAATAAATTAGGC	AGAAGGGCATTGAAGCACAC	Passerine and shorebird species	[25]
BbuS027	TCATGAGGAACTTTCAGTGCTC	GAAGAAAGGCAGCTCTCACC	*Bubo bubo*	[28]
BbuS102	AACTGATTTGGAAACCACCATC	CTGGAACACCCAGTGTTTGTC	*Bubo bubo*	[28]
Bbus116	GTTTCTGCAGCTGGGTCAG	AAACAGTTTCCATGCCTTACG	*Bubo bubo*	[28]
BbuS132	TCATTGTAGGTCCCATCCAAC	CCATATCTATCAAGCAACCTTGG	*Bubo bubo*	[28]
Oe050	AGAGTTGTCCTTGGTTGG	TTCTAGTAACCTCATCTGC	*Otus elegans*	[29]
Oe054	TCAGAAAGAAAACTTCAGCAACC	CATATATGTATACACAGGCACATGC	*Otus elegans*	[29]
15A6	ACCTCAGAAGCAGACAGAACC	CCTTTGCGATTGCTGTAAC	*Strix occidentalis*	[30]
Age5	ACGTTACAGACACCGATTACTTCC	AGCCACGCGTCTGATACTTT	*Accipiter gentilis*	[31]
Bb101	AATAACCCCAATAGAAGC	ACCAGAAGGAGATGAGACC	*Bubo bubo*	[21]
Bb126	TCTCCAGAAGGGTTGTCATC	TGCTAAAACCTTACAGAATAACAG	*Bubo bubo*	[21]
Oe142	TGAATCAGCAAACCTGTGCCTG	AGCTAACCTAGAGTCAGCCAGC	*Otus elegans*	[29]
Oe171	TTTTACAAACTACTAGTGCATGTCTCC	AGATGTTGTATTTCAGTGTCAG	*Otus elegans*	[29]
Oe3-21	GATTAGAGACCCGATTCCACA	TTATCTGAGTGGAAGGGTAGTGC	*Otus elegans*	[29]
Oe3-7	GTGGGTTTATTGCCCCCTCG	CAGATGAATGAATGGATAGATGG	*Otus elegans*	[22]

Primer ID	Forward Primer Sequence	Reverse Primer Sequence	Original Species	References
Oe149	CACACATCCATTTTGGGGTGC	GGATGCTGGAACTGACCTGC	*Otus elegans*	[22]
Oe081	GTAAGGGAAGTAGACGTCTGTTGG	CAACTTTGTGTCATCTGAAAGG	*Otus elegans*	[22]
Oe084	GGGCATAGTTAGACCTTTGCAG	CACATCTGTTGTTCTCCGTGTTACC	*Otus elegans*	[22]
SneD105	AGCCTTGGGAGGTTAAGTCCT	ACACGCAATCACTGAAGCAGT	*Strixnebulosa*	[24]
SneD113	AGCTCTTTCCCGACAGTGTCTA	GCCAAAAAACCCCATATCCTAG	*Strixnebulosa*	[24]
SneD202	GCTGGGCTTAGAAATTACATGG	CTCTGCCCTAATCAGGAACACT	*Strixnebulosa*	[24]
SneD211	TAGCCCTGTGGATTTGCATTAG	TCACCAGAAGTTCAAAGCAGGTAG	*Strixnebulosa*	[24]
SneD218	GGGTGTGAGAATGACCTACCTG	AGCAGACAGGAGATGGCTTTTA	*Strixnebulosa*	[24]
Fepo42	CGTATACATCGAAATAAATACC	CGAATAAAACATCCCTAACC	*Glaucidium brasilianum*	[23]
Oe053	CTCTGCATCTTAACGCACAGGAC	CCTCCAAGTGGACAGGAAAAGC	*Otus elegans*	[29]
Oe128	CGTTGTAAATGATGAATCGCCTAGTGC	ATGCATGTATACATACAAACCTGG	*Otus elegans*	[29]
Oe129	GTCACCTCTTGACATCCGAGTAGC	GCTAAGAGTCCATTTGCCCATCTG	*Otus elegans*	[29]

Parentage relationships among individuals within the family cluster were reconstructed in Colony 2.0 [**35**]. This analysis allows to check if an individual that is claimed to be bred in captivity truly descends from the declared parents. A single analysis was performed including all genotypes of all species together. We used non-inbreeding data and monogamy models, and set the genotyping error rate value 0.0001. The same analysis was conducted in Cervus [**36,37**] and the LOD score was computed using a proportion of loci typed = 0.60 and mistyped = 0.01. A supplementary calculation was conducted using the Delta score, which is defined as the disparity in LOD scores between the most probable candidate parents and the second most probable candidate parents. The analyses were conducted with confidence levels of 80% (relaxed) and 95% (strict).

3. RESULTS

3.1. Preliminary Screening of STR Polymorphic Loci

Thirty-two microsatellite markers were chosen to evaluate their cross-species amplification potential on three species (*Athene noctua, Strixaluco, Bubo bubo*) (**Table 1**), belonging to two different sub-families, Striginae and Surninae, and three tribes (Strigini, Bubonini, Athenini) respectively [**14**].

Two unrelated individuals per species were selected from the CITES sample database (managed since 1995 by the Italian Institute of Environmental Protection and Research –ISPRA-on behalf of the Italian Environmental Ministry). Data obtained revealed that twenty-one out of 32 markers used (15a6, Age5, Bb101, Bb126, BbuS027, Bbus116, Calex05, Fepo42, Oe054, Oe128, Oe129, Oe142, Oe149, Oe321, Oe53, Oe81, Oe84, SneD113, SneD202, SneD218, Tgu06) gave amplicons in *A. noctua*, *S. aluco* and *B. bubo* DNA samples. The remaining 11 microsatellites were discarded because they either did not give any amplification product or led to unreliable PCR amplicons.

3.2. Evaluation of the STR Potential in Family Groups

Three family groups (father, mother and one offspring) for each species (n = 9, **Table 1**) were chosen from the CITES database and analyzed at the twenty-one loci selected in the previous section.

Data obtained revealed that 12 of them resulted polymorphic in at least 2 species (15a6, Fepo42, Oe053, Oe054, Oe128, Oe129, Oe142, Oe149, Oe321, SneD113, SneD218, Tgu06) and were retained for further analyses. Fepo42 and Oe054 resulted monomorphic in *S. aluco*, and Oe129 did not give any amplification product in *B. bubo* but were retained.

3.3. Evaluation of the STR Potential in Other Subfamilies of Strigidae

In order to evaluate the potential application of the STR panel for forensic purposes, the 12 microsatellites were used on a larger panel of samples (**Table 1**), which included:

1. Additional species belonging to the already analyzed tribes of Bubonini and Strigini (*Bubo scandiacus, Strixuralensis, Strixnebulosa*);
2. Two species belonging to the tribes of Asionini (*Asiootus*) and Otini (*Otus scops*);
3. Two species belonging to the Surninae subfamily, Tribes Surnini (*Surniaulula, Glaucidium passerinum*).
4. *Tyto alba*, belonging to the Tytonidae family, subfamily Tytoninae.

Three family groups were chosen for each species except for *A. otus, S. uralensis* and *G. passerinum* because only two confirmed parental nuclei were available in the database. The final dataset consisted of 81 individuals belonging to 27 families.

The DNA of each of the 81 individuals was amplified using the twelve primer sets. Data obtained revealed that the DNA from the additional species included in the analysis (*O. scops, A. otus, B. scandiacus, S. ulula, S. uralensis, S. nebulosa, G. passerinum, T. alba*) was amplified at the examined loci with the following exceptions:

1. *S. ulula* showed fixed genotypes at four loci (Oe054, Oe129, Oe053, Oe321);
2. Ten loci were amplified in *G. passerinum*. However, only four loci were polymorphic. Because of lack of variable loci, *G. passerinum* was removed from the further analysis;
3. *S. uralensis* and *S. nebulosa* showed no polymorphisms at Oe321 and Oe054 loci;
4. In *A. otus*, a unique fixed allele was recorded at locus Oe142;
5. Six loci were amplified in *T. alba* (FePo42, Oe54, Oe128, Oe129, Oe321, Tgu06), but only 2 were variable (FePo42, Tgu06). Thus, this species was discarded from further analyses because two loci are too few to be able to distinguish individuals reliably.

All the percentage values of polymorphic loci per species varied between 100% (n = 12 in *A. noctua* and *O. scops*) and 66.67% (n = 8 in *S. ulula*), if we do not take into account the value in *G. passerinum* and *T. alba* (**Table 3**).

Table 3. Variability indices: Allele number (NA), effective allele number (NE); observed heterozygosity (HO), expected heterozygosity (HE), unbalanced expected heterozygosity (uHE), percentage of polymorphic loci (P%). Standard errors are shown in brackets.

Species	NA	NE	HO	HE	uHE	P%
Strixaluco	3.6 (±0.5)	2.8 (±0.4)	0.528 (±0.091)	0.525 (±0.083)	0.556 (±0.087)	83.33%
Strixnebulosa	3.0 (±0.3)	2.2 (±0.2)	0.565 (±0.096)	0.471 (±0.074)	0.499 (±0.079)	83.33%
Strixuralensis	3.0 (±0.5)	2.3 (±0.4)	0.528 (±0.119)	0.429 (±0.092	0.468 (±0.100)	75.00%

Species	NA	NE	HO	HE	uHE	P%
Bubo bubo	4.2 (±0.6)	2.6 (±0.4)	0.539 (±0.090)	0.545 (±0.068)	0.578 (±0.072)	91.67%
Bubo scandiacus	3.3 (±0.4)	2.6 (±0.3)	0.515 (±0.092)	0.550 (±0.061)	0.584 (±0.065)	91.67%
Otus scops	6.3 (±0.6)	4.6 (±0.5)	0.733 (±0.049)	0.750 (±0.027)	0.798 (±0.030)	100.00%
Athene noctua	4.1 (±0.5)	3.3 (±0.5)	0.620 (±0.061)	0.604 (±0.060)	0.641 (±0.064)	100.00%
Asiootus	3.2 (±0.4)	2.5 (±0.3)	0.681 (±0.097)	0.529 (±0.062)	0.590 (±0.070)	91.67%
Surniaulula	2.2 (±0.3)	1.70 (±0.21)	0.356 (±0.088)	0.318 (±0.074)	0.338 (±0.079)	66.67%
Mean	3.7 (±0.2)	2.8 (±0.14)	0.566 (±0.030)	0.525 (±0.025)	0.561 (±0.026)	87.04 (±3.70)%

3.4. Genetic Variability between and within Species

Statistical analysis was carried out on 9 species at 12 loci (**Table 3**). The mean allele number (NA) was 3.7 (±0.2), with the highest values recorded respectively in *O. scops* (0.6 ± 6.3) and the lowest one in *S. ulula* (0.3 ± 2.2). Mean expected and observed heterozygosity (He and Ho) ranged from 0.027 ± 0.750 and 0.049 ± 0.733 in *O. scops* to 0.074 ± 0.318 and 0.088 ± 0.356 in *S. ulula* with an average value of 0.025 ± 0.525 and 0.030 ± 0.566, respectively. The probability of identity resulted in different thresholds depending on the species and if it was estimated for unrelated or related individuals (**Figure 1**). A PID value lower than 0.001 was reached in all the species using an average of 3.9 markers, while the same value was achieved with $PID_{\_sib}$ using at least an average of 9.2 loci. *S. ulula*, *S. nebulosa* and *S. uralensis* were not included in this last computation because their threshold resulted higher using all the loci (minimum values respectively of 0.00162 ,0.0155 and 0.00246) (**Table 4**).

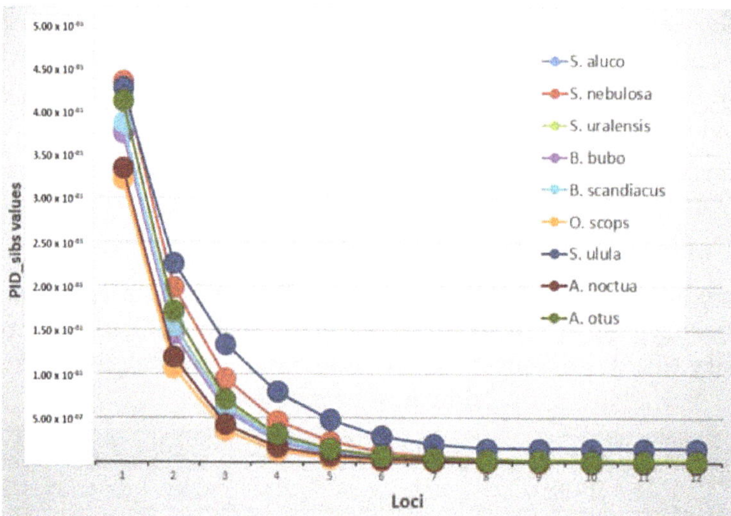

Figure 1. Graph of the $PID_{\_sib}$ trend in the species.

Table 4. PID and PID_sib values: In the table, values lower than 0.001 were bold-evidenced. In the PID_sib columns, this threshold wasindicated with (*). The PID_sib never reached this value in *Surniaulula* (0.0155), *Strixnebulosa* (0.00162) and *Strixuralensis* (0.0246). When PID or PID_sib values stop decreasing it means that the markers are not variable and informative (underlined).

Strixaluco			**Strixnebulosa**			**Strixuralensis**		
locus	**PID**	**PID_sib**	**locus**	**PID**		locus	**PID**	**PID_sib**
Oe149	8.16×10^{-02}	3.78×10^{-01}	15a6	1.40×10^{-01}	4.33×10^{-01}	Oe128	8.16×10^{-02}	3.82×10^{-01}
15a6	6.98×10^{-03}	1.46×10^{-01}	Oe53	2.38×10^{-02}	1.99×10^{-01}	Oe53	9.28×10^{-03}	1.57×10^{-01}
SneD218	**7.10×10^{-04}**	5.87×10^{-02}	SneD218	4.82×10^{-03}	9.54×10^{-02}	Oe142	1.25×10^{-03}	6.72×10^{-02}
Oe53	7.22×10^{-05}	2.36×10^{-02}	Oe149	1.01×10^{-03}	4.68×10^{-02}	Oe149	**1.86×10^{-04}**	2.96×10^{-02}
Oe142	7.77×10^{-06}	9.59×10^{-03}	Fepo42	**2.28×10^{-04}**	2.35×10^{-02}	SneD218	3.23×10^{-05}	1.38×10^{-02}
Oe128	1.45×10^{-06}	4.56×10^{-03}	Oe142	5.18×10^{-05}	1.18×10^{-02}	15a6	7.33×10^{-06}	6.93×10^{-03}
Oe129	3.52×10^{-07}	2.42×10^{-03}	SneD113	1.24×10^{-05}	6.07×10^{-03}	Oe129	2.14×10^{-06}	3.82×10^{-03}
SneD113	9.97×10^{-08}	1.33×10^{-03}	Oe129	3.17×10^{-06}	3.20×10^{-03}	Fepo42	1.20×10^{-06}	2.87×10^{-03}
Oe321	3.77×10^{-08}	7.97×10^{-04}*	Oe128	1.14×10^{-06}	1.99×10^{-03}	Tgu06	8.74×10^{-07}	2.46×10^{-03}
Tgu06	2.11×10^{-08}	5.99×10^{-04}	Tgu06	7.53×10^{-07}	1.62×10^{-03}	Oe054	8.74×10^{-07}	2.46×10^{-03}
Fepo42	2.11×10^{-08}	5.99×10^{-04}	Oe054	7.53×10^{-07}	1.62×10^{-03}	Oe321	8.74×10^{-07}	2.46×10^{-03}
Oe054	2.11×10^{-08}	5.99×10^{-04}	Oe321	7.53×10^{-07}	1.62×10^{-03}	SneD113	8.74×10^{-07}	2.46×10^{-03}
Bubo bubo			***Bubo scandiacus***			***Otus scops***		
locus	**PID**	**PID_sib**	**locus**	**PID**	**PID_sib**	**locus**	**PID**	**PID_sib**
SneD218	7.65×10^{-02}	3.74×10^{-01}	Oe53	8.79×10^{-02}	3.86×10^{-01}	Oe054	2.99×10^{-02}	3.22×10^{-01}
Oe53	7.26×10^{-03}	1.46×10^{-01}	SneD218	8.94×10^{-03}	1.55×10^{-01}	Oe149	1.18×10^{-03}	1.08×10^{-01}
15a6	**8.09×10^{-04}**	6.01×10^{-02}	SneD113	1.01×10^{-03}	6.48×10^{-02}	SneD113	**4.70×10^{-05}**	3.60×10^{-02}
Oe142	9.11×10^{-05}	2.56×10^{-02}	Oe142	**1.36×10^{-04}**	2.76×10^{-02}	15a6	2.73×10^{-06}	1.28×10^{-02}
Fepo42	1.48×10^{-05}	1.16×10^{-02}	Oe128	2.52×10^{-05}	1.31×10^{-02}	Fepo42	1.85×10^{-07}	4.70×10^{-03}
SneD113	2.70×10^{-06}	5.77×10^{-03}	Oe054	5.84×10^{-06}	6.70×10^{-03}	Oe142	1.48×10^{-08}	1.79×10^{-03}
Oe128	8.03×10^{-07}	3.16×10^{-03}	Tgu06	1.50×10^{-06}	3.49×10^{-03}	SneD218	1.47×10^{-09}	7.19×10^{-04}*
Oe149	2.23×10^{-07}	1.78×10^{-03}	Oe149	4.68×10^{-07}	2.03×10^{-03}	Oe128	1.48×10^{-10}	2.93×10^{-04}
Oe054	6.17×10^{-08}	1.00×10^{-03}	Fepo42	1.68×10^{-07}	1.26×10^{-03}	Oe53	2.49×10^{-11}	1.34×10^{-04}
Tgu06	2.33×10^{-08}	5.98×10^{-04}*	15a6	6.83×10^{-08}	7.92×10^{-04} *	Oe321	4.46×10^{-12}	6.45×10^{-05}
Oe321	1.55×10^{-08}	4.89×10^{-04}	Oe129	2.78×10^{-08}	4.99×10^{-04}	Tgu06	8.29×10^{-13}	3.13×10^{-05}
Oe129	\	\	Oe321	2.78×10^{-08}	4.99×10^{-04}	Oe129	1.76×10^{-13}	1.54×10^{-05}
Surniaulula			***Athene noctua***			***Asiootus***		
locus	**PID**	**PID_sib**	**locus**	**PID**	**PID_sib**	**locus**	**PID**	**PID_sib**
SneD218	1.34×10^{-01}	4.26×10^{-01}	SneD113	3.94×10^{-02}	3.34×10^{-01}	Tgu06	1.14×10^{-01}	4.10×10^{-01}
Oe142	3.56×10^{-02}	2.27×10^{-01}	Oe149	2.42×10^{-03}	1.20×10^{-01}	SneD218	1.36×10^{-02}	1.72×10^{-01}
15a6	1.34×10^{-02}	1.35×10^{-01}	Oe129	**1.51×10^{-04}**	4.30×10^{-02}	Fepo42	1.62×10^{-03}	7.20×10^{-02}

Strixaluco			Strixnebulosa			Strixuralensis		
Fepo42	4.63×10^{-03}	8.06×10^{-02}	SneD218	1.46×10^{-05}	1.71×10^{-02}	Oe129	$\mathbf{2.41 \times 10^{-04}}$	3.17×10^{-02}
Oe128	1.77×10^{-03}	4.86×10^{-02}	Oe142	1.96×10^{-06}	7.27×10^{-03}	Oe128	4.70×10^{-05}	1.50×10^{-02}
SneD113	$\mathbf{5.87 \times 10^{-04}}$	2.95×10^{-02}	15a6	3.66×10^{-07}	3.38×10^{-03}	15a6	9.91×10^{-06}	7.33×10^{-03}
Tgu06	2.86×10^{-04}	2.06×10^{-02}	Fepo42	6.59×10^{-08}	1.59×10^{-03}	Oe149	3.24×10^{-06}	4.31×10^{-03}
Oe149	1.60×10^{-04}	1.55×10^{-02}	Oe321	1.34×10^{-08}	7.88×10^{-04} *	SneD113	1.24×10^{-06}	2.60×10^{-03}
Oe054	1.60×10^{-04}	1.55×10^{-02}	Oe054	5.22×10^{-09}	4.80×10^{-04}	Oe321	5.04×10^{-07}	1.64×10^{-03}
Oe129	1.60×10^{-04}	1.55×10^{-02}	Oe128	2.59×10^{-09}	3.39×10^{-04}	Oe054	2.00×10^{-07}	1.06×10^{-03}
Oe321	1.60×10^{-04}	1.55×10^{-02}	Tgu06	1.36×10^{-09}	2.50×10^{-04}	Oe53	9.20×10^{-08}	7.18×10^{-04} *
Oe53	1.60×10^{-04}	1.55×10^{-02}	Oe53	7.63×10^{-10}	1.87×10^{-04}	Oe142	9.20×10^{-08}	7.18×10^{-04}

A principal component analysis was then carried out, revealing that, as expected, the three species of the genus *Strix* (*S. uralensis*, *S. nebulosa*, *S. aluco*) overlapped in the PCA graphic (**Figure 2**). *A. otus* and *O. scops* showed a low distance from each other and resulted not very distant from the *Strix* species. *S. ulula* exhibited the greatest genetic distance from the other species while *A. noctua* was in a mid-range position (**Figure 2**). *B. scandiacus* and *B. bubo* did not overlap, but the genetic distance of the two species is low.

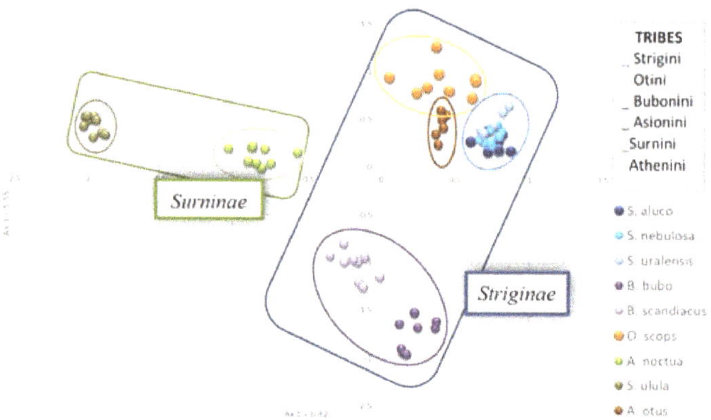

Figure 2. Principal Component Analysis plot obtained using Genetix and visualized in Excel. *Strix* sp. overlapped while *Bubo bubo* and *Bubo scandiacus* plotted very closed each other. *Surniaulula* individuals diverged consistently from other species. Ovals in different colors represent the distribution of individuals belonging to six tribes of Striginae and Surninae, in the blue and green squares, respectively.

3.5. Evaluation of the STR Potential Panel for Parentage Analysis in Family Groups

The paternity test yielded inconsistent results and different reliability values depending on the species (**Table 5** and **Table 6**). In Colony, the correct parent pair was assigned with the maximum probability value (1.00) but decreased in *S.*

nebulosa (0.992 and 0.907), *S. uralensis* (0.997), and *S. ulula* (0.983 and 0.753) when the computation has been limited to association with putative mother or father. In two individuals, respectively, of *S. uralensis* (S_ur6) and *S. ulula* (S_ul3), the probability of assignment to the right parent is reduced to 0.500.

Table 5. Parental assignment values from Colony: the lowest values were recovered when the one-parent assignment was tested. In the table, misallocation or low assignment probabilities are highlighted in bold.

Parent Pair Probability				Single Parent Probability		
Offspring	Father	Mother	Probability Value	Main Probability		Second Probability
S_al 3	S_al 1	S_al 2	1.00			
S_al 6	S_al 4	S_al 5	1.00			
S_al 9	S_al 7	S_al 8	1.00			
S_ne 3	S_ne 1	S_ne 2	1.00	S_ne 2	0.907	S_ne 8
S_ne 6	S_ne 4	S_ne 5	1.00	S_ne 4	0.992	S_ne 1
S_ne 9	S_ne 7	S_ne 8	1.00			
S_ur 3	S_ur 1	S_ur 2	1.00	S_ur 2	0.997	S_ur 5
S_ur 6	S_ur 4	S_ur 5	1.00	S_ur 5	**0.500**	S_ur 2
B_bu 3	B_bu 1	B_bu 2	1.00			
B_bu 6	B_bu 4	B_bu 5	1.00			
B_bu 9	B_bu 7	B_bu 8	1.00			
B_sc 3	B_sc 1	B_sc 2	1,00			
B_sc 6	B_sc 4	B_sc 5	1.00	B_sc 6	0.999	
B_sc 9	B_sc 7	B_sc 8	1,00			
O_sc 3	O_sc 1	O_sc 2	1.00			
O_sc 6	O_sc 4	O_sc 5	1.00			
O_sc 9	O_sc 7	O_sc 8	1.00			
A_no 3	A_no 1	A_no 2	1.00			
A_no 6	A_no 4	A_no 5	1.00			
A_no 9	A_no 7	A_no 8	1.00			
S_ul 3	S_ul 1	S_ul 2	1.00	S_ul 1	**0.500**	S_ul 4
S_ul 6	S_ul 4	S_ul 5	1.00	S_ul 4	0.983	S_ul 7
				S_ul 5	**0.753**	S_ul 2
S_ul 9	S_ul 7	S_ul 8	1.00			
A_ot 3	A_ot 1	A_ot 2	1.00			
A_ot 6	A_ot 7	A_ot 8	1.00			

Table 6. Parentage analysis in Cervus using the LOD and Delta computation. The different confidence values are indicated by the following codes: * = confidence level higher than %95. In the table, misallocation or low assignment probabilities are highlighted in bold.

LOD						Delta				
Offspring	Candidate Mother	Pair Confidence	Candidate Father	Pair Confidence	Trio Confidence	Offspring	Candidate Mother	Pair Confidence	Candidate Father	Pair Confidence
S_al 3	S_al 1	*	S_al 2	*	*	S_al 3	S_al 1	*	S_al 2	*
S_al 6	S_al 3	*	S_al 4	*	*	S_al 6	S_al 3	*	S_al 4	*
S_al 9	S_al 7	*	S_al 8	*	*	S_al 9	S_al 7	*	S_al 8	*
S_ne 3	S_ne 1	*	S_ne 2	*	*	S_ne 3	S_ne 1	*	S_ne 2	*
S_ne 6	S_ne 4	*	S_ne 5	*	*	S_ne 6	S_ne 4	*	S_ne 5	*
S_ne 9	S_ne 7	*	S_ne 8	*	*	S_ne 9	S_ne 7	*	S_ne 8	*
S_ur 3	S_ur 1	*	S_ur 2	*	*	S_ur 3	S_ur 1	*	S_ur 2	*
S_ur 6	S_ur 4		S_ur 5	*	*	S_ur 6	S_ur 4		S_ur 5	*
B_bu 3	B_bu 1	*	B_bu 2	*	*	B_bu 3	B_bu 1	*	B_bu 2	*
B_bu 6	B_bu 4	*	B_bu 5	*	*	B_bu 6	B_bu 4	*	B_bu 5	*
B_bu 9	B_bu 7	*	B_bu 8	*	*	B_bu 9	B_bu 7	*	B_bu 8	*
B_sc 3	B_sc 1	*	B_sc 2	*	*	B_sc 3	B_sc 1	*	B_sc 2	*
B_sc 6	B_sc 4	*	B_sc 5	*	*	B_sc 6	B_sc 4	*	B_sc 5	*
B_sc 9	B_sc 7	*	B_sc 8	*	*	B_sc 9	B_sc 7	*	B_sc 8	*
O_sc 3	O_sc 1	*	O_sc 2	*	*	O_sc 3	O_sc 1	*	O_sc 2	*
O_sc 6	O_sc 4	*	O_sc 5	*	*	O_sc 6	O_sc 4	*	O_sc 5	*
O_sc 9	O_sc 7	*	O_sc 8	*	*	O_sc 9	O_sc 7	*	O_sc 8	*

LOD						Delta				
Offspring	Candidate Mother	Pair Confidence	Candidate Father	Pair Confidence	Trio Confidence	Offspring	Candidate Mother	Pair Confidence	Candidate Father	Pair Confidence
A_no 3	A_no 1	*	A_no 2	*	*	A_no 3	A_no 1	*	A_no 2	*
A_no 6	A_no 4	*	A_no 5	*	*	A_no 6	A_no 4	*	A_no 5	*
A_no 9	A_no 7	*	A_no 8	*	*	A_no 9	A_no 7	*	A_no 8	*
S_ul 3	S_ul 1	*	S_ul 5	*	*	S_ul 3	S_ul 1	*	**S_ul 5**	*
S_ul 6	S_ul 4	*	**S_ul 5**	*	*	S_ul 6	S_ul 4	*	S_ul 5	*
S_ul 9	S_ul 7	*	S_ul 8	*	*	S_ul 9	S_ul 7	*	S_ul 8	*
A_ot 3	A_ot 1	*	A_ot 2	*	*	A_ot 3	A_ot 1	*	A_ot 2	*
A_ot 6	A_ot 7	*	A_ot 8	*	*	A_ot 6	A_ot 7	*	A_ot 8	*

In Cervus, using the LOD computation, all the individuals have been correctly associated with the right parents with trio confidence values higher than 95%, except for S_ul3 that has been associated with the wrong father (S_ul5 instead of S_ul2). In *S. uralensis* (S_ur6), the assignment to the right mother did not reach a significant value. Using the Delta calculation, the probability values decreased in *S. nebulosa*, *S. uralensis* and *S. ulula* (see **Table 6** for more details). Again, S_ul3 was associated with the wrong father.

4. DISCUSSION

Presently, international wildlife trafficking is acknowledged as one of the most significant organized transnational offenses [38], on par with the trafficking of narcotics, weapons, and individuals (World Wildlife Report, United Nations: Office on Drugs and Crime).Genotyping assays through microsatellites are a quick, informative and low-cost approach for linking items of evidence to crimes in forensic investigations [**39**]. Since microsatellites have high mutation rates, they are used within the context of monitoring illegal wildlife trade primarily to identify individuals, assign individuals to specific populations or for relatedness testing. Oklander et al. [**39**] generated a multilocus microsatellite genotype reference database of the black and gold howler monkey (*Alouatta caraya*), a neotropical primate threatened by habitat loss and capture for illegal trade in Argentina, to assign confiscated individuals to localities of origin, illustrating the applicability of genotype databases for inferring hotspots of illegal capture. Potoczniak et al. [**39**] developed a STR genotyping assay able to associate a biological sample to Asian elephant (*Elephas maximus*) or African elephant (*Loxodonta africana*). Fitzsimmons et al. [**40**] designed a panel of 26 microsatellite loci for *Crocodylus* spp. to answer questions regarding population assignment, mating system and geneflow. Jan and Fumagalli [**41**] isolated DNA microsatellite markers in seven parrot species threatened with extinction and subjected to illegal trafficking, characterized a total of 106 polymorphic microsatellite markers and tested them for individual identification and parental analyses. Mucci and colleagues [**42**] developed a panel of 16 de novo sequenced microsatellites (STRs) for *Testudo graeca* and tested its effectiveness for parentage analysis in two other species of endangered tortoises, *T. hermanni* and *T. marginata*.

Given the utility of STR-based approaches to answer questions related to wildlife crime investigations as forensic genetics, efforts should be devoted to the characterization of microsatellite primers for species threatened by illegal trafficking. However, while in human forensics, the selection of around 20 core STR loci allowed the standardization around the globe for human identity testing [**43,44**], accomplishing this same achievement is more challenging for the hyper-diverse animal assemblage encountered in wildlife forensics [**38**]. The lack of species-specific molecular markers or their inadequate representation in genetic databases is major limitation in wildlife forensics [**45**].

Cross-amplification is a widely used approach permitting to avoid investing time and money in the development of new markers, and many studies have proven the efficiency of microsatellite loci developed in closely related species [**42,46,47,48,49**].

The first aim of this study was to test for the presence of a minimum number of microsatellite loci reliable for individual identification and parentage analysis for forensic purposes in 11 species of Strigiformes listed in the CITES Appendix II and regularly traded in the Italian national market.

Literature data show that cross-species transferability is unevenly distributed across taxa. Barbará et al. [**50**] reviewed 64 primer notes and found more than 40% transfer success in mammals, more than 25% in fishes and more than 10% in birds. Our results permitted to define a unique panel of 12 out of 32 highly polymorphic microsatellites (37%), able to identify individuals in nine species of two subfamilies of Strigiformes (Striginae and Surninae) belonging to Family Strigidae (*A. otus*, *A. noctua*, *B. bubo*, *B. scandiacus*, *O. scops*, *S. aluco*, *S. nebulosa*, *S. uralensis*, *S. ulula*).

The test was non-efficient in individuals of *T. alba* in which only six markers yield a positive result but four of them resulted monomorphic. Such results could rely on the fact that *T. alba* is the only species of this study belonging to a different family [**51**]. The two monophyletic families of Strigiformes, Tytonidae and Strigidae, diverged in the middle of the Eocene [**52**]. The phylogenetic divergence of Tytonidae from Strigidae could be justified by the retrieved inefficiency of markers panel tested on species. Since only two subfamilies represent Tytonidae family with a single genus each [**53**]—Tytoninae with *Tyto* and Phodilinae with *Phodilus*—it is not possible to verify the discriminant power of tested markers on other species of the same taxon.

The common barn owl *T. alba* is one of the most cosmopolitan species and represents a taxon-rich species complex with several subspecies [**54**]. The interest in its conservation status clears the need for integrating this set with more polymorphic loci for this species. Twenty-one microsatellite loci already isolated and characterized in *T. alba* [**26**] could be tested for cross-amplification in CITES species to implement this panel.

The principal component analysis results reflect the most recent knowledge about taxonomy and systematics of Strigiformes [**53**].

A slightly better but similar result was obtained for *G. passerinum*, for which only 4 markers were polymorphic. The high number of monomorphic loci in this species could be due to several cumulative factors: the low number of related individuals analyzed, the high inbreeding of captive-bred individuals, or the inefficiency of the selected markers for this species. Unlike *T. alba*, microsatellites were developed only for a species of the same genera: *Glaucidium brasilianum* [**23**].

Evaluating the STR panel potential in identifying family groups, we found that in two out of the nine species analyzed (*S. uralensis* and *S. ulula*), the probability value associated with parent pairs was reduced. The $PID_{\_sib}$ almost reached the 0.001 threshold value in *S. ulula*, *S. nebulosa* and *S. uralensis* (see **Table 4**: Probability of identity values, PID and $PID_{\_sibs}$). However, PID and $PID_{\_sib}$ values are subjected to bias due to the low number of tested individuals per species and bottleneck in captive breeding facilities. For these three species, it could be helpful to increase the number of samples and markers to obtain a more confident individual identification and association to parent pairs.

Besides reducing time and costs when adopting a cross-amplification approach, another advantage of using a shared panel among several species is the possibility of comparing genetic variability values among species. Nevertheless, even if microsatellites loci are very useful genetic markers in studying the mating system, population genetics, and conservation of owls, many studies focus on species belonging to the same genus. Dial et al. [**55**] screened many markers developed in strigids but found only four polymorphic pairs in the great horned owl (*Bubo virginianus*), short-eared owl (*Asioflammeus*) and the snowy owl (*Bubo scandiacus*). In addition, another eight reliably amplified polymorphic fragments only in the great horned owl, eleven in the short-eared owl, and ten in the snowy owl. Hsu et al. [**22**] developed six new microsatellite markers containing tetranucleotide repeat motifs (GATA/CTAT) for Lanyu scops owl (*Otus elegans botelensis*). They tested them, and additional further microsatellite primer pairs previously developed from *O. elegans* on four other species of owls (*O. lettia*, *O. spilocephalus*, *O. scops* and *Ninoxscutulata*). Data obtained showed a reduced degree of polymorphism with most of the loci resulting polymorphic in the three *Otus* owls but only five in *N. scutulata*.

Our study gives a valuable tool to implement research involving most Strigidae threatened by illegal trafficking, habitat loss and fragmentation in Italy and other countries.

Delport et al. [**56**] tested 19 loci originally developed for Vidua and Geospiza for cross-amplification in Nesospiza buntings. They detected a degree of polymorphism and heterozygosity lower in loci developed for Vidua than those explicitly developed for Nesospiza. These data demonstrate that microsatellite markers isolated in the reference species are frequently less variable in related species. Moreover, cross-species amplification is usually limited to the loci that were found polymorphic in the referent species. In this study, we selected from the literature the most variable loci in each reference species: most polymorphic loci were discarded only when they were found monomorphic or did not give amplification products in all the target species, with only three exceptions as mentioned in Results (Paragraph ii), because of their utility in other species. We are aware that this a priori selection can cause an ascertainment bias, as suggested by Delport et al. [**56**]; however, we found high levels of polymorphisms in nine out 11 analyzed species.

Variability indices found in different species using this panel were not discordant from the ones found in the natural populations with different markers: Pellegrino et al. [**57**] found in the *A. noctua* European populations an average and an effective number of alleles = 5.6 and 3.5, respectively, and observed and expected heterozygosities equal to 0.59 and 0.61, respectively. Pertoldi et al. [**58**] found lower alleles in the Danish population (effective number of alleles = 2.8; Ho = 0.51 and He = 0.60), probably caused by a population bottleneck in the last decades.

Microsatellite loci represent reliable molecular markers to describe genetic variability or its drastic reduction, as demonstrated by Macías-Duarte et al. [**59**] that found different values in three different populations of *Athene cunicularia,* with the average number of alleles varying from 2.7 in Clarion Islands to 5.1 in Florida and 22.5 in Western North America.

Though their high polymorphism makes them adequate for conservation and forensic genetics purposes, the main difficulties are represented by comparing samples between laboratories.

This problem that has been resolved through the exchange of reference samples has been recently fixed by the set-up of an allelic ladder [**60**]. According to these authors, we constructed an allelic ladder for each locus to standardize a protocol between laboratories for conservation and forensic purposes.

Comparability between laboratories is now also possible thanks to high-throughput sequencing (HTS) technologies [**61**]. This application that has been developed and used in human forensics [**62,63,64**] has been already applied also in conservation genetics [**65,66,67,68**]. This method permits the sequencing of STRs, allowing the identification of the correct number of repeats. The possibility of multiplexing several dozen of markers from a single individual will allow cost and time reduction. De Barba et al. [**69**] used this protocol in the study of a brown bear population yielded reliable results of parentage analysis also from low quality DNA, confirming a broader application in conservation genetics and forensics.

5. AT A GLANCE

The recently enhanced set of 12 genetic markers enables authorities to effectively probe potential smugglers and fraudulent parental claims, as well as establish a connection between evidence and an individual. For instance, it can be used to identify instances of illegal transfer of a bird between different breeding facilities.

Moreover, the chosen markers can be utilized to evaluate the degree of relatedness among captive-bred individuals, a critical factor in formulating effective breeding protocols to prevent loss of genetic diversity and reduce inbreeding. Furthermore, these markers are valuable for research studies on wild populations, since they enable the identification of individual genotypes, facilitating the description of the mating system, genetic variability, population structure, and gene flow across nine distinct species of Strigiformes that are of conservation significance.

6. CONCISE SUMMARY

The population of Strigiformes is experiencing a significant decrease primarily due to the loss and destruction of their natural habitats, as well as illegal hunting and trapping. Furthermore, the rise in bird commerce and the surging preference for wild-caught birds over captive-bred ones are anticipated to foster illicit trafficking. Biomolecular inquiry is a useful method for monitoring illegal trade and studying genetic diversity for the purpose of biodiversity conservation. Microsatellite loci, often known as STRs (short tandem repeats), are the primary markers employed for investigating genetic variability. Although microsatellite loci specific to Strigiformes are available, a distinct set of markers that allows for the characterization of genetic diversity across species has not yet been found. We conducted a comprehensive assessment of 32 microsatellite markers with significant levels of genetic variation. The objective was to determine the effectiveness of a specific microsatellite panel in several species of Strigiformes, and its potential use in conservation and forensic contexts. The study encompassed 84 individuals from 28 parental groups and 11 species of Strigiformes. Following the analysis of polymorphic microsatellite loci, assessment of genetic diversity, and determination of kinship, we established a final set of 12 microsatellite loci capable of distinguishing individuals among 9 species of Strigiformes. The STR panel can assist authorities in forensic investigations involving suspected smugglers and false parental claims. Additionally, it can be valuable in assessing relatedness among individuals in captive-bred populations and conducting research projects aimed at describing genetic variability in wild populations.

REFERENCES

1. Anderson, S.C.; Elsen, P.R.; Hughes, B.B.; Tonietto, R.K.; Bletz, M.C.; Gill, D.A.; Holgerson, M.A.; Kuebbing, S.E.; McDonough MacKenzie, C.; Meek, M.H.; et al. Trends in ecology and conservation over eight decades. *Front. Ecol. Environ.* **2021**, *19*, 274–282. **[Google Scholar] [CrossRef]**
2. Bishop, J.M.; Leslie, A.J.; Bourquin, S.L.; O'Ryan, C. Reduced effective population size in an overexploited population of the Nile crocodile (*Crocodylusniloticus*). *Biol. Conserv.* **2009**, *142*, 2335–2341. **[Google Scholar] [CrossRef]**

3. Betancourth-Cundar, M.; Palacios-Rodríguez, P.; Mejía-Vargas, D.; Paz, A.; Amézquita, A. Genetic differentiation and overexploitation history of the critically endangered Lehmann's Poison Frog: Oophagalehmanni. *Conserv. Genet.* **2020**, *21*, 453–465. **[Google Scholar] [CrossRef]**
4. Righi, T.; Splendiani, A.; Fioravanti, T.; Casoni, E.; Gioacchini, G.; Carnevali, O.; Barucchi, V.C. Loss of mitochondrial genetic diversity in overexploited mediterranean swordfish (*Xiphias gladius*, 1759) population. *Diversity* **2020**, *12*, 170. **[Google Scholar] [CrossRef]**
5. Baker, S.E.; Cain, R.; Kesteren, F.V.; Zommers, Z.A.; D'Cruze, N.; MacDonald, D.W. Rough trade: Animal welfare in the global wildlife trade. *BioScience* **2013**, *63*, 928–938. **[Google Scholar] [CrossRef] [Green Version]**
6. Tingley, M.W.; Harris, J.B.C.; Hua, F.; Wilcove, D.S. The pet trade's role in defaunation. *Science* **2017**, *356*, 1–916. **[Google Scholar] [CrossRef] [PubMed]**
7. Panter, C.T.; Atkinson, E.D.; White, R.L. Retraction: Quantifying the global legal trade in live CITES-listed raptors and owls for commercial purposes over a 40-year period. *Avocetta* **2019**, *43*, 23–36. **[Google Scholar] [CrossRef]**
8. Vall-Llosera, M.; Su, S. Trends and characteristics of imports of live CITES-listed bird species into Japan. *Ibis* **2019**, *161*, 590–604. **[Google Scholar] [CrossRef]**
9. Brochet, A.L.; Van Den Bossche, W.; Jbour, S.; Ndang'Ang'A, P.K.; Jones, V.R.; Abdou, W.A.L.I.; Al-Hmoud, A.R.; Asswad, N.G.; Atienza, J.C.; Atrash, I.; et al. Preliminary assessment of the scope and scale of illegal killing and taking of birds in the Mediterranean. *Bird Conserv. Int.* **2016**, *26*, 1–28. **[Google Scholar] [CrossRef] [Green Version]**
10. Domínguez, M.; Tiedemann, R.; Reboreda, J.C.; Segura, L.; Tittarelli, F.; Mahler, B. Genetic structure reveals management units for the yellow cardinal (*Gubernatrix cristata*), endangered by habitat loss and illegal trapping. *Conserv. Genet.* **2017**, *18*, 1131–1140. **[Google Scholar] [CrossRef]**
11. Harris, J.B.C.; Green, J.M.H.; Prawiradilaga, D.M.; Giam, X.; Giyanto; Hikmatullah, D.; Putra, C.A.; Wilcove, D.S. Using market data and expert opinion to identify overexploited species in the wild bird trade. *Biol. Conserv.* **2015**, *187*, 51–60. **[Google Scholar] [CrossRef]**
12. Evans, S.R.; Sheldon, B.C. Interspecific patterns of genetic diversity in birds: Correlations with extinction risk. *Conserv. Biol.* **2008**, *22*, 1016–1025. **[Google Scholar] [CrossRef] [PubMed]**
13. Enríquez, P.L.; Eisermann, K.; Mikkola, H.; Motta-Junior, J.C. A review of the systematics of Neotropical owls (Strigiformes). *Neotrop. Owls Divers. Conserv.* **2017**, 1–670. **[Google Scholar] [CrossRef]**

14. Wink, M.; El-Sayed, A.A.; Sauer-Gürth, H.; Gonzalez, J. Molecular phylogeny of owls (Strigiformes) inferred from DNA sequences of the mitochondrial cytochrome b and the nuclear RAG-1 gene. *Ardea* **2009**, *97*, 581–591. **[Google Scholar] [CrossRef] [Green Version]**

15. Abdel Rabou, A.F.N. On the Owls (Order Strigiformes) Inhabiting the Gaza Strip—Palestine. *JOJ Wildl. Biodivers.* **2020**, *3*, 1–11. **[Google Scholar] [CrossRef]**

16. Wan, H.Y.; Ganey, J.L.; Vojta, C.D.; Cushman, S.A. Managing emerging threats to spotted owls. *J. Wildl. Manag.* **2018**, *82*, 682–697. **[Google Scholar] [CrossRef]**

17. Li, L.; Jiang, Z. International trade of CITES listed bird species in China. *PLoS ONE* **2014**, *9*, e85012. **[Google Scholar] [CrossRef]**

18. Ribeiro, J.; Reino, L.; Schindler, S.; Strubbe, D.; Vall-llosera, M.; Araújo, M.B.; Capinha, C.; Carrete, M.; Mazzoni, S.; Monteiro, M.; et al. Trends in legal and illegal trade of wild birds: A global assessment based on expert knowledge. *Biodivers. Conserv.* **2019**, *28*, 3343–3369. **[Google Scholar] [CrossRef]**

19. Nijman, V.; Nekaris, K.A.I. The Harry Potter effect: The rise in trade of owls as pets in Java and Bali, Indonesia. *Glob. Ecol. Conserv.* **2017**, *11*, 84–94. **[Google Scholar] [CrossRef]**

20. Siriwat, P.; Nekaris, K.A.I.; Nijman, V. Digital media and the modern-day pet trade: A test of the 'Harry Potter effect' and the owl trade in Thailand. *Endanger. Species Res.* **2020**, *41*, 7–16. **[Google Scholar] [CrossRef] [Green Version]**

21. Isaksson, M.; Tegelström, H. Characterization of polymorphic microsatellite markers in a captive population of the eagle owl (*Bubo bubo*) used for supportive breeding. *Mol. Ecol. Notes* **2002**, *2*, 91–93. **[Google Scholar] [CrossRef]**

22. Hsu, Y.C.; Li, S.H.; Lin, Y.S.; Severinghaus, L.L. Microsatellite loci from Lanyu scops owl (*Otus elegans botelensis*) and their cross-species application in four species of strigidae. *Conserv. Genet.* **2006**, *7*, 161–165. **[Google Scholar] [CrossRef]**

23. Proudfoot, G.; Honeycutt, R.; Douglas Slack, R. Development and characterization of microsatellite DNA primers for ferruginous pygmy-owls (*Glaucidium brasilianum*). *Mol. Ecol. Notes* **2005**, *5*, 90–92. **[Google Scholar] [CrossRef]**

24. Hull, J.M.; Keane, J.J.; Tell, L.A.; Ernest, H.B. Development of 37 microsatellite loci for the great gray owl (*Strixnebulosa*) and other *Strix* spp. owls. *Conserv. Genet.* **2008**, *9*, 1357–1361. **[Google Scholar] [CrossRef]**

25. Klein, A.; Horsburgh, G.J.; Kupper, C.; Major, A.; Lee, P.L.M.; Hoffmann, G.; Matics, R.; Dawson, D.A. Microsatellite markers characterized in the barn owl (*Tyto alba*) and high utility in the other owls (Strigiformes: AVES). *Mol. Ecol. Resour.* **2009**, *9*, 1513–1519. **[Google Scholar] [CrossRef]**

26. Burri, R.; Antoniazza, S.; Siverio, F.; Klein, Á.; Roulin, A.; Fumagalli, L. Isolation and characterization of 21 microsatellite markers in the barn owl (*Tyto alba*). *Mol. Ecol. Resour.* **2008**, *8*, 977–979. **[Google Scholar] [CrossRef] [PubMed] [Green Version]**

27. Hogan, F.E.; Cooke, R.; Norman, J.A. Reverse ascertainment bias in microsatellite allelic diversity in owls (Aves, Strigiformes). *Conserv. Genet.* **2009**, *10*, 635–638. **[Google Scholar] [CrossRef]**

28. Kleven, O.; Dawson, D.A.; Gjershaug, J.O.; Horsburgh, G.J.; Jacobsen, K.O.; Wabakken, P. Isolation, characterization and predicted genome locations of Eurasian eagle-owl (*Bubo bubo*) microsatellite loci. *Conserv. Genet. Resour.* **2013**, *5*, 723–727. **[Google Scholar] [CrossRef] [Green Version]**

29. Hsu, Y.C.; Severinghaus, L.L.; Lin, Y.S.; Li, S.H. Isolation and characterization of microsatellite DNA markers from the Lanyu scops owl (*Otus elegans botelensis*). *Mol. Ecol. Notes* **2003**, *3*, 595–597. **[Google Scholar] [CrossRef] [Green Version]**

30. Thode, A.B.; Maltbie, M.; Hansen, L.A.; Green, L.D.; Longmire, J.L. Microsatellite markers for the Mexican spotted owl (*Strix occidentalis* lucida). *Mol. Ecol. Notes* **2002**, *2*, 446–448. **[Google Scholar] [CrossRef]**

31. Topinka, J.R.; May, B. Development of polymorphic microsatellite loci in the northern goshawk (*Accipiter gentilis*) and cross-amplification in other raptor species. *Conserv.Genet.* **2004**, *5*, 861–864. **[Google Scholar] [CrossRef]**

32. Peakall, R.; Smouse, P.E. GenAlEx 6.5: Genetic analysis in Excel. Population genetic software for teaching and research--an update. *Bioinformatics* **2012**, *28*, 2537–2539. **[Google Scholar] [CrossRef] [PubMed] [Green Version]**

33. Valière, N. GIMLET: A computer program for analysing genetic individual identification data. *Mol. Ecol. Notes* **2002**, *2*, 377–379. **[Google Scholar] [CrossRef]**

34. Jones, O.R.; Wang, J. COLONY: A program for parentage and sibship inference from multilocus genotype data. *Mol. Ecol. Resour.* **2010**, *10*, 551–555. **[Google Scholar] [CrossRef]**

35. Marshall, T.C.; Slate, J.; Kruuk, L.E.B.; Pemberton, J.M. Statistical confidence for likelihood-based paternity inference in natural populations. *Mol. Ecol.* **1998**, *7*, 639–655. **[Google Scholar] [CrossRef] [Green Version]**

36. Kalinowski, S.T.; Taper, M.L.; Marshall, T.C. Revising how the computer program CERVUS accommodates genotyping error increases success in paternity assignment. *Mol. Ecol.* **2007**, *16*, 1099–1106. **[Google Scholar] [CrossRef] [PubMed]**

37. Belkhir, K.; Borsa, P.; Chikhi, L.; Raufaste, N.; Bonhomme, F. *Genetix 4.05, Logiciel Sous Windows Tm Pour La GénétiqueDes Populations*; Laboratoire Genome, Populations, Interactions, CNRS UMR 5000, Universite de Montpellier II: Montpellier, France, 1996–2004.
38. Smart, U.; Cihlar, J.C.; Budowle, B. International Wildlife Trafficking: A perspective on the challenges and potential forensic genetics solutions. *Forensic Sci. Int. Genet.* **2021**, *54*, 102551. **[Google Scholar] [CrossRef] [PubMed]**
39. Potoczniak, M.J.; Chermak, M.; Quarino, L.; Tobe, S.S.; Conte, J. Development of a multiplex, PCR-based genotyping assay for African and Asian elephants for forensic purposes. *Int. J. Leg. Med.* **2020**, *134*, 55–62. **[Google Scholar] [CrossRef]**
40. Fitzsimmons, N.N.; Tanksley, S.; Forstner, M.R.J.; Louis, E.E.; Daglish, R.; Gratten, J.; Davis, S. Microsatellite markers for Crocodylus: New genetic tools for population genetics, mating system studies and forensics. *Crocodilian Biol. Evol.* **2001**, 51–57. **[Google Scholar]**
41. Jan, C.; Fumagalli, L. Polymorphic DNA microsatellite markers for forensic individual identification and parentage analyses of seven threatened species of parrots (family Psittacidae). *PeerJ* **2016**, *2016*, 1–7. **[Google Scholar] [CrossRef] [Green Version]**
42. Mucci, N.; Giangregorio, P.; Cirasella, L.; Isani, G.; Mengoni, C. A new STR panel for parentage analysis in endangered tortoises. *Conserv. Genet. Resour.* **2020**, *12*, 67–75. **[Google Scholar] [CrossRef]**
43. Zhou, Z.; Shao, C.; Xie, J.; Xu, H.; Liu, Y.; Zhou, Y.; Liu, Z.; Zhao, Z.; Tang, Q.; Sun, K. Genetic polymorphism and phylogenetic analyses of 21 non-CODIS STR loci in a Chinese Han population from Shanghai. *Mol. Genet. Genom. Med.* **2020**, *8*, 1–8. **[Google Scholar] [CrossRef] [PubMed]**
44. Kumar, A.; Kumar, R.; Kumawat, R.K.; Shrivastava, P.; Chaubey, G. Genomic diversity at 22 STR loci (extended CODIS STR) in the population of Rajasthan, India. *Gene Rep.* **2021**, *23*, 101150. **[Google Scholar] [CrossRef]**
45. Rocco, F.D.; Anello, M. The Use of Forensic DNA on the Conservation of Neotropical Mammals. In *Molecular Ecology and Conservation Genetics of Neotropical Mammals*; Springer Nature Switzerland AG: Cham, Switzerland, 2021; pp. 85–98. **[Google Scholar]**
46. Gebhardt, K.J.; Waits, L.P. Cross-species amplification and optimization of microsatellite markers for use in six Neotropical parrots. *Mol. Ecol. Resour.* **2008**, *8*, 835–839. **[Google Scholar] [CrossRef] [PubMed]**

47. Dawson, D.A.; Horsburgh, G.J.; Küpper, C.; Stewart, I.R.K.; Ball, A.D.; Durrant, K.L.; Hansson, B.; Bacon, I.; Bird, S.; Klein, Á.; et al. New methods to identify conserved microsatellite loci and develop primer sets of high cross-species utility—As demonstrated for birds. *Mol. Ecol. Resour.* **2010**, *10*, 475–494. **[Google Scholar] [CrossRef] [PubMed]**

48. Maulidi, A.; Fatchiyah, F.; Hamidy, A.; Kurniawan, N. Microsatellite Marker for Cross-Species Amplification: Study Case for Indonesian Sundaland Python (Serpentes: Pythonidae). *J. Exp. Life Sci.* **2018**, *8*, 61–65. **[Google Scholar] [CrossRef] [Green Version]**

49. Du Toit, Z.; Dalton, D.L.; du Plessis, M.; Jansen, R.; Grobler, J.P.; Kotzé, A. Isolation and characterization of 30 STRs in Temminck's ground pangolin (*Smutsiatemminckii*) and potential for cross amplification in other African species. *J. Genet.* **2020**, *99*. **[Google Scholar] [CrossRef]**

50. Barbará, T.; Palma-Silva, C.; Paggi, G.M.; Bered, F.; Fay, M.F.; Lexer, C. Cross-species transfer of nuclear microsatellite markers: Potential and limitations. *Mol. Ecol.* **2007**, *16*, 3759–3767. **[Google Scholar] [CrossRef]**

51. König, C.; Weick, F. *Owls of the World*; AandC Black Publishers Ltd.: London, UK, 2008. **[Google Scholar]**

52. Prum, R.O.; Berv, J.S.; Dornburg, A.; Field, D.J.; Townsend, J.P.; Lemmon, E.M.; Lemmon, A.R. A comprehensive phylogeny of birds (Aves) using targeted next-generation DNA sequencing. *Nature* **2015**, *526*, 569–573. **[Google Scholar] [CrossRef] [PubMed]**

53. Wink, M.; Sauer-Gurth, H. Molecular taxonomy and systematics of owls (Strigiformes)—An update. *Airo* **2021**, *29*, 487–500. **[Google Scholar]**

54. Aliabadian, M.; Alaei-Kakhki, N.; Mirshamsi, O.; Nijman, V.; Roulin, A. Phylogeny, biogeography, and diversification of barn owls (Aves: Strigiformes). *Biol. J. Linn. Soc.* **2016**, *119*, 904–918. **[Google Scholar] [CrossRef]**

55. Dial, C.R.; Talbot, S.L.; Sage, G.K.; Seidensticker, M.T.; Holt, D.W. Cross-species Amplification of Microsatellite Markers in the Great Horned Owl *Bubo virginianus*, Short-eared Owl *Asioflammeus* and Snowy Owl *B. scandiacus* for Use in Population Genetics, Individual Identification and Parentage Studies. *J. Yamashina Inst. Ornithol.* **2012**, *44*, 1–12. **[Google Scholar] [CrossRef]**

56. Delport, W.; Grant, T.J.; Ryan, P.G.; Bloomer, P. Ten microsatellite loci for evolutionary research on Nesospiza buntings. *Mol. Ecol. Notes* **2006**, *6*, 1180–1183. **[Google Scholar] [CrossRef]**

57. Pellegrino, I.; Negri, A.; Boano, G.; Cucco, M.; Kristensen, T.N.; Pertoldi, C.; Randi, E.; Šálek, M.; Mucci, N. Evidence for strong genetic structure in European populations of the little owl Athene noctua. *J. Avian Biol.* **2015**, *46*, 462–475. **[Google Scholar] [CrossRef]**

58. Pertoldi, C.; Pellegrino, I.; Cucco, M.; Mucci, N.; Randi, E.; Laursen, J.T.; Sunde, P.; Loeschcke, V.; Kristensen, T.N.; du Toit, Z.; et al. Genetic consequences of population decline in the Danish population of the little owl (*Athene noctua*). *Mol. Ecol.* **2020**, *14*, 358–367. **[Google Scholar] [CrossRef] [Green Version]**

59. Macías-Duarte, A.; Conway, C.J.; Holroyd, G.L.; Valdez-Gómez, H.E.; Culver, M. Genetic Variation among Island and Continental Populations of Burrowing Owl (*Athene cunicularia*) Subspecies in North America. *J. Raptor Res.* **2019**, *53*, 127–133. **[Google Scholar] [CrossRef] [Green Version]**

60. Biello, R.; Zampiglia, M.; Corti, C.; Deli, G.; Biaggini, M.; Crestanello, B.; Delaugerre, M.; Di Tizio, L.; Leonetti Francesco, L.; Stefano, C.; et al. Mapping the geographic origin of captive and confiscated Hermann's tortoises: A genetic toolkit for conservation and forensic analyses. *Forensic Sci. Int. Genet.* **2021**, *51*, 102447. **[Google Scholar] [CrossRef] [PubMed]**

61. Glenn, T.C. Field guide to next-generation DNA sequencers. *Mol. Ecol. Resour.* **2011**, *11*, 759–769. **[Google Scholar] [CrossRef]**

62. Fordyce, S.L.; Ávila-Arcos, M.C.; Rockenbauer, E.; Børsting, C.; Frank-Hansen, R.; Petersen, F.T.; Willerslev, E.; Hansen, A.J.; Morling, N.; Gilbert, T.P. High-throughput sequencing of core STR loci for forensic genetic investigations using the Roche Genome Sequencer FLX platform. *BioTechniques* **2011**, *51*, 127–133. **[Google Scholar] [CrossRef] [PubMed]**

63. Bornman, D.M.; Hester, M.E.; Schuetter, J.M.; Kasoji, M.D.; Minard-Smith, A.; Barden, C.A.; Faith, S.A. Short-read, high-throughput sequencing technology for STR genotyping. *BioTechniques. Rapid Dispatches* **2012**, *23*, 1–7. **[Google Scholar] [CrossRef]**

64. Van Neste, C.; Van Nieuwerburgh, F.; Van Hoofstat, D.; Deforce, D. Forensic STR analysis using massive parallel sequencing. *Forensic Sci. Int. Genet.* **2012**, *6*, 810–818. **[Google Scholar] [CrossRef]**

65. Stoeckle, B.C.; Theuerkauf, J.; Rouys, S.; Gula, R.; Lorenzo, A.; Lambert, C.; Kaeser, T.; Kuehn, R. Identification of polymorphic microsatellite loci for the endangered Kagu (*Rhynochetos jubatus*) by high-throughput sequencing. *J. Ornithol.* **2012**, *153*, 249–253. **[Google Scholar] [CrossRef] [Green Version]**

66. Pimentel, J.S.M.; Carmo, A.O.; Rosse, I.C.; Martins, A.P.V.; Ludwig, S.; Facchin, S.; Pereira, A.H.; Brandão-Dias, P.F.P.; Abreu, N.L.; Kalapothakis, E. High-throughput sequencing strategy for microsatellite genotyping using neotropical fish as a model. *Front. Genet.* **2018**, *9*, 1–8. **[Google Scholar] [CrossRef] [Green Version]**

67. Qi, W.H.; Lu, T.; Zheng, C.L.; Jiang, X.M.; Jie, H.; Zhang, X.Y.; Yue, B.S.; Zhao, G.J. Distribution patterns of microsatellites and development of its marker in different genomic regions of forest musk deer genome based on high throughput sequencing. *Aging* **2020**, *12*, 4445–4462. **[Google Scholar] [CrossRef] [PubMed]**

68. Song, C.; Feng, Z.; Li, C.; Sun, Z.; Gao, T.; Song, N.; Liu, L. Profile and development of microsatellite primers for Acanthogobiusommaturus based on high-throughput sequencing technology. *J. Oceanol. Limnol.* **2020**, *38*, 1880–1890. **[Google Scholar] [CrossRef]**

69. De Barba, M.; Miquel, C.; Lobréaux, S.; Quenette, P.Y.; Swenson, J.E.; Taberlet, P. High-throughput microsatellite genotyping in ecology: Improved accuracy, efficiency, standardization and success with low-quantity and degraded DNA. *Mol. Ecol. Resour.* **2017**, *17*, 492–507. **[Google Scholar] [CrossRef] [PubMed]**

17

INDIVIDUAL IDENTIFICATION USING SHORT TANDEM REPEATS REITERATE THE PROCESS OF ANALYZING AND GATHERING SECONDARY DATA THROUGH MICROBIOME ANALYSIS

1. INTRODUCTION: AN OVERVIEW

Evidence analysis plays a vital role in substantiating the alibis of individuals implicated in legal proceedings and is of utmost importance in the courtroom for the purpose of reconstructing crime scenes and tracking suspects. Obtaining evidence pertaining to the victim or suspect is crucial in an investigation. Identifying the presence of an individual who fulfills specific criteria at the location is crucial in forensic inquiry as it establishes a link between the suspect or victim and aids in resolving the case. Important trace evidence possesses several distinct qualities, with evidence of human origin being the most conspicuous. The majority of this evidence comprises human DNA, which is valuable for forensic assessment. Furthermore, many forms of evidence, including unidentifiable and post-mortem residues, bodily hair, and behavioral indicators found on the victim's body or human-derived substances, might be gathered. The study yields information that can be utilized for crime scene identification and as evidence of an alibi. DNA evidence establishes a direct link between the case, the victim, and the suspect [1]. STR analysis is more efficient when there is unequivocal indication of human DNA. Human genetic material can be recognized using DNA typing, which allows for the identification of persons involved in the case [2,3]. Ensuring the protection of human DNA and conducting analysis of Short Tandem Repeats (STRs) are crucial in order to accurately identify potential suspects and victims who may be compared. Consequently, laboratories around the world are researching the optimal methods for obtaining DNA evidence with maximum efficiency and effectiveness. Moreover, research on evidence derived from low-level DNA profiles obtained from trace DNA indicates that trace DNA can contribute to the effective resolution of cases [4,5]. Nevertheless, several laboratories refrain from analyzing trace DNA evidence due to the limited amount of biomass in trace DNA or touch DNA samples [6].

STR analysis cannot guarantee a 100% success rate, and the assay success varies with the sample quantity and quality. Human DNA is considerably affected by environmental factors. The STR analysis may not be possible because DNA is damaged by high temperature, humidity, UV light, and microbial activation [7]. Therefore, low copy number (LCN) samples subjected to STR analysis sometimes render uninterpretable or inconclusive results [**8**]. Additionally, other results cannot be inferred from the results of LCN samples alone. Even if a sample contains non-existent or non-identifiable human DNA, it is likely that other substances or genes are present, as samples are collected from various sites and situations. Thus, other analytical methods may supplement information from low-quality STR results.

The human body hosts numerous microorganisms, and the human microbiome is influenced by the human body. For example, the microbiome is affected by various factors such as living environment [**9**], eating habits [**10**], age, sex [**11,12**], race [**13**], and the existence of a distinct microbiome field [**14**]. Thus, studies investigating whether individuals can be identified using microbiome characteristics that differ from one individual to another are being actively conducted [**15,16,17**]. Analysis of the microbiome has progressed to the point that the microbiome can be traced using biodiversity analysis. Furthermore, the microbiome is unique to an individual or a group and can reveal characteristic features known as the "microbial fingerprint." Individual microorganisms are used in the forensic field and in research related to individual identification during the progression of a case. Studies have included object and owner tracking [**17,18**], environmental tracking using soil microbes [**19**], bloodstain tracking [**20**], post-mortem interval tracking [**14**], and oral saliva identification [**21,22,23**]. By examining the characteristics of microorganisms (bacteria, fungi, and viruses) and linking them to individuals and case evidence, forensic microbiology can help obtain clues that can help resolve the case. Additionally, analyzing the microbiome makes it possible to montage trace DNA from suspects or victims and provides information on background tracing and scene reconstruction. Microorganisms are ubiquitous, and their quantity is considerably greater than that of human DNA. Proofs found at the scene may include those that do not have visible features, such as blood. The discovery of human DNA at crime scenes alone may not lead to prosecution based on genetic evidence. In fact, the prosecution of a case in Korea was canceled, despite the fact that the suspect was identified using Y-STR (STR on the Y chromosome) analysis [**24**]. Therefore, additional analysis is needed to supplement human genetic information and reveal the case background, proof of alibi, and connection of evidence.

Herein, we propose a human bacterial profiling method using next-generation sequencing to collect secondary information. In this study, using bacterial profiling, we investigated whether additional information can be obtained from samples of failed STR assays (**Figure 1**). Furthermore, we devised a method to classify and identify samples of unknown origin.

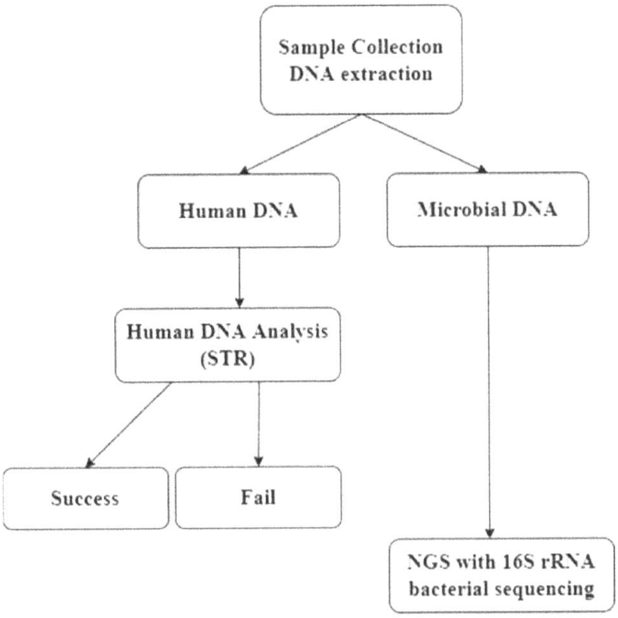

Figure 1. Schematic of the application of Bacterial profiling. Human DNA analysis and microbial DNA analysis are possible simultaneously through one DNA extraction.

2. MATERIALS AND METHODS

2.1. Sample Collection and DNA Extraction

Samples were collected from student volunteers of Eulji University (Republic of Korea, Daejeon). The study protocol was approved by the internal review board of Eulji University (IRB No. EUIRB 2020-13). All volunteers were healthy adults in their 20s, with one male and nine female university students, and samples were collected from each individual during the same visit. A questionnaire was used to determine if the participant was under treatment with antibiotics or if there was environmental contamination, such as alcoholic disinfection of the hand or mobile phone. If a volunteer was administered antibiotics or had disinfected the hand or mobile phone within 2 h before sampling, sampling was discontinued.

We used sterile, DNA-free cotton-tipped applicators (Puritan Medical Products Co., Guilford, ME, USA) to swab individual mobile phones and surface skin of the fingertip of the ventral joint. The tip of the applicator was then cut with sterile pair of scissors and collected in a 1.5 mL collection tube. Oral gargle and urine samples were collected in sterile 8 mL collection tubes. To obtain gargle samples, the volunteers were provided drinking water to swirl the floor of their mouths 10 times. The samples were stored at -80 °C until DNA extraction. Each sample swab was mixed with 700 μL of phosphate-buffered saline and 12 glass beads (2 mm) in a 1.5 mL Eppendorf (Ep) tube. The mixture was vortexed for 15 min and centrifuged at 8000 rpm for 5 min. The supernatant was removed, and the pellet was mixed with 20 μL of egg white lysozyme (Amresco, Solon, OH, USA) and incubated at 37 °C for 1 h. Subsequently, the total DNA was extracted using a QIAamp® DNA Mini Kit (Qiagen, Hilden, Germany). The DNA was eluted with 80 μL of elution buffer. The extracted DNA samples were stored at -80 °C until library preparation and sequencing.

2.2. STR Analysis

The extracted DNA samples were quantified using a Quantifiler Trio Quantification kit (Thermo Fisher Scientific, Waltham, MA, USA) and an Applied Biosystems 7500 Real-Time Polymerase Chain Reaction (PCR) System (Thermo Fisher Scientific, Waltham, MA, USA). The cycling parameters were set and data analysis was performed in accordance with the recommendations of the manufacturer. The samples were amplified in triplicate. The DNA concentration was adjusted to 1.0 ng/μL in a final PCR mixture volume of 25 μL. The samples with higher DNA concentrations were diluted to 1 ng/μL.

The PCR product (1 μL) was added to 8.5 μL of highly deionized (Hi-Di) formamide (Applied Biosystems, Zug, Switzerland) and 0.5 μL of 600 LIZ (20–600 nucleotide range, Applied Biosystems) Size Standard (Applied Biosystems). STR amplification was performed using 1 ng of template DNA and an AmpFLSTRIdentifiler PCR Amplification Kit (Thermo Fisher Scientific). Amplification was performed using an ABI Prism 310 Genetic Analyzer, and the data were analyzed with GeneMapper ID software v3.2.1 (Thermo Fisher Scientific, MA, USA).

2.3. Library Preparation of 16S rRNA Amplicons

Forty samples were analyzed by high-throughput 16S rRNA gene amplicon sequencing on an Ion GeneStudio S5 platform (Thermo Fisher Scientific). The V3–V4 regions of the 16S rRNA gene from each sample were amplified using the following primers: 341F (5′-CCTACGGGNGGCWGCAG-3′), which contained a sample-specific 6–8-base pair tag sequence, and 805R (5′-GACTACHVGGGTATCTAATCC-3′). PCR amplification was performed using Platinum PCR SuperMix High Fidelity master mix (Invitrogen, Carlsbad, CA, USA) with 2.5 ng of template DNA and 50 nmol of each primer in a 27-μL reaction mixture volume. The PCR conditions were as follows: 94 °C for 3 min, followed by 30 cycles at 94 °C for 30 s, 50 °C for 30 s, and 72 °C for 30 s. To remove residual primer dimers and contaminants, the amplicon libraries were purified using an AgencourtAMPure XP DNA purification kit (Beckman Coulter

Life Sciences, Indianapolis, IN, USA). The samples were eluted with 15 μL of low Tris-EDTA (TE) buffer. The DNA concentration and quality were assessed using the dsDNA High Sensitivity Assay Kit on a Qubit 4 Fluorometer (Thermo Fisher Scientific, Waltham, MA, USA). The fragment size and quality of the pooled DNA were assessed using an Agilent 2100 Bioanalyzer (Agilent Technologies, Palo Alto, CA, USA). The enriched particles were loaded on to an Ion 530 chip (Thermo Fisher Scientific) and sequenced using an Ion GeneStudio S5 instrument [**25**].

2.4. Analysis of 16S rRNA Amplicon Sequences

Reads were excluded from the analysis if they were shorter than 500 bp or were inappropriately paired. The sequence data were analyzed using EzBioCloud 16S rRNA gene-based microbiome taxonomic profiling (MTP) and the PICRUST algorithm (ChunLab, Seoul, Korea) with "Bacteria" as a target taxon in the prokaryotic 16S rRNA gene database PKSSU4.0. Sequences processed using the EzBio Cloud 16S rRNA gene-based MTP pipeline were clustered into operational taxonomic units (OTUs) using a 97%-similarity cut-off and identified using QIIME-MOTHUR algorithms.

2.5. Statistical Analysis

To examine differences in the bacterial community diversity, Microbiome Analyst software (**www.microbiomeanalyst.ca**, accessed on 25 January 2021,) [**26**] was used to evaluate α and β diversities and to conduct group comparison and classification. The Shannon index was used to evaluate α diversity. Principal coordinate analysis (PCoA) was conducted using Jensen–Shannon divergence and evaluated using the permutational multivariate analysis of variance (PERMANOVA). The microbial composition among groups was compared using linear discriminant analysis (LDA) effect size (LEfSe), and the relative abundance of the core microbiome taxa was assessed at the genus level. Kruskal–Wallis H test correction was performed using SPSS ver. 20.0 to evaluate the inter-group significance at the genus and species levels.

3. RESULTS

3.1. STR Analysis

The samples underwent STR analysis. The DNA concentration obtained from the gargle, urine, finger, and mobile phone samples was 16.67, 2.16, 0.85, and 0.42 ng/μL, respectively. Four DNA concentrations that were not initially discovered (below 0 ng/μL) were examined in the fingertip and mobile phone samples.

The STR analysis results are shown in **Table 1**. For actual individual identification, statistical analysis should be performed for each gene peak. However, in this study, the number of samples to be confirmed was not large enough for a statistical approach, and the general approach of simply counting the number of matching loci was able to distinguish them from each other [27]. The urine and gargle fluid samples provided a full profile that could identify individuals. For both fingertip and mobile phone samples, 8 of the 10 samples were confirmed identifiable full profiles. Two fingertip and two mobile phone samples from two subjects could not be used to identify individuals.

Table 1. Results of STR analysis.

	Sample (n = 40)	Full Profile (Loci = 24)	Partial Profile			Mixed	Full + Minor
			1~10	11~19	20~23		
Fingertip	10	4	1	1	1	1	2
Mobile phone	10	7	1	1	0	0	1
Urine	10	9	0	0	0	0	1
Gargle	10	9	0	0	0	0	1

Mixed = although mixed with other genes, the degree to which an individual profile can be identified.

3.2. Microbiome Analysis

3.2.1. Overview of Taxonomic Diversity

The samples were classified into gargle, urine, mobile phone, and fingertip according to the collection site. Among the 40 samples, 5 were discarded during quality control processing. Thus, 35 samples were used in the study. We obtained 660,095 valid 16S rRNA reads after quality filtering. For each group, 340,525 (gargle), 103,510 (urine), 112,877 (fingers), and 103,183 (mobile phone) reads were obtained.

The composition of the bacterial community of each sample was examined for relative abundance at the genus level. The genera that accounted for the highest proportion in the gargle samples were *Streptococcus*, *Rothia*, *Gemella*, *Heamophillus*, *Neisseria*, and *Granulicatella*. The genera that accounted for the highest proportion in the urine samples were *Lactobacillus*, *Gardnerella*, *Streptococcus*, and *Prevotella*. The genera with the highest proportion in the fingertip samples were *Streptococcus*, *Rothia*, *Cutibacterium*, and *Staphylococcus*. Finally, the genera that accounted for the highest proportion in the mobile phone samples were *Streptococcus*, *Rothia*, *Porphyromonas*, and *Neisseria*. The average number of OTUs in each sample was 643 for gargle, 229 for urine, 520 for finger, and 400 for mobile phone.

3.2.2. α and β Diversities

α-Diversity analysis was performed to determine the abundance of each species (**Figure 2**). The analysis was performed at the species level and tested using the Kruskal–Wallis method. The overall p-value was less than 0.05, which indicated significant diversity of all species. In the α-diversity analysis, the samples with high abundance, diversity, and uniformity were the gargle, fingertip, and mobile phone samples. Urine had lower diversity than the other three sample types.

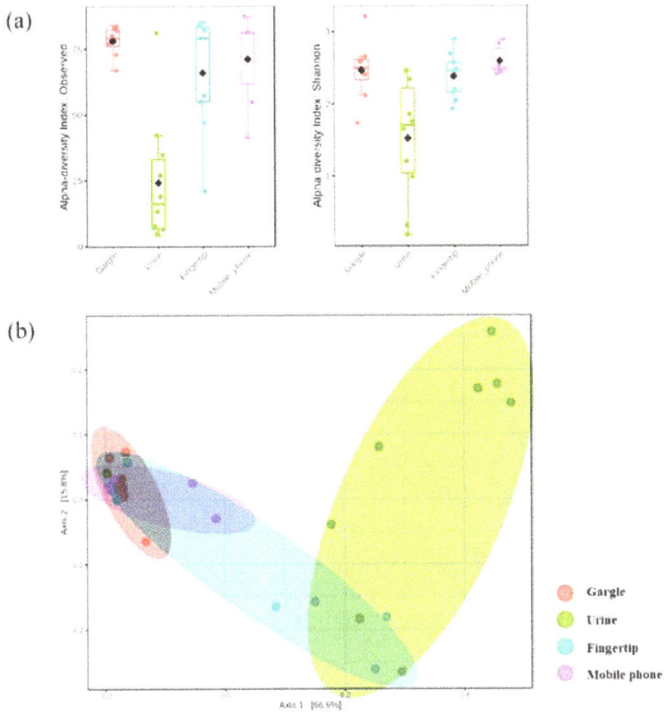

Figure 2. Results of α and β diversity, which reveal the total strain abundance and type (a) α diversity with Observed (left) and Shannon (right) index and (b) β diversity with Jensen-Shannon divergence index.

β-Diversity analysis using Jensen–Shannon divergence demonstrated bacterial community clustering for each sample type. Urine samples had a few overlapping parts with the other sample types, indicating differences in the bacterial community composition from other sample types. Several communities in the fingertip samples were similar to the bacterial communities from the gargle and mobile phone samples, suggesting that the fingers share bacterial species through contact with other body parts.

3.2.3. Creation of a List of Strains and Matching Assessment

Next, a list was prepared using the representative markers reported for each site and the sample analysis method used in the study (**Table 2**). This list was prepared at the genus level using LEfSe, the core microbiome, and the reported general microbiome from previous studies. LEfSe was used to analyze gargle, urine, and finger metagenomics data. The linear discriminant analysis (LDA) score was derived using a Kruskal–Wallis rank sum test, where the significant LDA scores were more than 2.0. The false discovery rate (FDR) was set to 0.05. In the core microbiomes, the relative abundance of the bacteria in each group was determined using a relative abundance cut-off value of 0.01. The LEfSe and core microbiome data were analyzed by selecting the most abundant strain present in each body part and a representative species present in the urine and gargle samples.

Table 2. Ten representative strains at the Genus level for each region, constructed using the microbiome or indicative bacteria reported in previous studies and sample analysis conducted within the study.

Gargle	Urine	Fingertip
Streptococcus	Escherichia	Corynebacterium
Veillonella	Staphylococcus	Streptococcus
Prevotella	Finegoldia	Staphylococcus
Neisseria	Atopobium	Micrococcus
Haemophilus	Lactobacillus	Veillonella
Porphyromonas	Corynebacterium	Dermacoccus
Rothia	Gardnerella	Cutibacterium
Actinomyces	Campylobacter	Enhydrobacter
Campylobacter	Peptoniphilus	Sphingomonas
Tannerella	Anaerococcus	Lawsonella

We observed that gargle samples had the highest probability of confirming the microbiome composition, followed by the fingertip and mobile phone samples. For the urine samples, the average agreement rate was 66%, which was lower than that of the gargle and finger samples. Unlike the microbiome of other body parts, the number of bacteria present in urine is small and is classified using strains originating from the genital tract according to sex [**28**]. Identification of the strains from mobile phones was confirmed using the list of fingertip strains. The average coincidence rate was approximately 60%. Among the eight target samples, six matched the fingertip strain by more than 60%. Thus, it was inferred that the corresponding sample is related to the skin.

By confirming 10 genera in the gargle samples, significant differences were observed in the gargle and urine samples compared, thereby making it possible to discriminate body fluids ($p < 0.05$). However, for *Streptococcus* and *Prevotella*, there was no significant difference because these strains are commonly found in urine and the mouth [**28,29**]. Among the remaining eight strains, *Tannerella* was the most different compared with the other strains. There was no significant difference in the gargle samples compared to the finger and mobile phone samples, which is consistent with the findings of a previous study, showing that a large number of oral bacteria are introduced via exposure to external environments such as fingers and mobile phones or by personal habits and behaviors [**30**].

In urine samples, *Staphylococcus, Finegoldia, Corynebacterium,* and *Campylobacter* showed significant differences compared with those in the other groups. For the fingertip samples, no significant differences were observed compared to the mobile phone samples. Similarly, no significant differences were found when compared with the gargle samples. However, there were strains with significant differences between the urine and gargle samples and the mobile phone samples. *Staphylococcus, Dermacoccus, Cutibacterium,* and *Enhydrobacter* were specifically identified on mobile phones. *Sphingomonas* was the most frequent in urine.

The possibility of final classification was investigated by applying our classification method to the selected sample using the strain reported as the major microbiome component at the species level. The strains observed in the gargle samples were *Streptococcus salivarius, S. sanguinis,* and *Neisseria subflava* [**31**], whereas those in the urine samples were *Lactobacillus* spp. and *Gardnerella vaginalis* [**32**]. There are three strains in the fingertip samples: *Cutibacterium acnes, Corynebacterium tuberculostearicum,* and *Micrococcus lutes* [**33**]. Using our classification method, three strains were identified in all gargle samples and *Lactobacillus* species were found in all urine samples. Among them, *G. vaginalis* was found in only 5 of 10 samples. *Gardnerella vaginalis* is found in females and is increased in the presence of bacterial vaginosis [**21**]. In fingertip and mobile phone samples, *C. acnes* was found in all samples on the fingers, *C. tuberculostearicum* was found in eight out of nine samples, and *M. luteus* was detected in six out of nine samples. In the mobile phone samples, *C. acnes* was found in all individuals, whereas the other two bacterial species were not detected in mobile phone samples. Thus, by identifying skin-related microorganisms, it can be inferred whether the sample is directly related to humans or was obtained from a surface where primary or secondary transfer occurred.

3.3. Personal Feature Tracking and Unique Bacterial Features

The bacterial strains present in each sample type were as follows. First, numerous strains derived from microorganisms in the human upper respiratory tract, oral cavity, feces, and intestines were found in the gargle samples [**22,29**]. Additionally, microorganisms related to the environment, such as plant roots, sea water, and soil were observed. *Avibacterium,* which is found in chicken beaks, and *Bombiscardovia,* which is found in the digestive tract of bees, were also detected [**23,34**]. Among environmental microorganisms, *Skermanella,* which is found in Korean aerial environments, was observed [**34**]. In the case of urine, microbes associated with the intestines, urethra, vagina, and cervix, as well as oral bacteria, were found

[21,31]. By matching samples with female subjects, we confirmed that the sex-related information was consistent.

3.4. Bacterial Analysis of Failed STR an Analysis Samples

We conducted Next-Generation Sequencing (NGS) to analyze the bacterial composition using 16S rRNA profiling. The samples used for this analysis were those that had previously failed in the Short Tandem Repeat (STR) study. This information is presented in Figure 3. The bacterial sequencing yielded valid results for all failed samples in the STR analysis. The bacterial compositions varied between samples collected from mobile phones and fingers. Thus, the bacterial sequencing findings were compared to the results of a previous investigation where the microorganism profiles of both mobile phones and individuals' hands exhibited resemblances [16]. Four samples, specifically No. 12, 25, 34, and 44, which were unsuccessful in STR analysis, were collected from finger tips and mobile phones. These samples were then mixed and analyzed for bacterial profiling. The samples were subjected to analysis utilizing Jensen–Shannon divergence in order to ascertain the level of similarity among them. It has been verified that the samples that did not pass the STR analysis can be differentiated based on their respective sample locations.

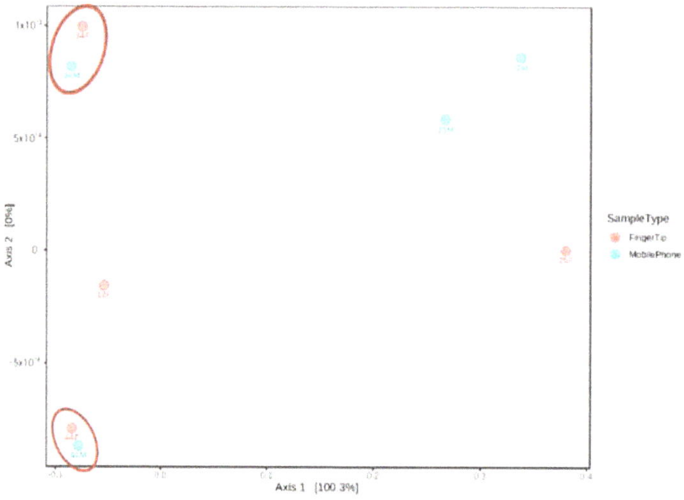

Figure 3. β **diversity and relationship of failed STR samples with fingertip-mobile phone (Jensen-Shannon divergence). The correlation is confirmed by measuring the similar distance between each sample using a two-dimensional distance measurement expression. The red circle means nearest sample).**

4. DISCUSSION

Forensic investigation frequently encounters contamination of DNA samples with non-human genetic material. The aim of the study was to validate the feasibility of conducting supplementary microbiome analysis in cases where STR analysis is unsuccessful. Furthermore, we demonstrated the feasibility of sample categorization and monitoring of individual traits by microbiome analysis. The source of the sample might be ascertained by analyzing the bacterial composition of each individual sample. Additional Next-Generation Sequencing (NGS) analysis can provide supplementary information regarding the relationship between experimental samples that yield invalid results in the Short Tandem Repeat (STR) study. Furthermore, by the examination of bacterial strains in each sample based on certain microbiological traits, it is feasible to acquire data pertaining to the gender, habitat, or physical state of the sample's origin.

Owing to the wide distribution of microorganisms, complete individual results were not obtained through microbial profiling, although tracked information could be obtained about the surrounding environment, sex, or physical condition of the subject using survey and subject information matching. In gargle samples, we observed strains derived from various communities such as those from feces and intestines, in addition to human oral microbes. Furthermore, the observed environmental microorganisms were derived from various environments, such as plant roots, sea water, soil, and deep mineral water. Therefore, we were able to observe whether bacteria were included because of respiration through the mouth. However, unlike oral bacteria, periodontal and upper respiratory tract bacteria were present in small proportions and excluded from the oral microbiome composition. Using these strains, environmental tracking was possible. Of the bacteria found in the oral cavity, *Skermanella* is found in the air of Korea [**35**], suggesting that individuals with this species in the gargle samples currently reside in Korea.

In urine, bacteria related to the intestines, urethra, cervix, vagina, soil or water environment, and outdoor air were found to be similarly diverse. Urine is particularly likely to be contaminated by contact with the hand of the subject at the time of collection, and the distribution of bacteria may vary depending on the collection environment and method. Additionally, urine is a body fluid that is linked to health status and sex-related information. In females, the perineum is shorter than that in men, and intestinal microbes found in the anus or large intestine are found in the female genitalia. Thus, not only urinary-tract-related bacteria but also anal-, fecal-, and intestinal-related bacteria are found in urine-related samples with high abundance. Thus, sex can be predicted using this method. Females may suffer from bacterial vaginosis, and the associated bacteria may be detected in large amounts [**36**]. In this study, many *Lactobacillus* species were found in the urinary tract. Additionally, *G. vaginalis*, the causative agent of bacterial vaginosis, was frequently observed. However, the volunteers were not questioned regarding vaginitis. Our results indicate that the biological environment or sex of the volunteers can be inferred by analyzing the bacterial profile.

Generally, specific strains were found related to the sample collection site, such as the mouth and urinary tract, although this was not the case for skin sites. The existence of skin microbes related to the genus *Cutibacterium*, as reported in other studies, was confirmed in this study [**19**]. However, as the fingers are most exposed to the external environment and often come into contact with other surfaces, the bacterial community composition is complicated by various environments [**37**]. Thus, it is possible to find a characteristic bacterial community related to the contact environment, although contrarily, it is appropriate to interpret skin microbiome analysis results with the possibility of contamination in mind. At the time of sample collection, it was recommended not to wash hands for 3 h before sample collection, although no specific precautions were taken. Thus, the volunteers may have been in contact with each other or other college students, for example, by touching or unconscious contact. DNA transfer can occur by two routes: primary and secondary; it has been reported that human DNA as well as bacterial DNA can be transferred [**38**]. These external factors may have attributed to the difficulty in identifying tracing microorganisms using skin samples.

Owing to the outbreak of COVID-19 at the time of sample collection, the Korean government recommended the regular use of disinfectants. Hence, hand sanitizers were actively used. Any subjects who disinfected and washed their hands more than a certain number of times a day were excluded. As bacteria are vulnerable to alcohol-mediated degradation, if disinfection is carried out several times a day there is a high possibility that characteristic bacteria will be killed [**39**]. As these habits are related to social changes, it may be helpful to consider the variables for changes according to social aspects when conducting skin microbiome studies of the hand. Additionally, although there are microbes on surfaces such as the skin and mobile phones, they do not necessarily share all strains because of different biological and nutritional conditions [**38**]. Therefore, when performing a microbiome-wide analysis, it is necessary to confirm the characteristics of the possession identified from the owner. The method by which information of the object surface is analyzed can also help the investigation.

In the case of sample information derived from a single person, profiling can be integrated with information of other samples. In the case of volunteer 7, *Mobiluncus*, a bacterium found in the male partner of a woman with bacterial vaginosis, was found in the gargle sample [**40**]. Additionally, numerous *Lactobacillus* species were found in the urine of volunteer 7, and it was confirmed through the questionnaire that the subject was a man with a girlfriend. In the case of volunteer 21, *Neiserria sicca* was found in all collection sites. This bacterium is found in patients with weakened immunity or atopy [**41**]. Volunteer 21 was confirmed to have atopic skin disease using the questionnaire. In the case of volunteer 2, *Massiliaaerilata* and *Corynebacterium kreppenstedtii*, which are commonly found on the hands, mobile phones, plant-related substances, or plant roots, were observed [**42,43**]. Volunteer 2 confirmed the presence of companion plants at home in the survey. Volunteer 20 had *Lactobacillus iners*, *Gardnella vaginalis*, and *Atopobiumvaginae* on the mobile phone. These bacteria are found in the vagina and urinary tract of females, suggesting that the owner of the mobile phone was a woman [**16**]. Additionally, as volunteer 20 did not disinfect the mobile phone, there is a high probability that the microbiological composition is associated with the owner. Our results suggest that individual profiles can be obtained by microbial analysis through information matching between multiple samples and confirmation using questionnaires.

Microbiome analysis of samples from human sources has potential for forensic applications by providing information on the identification of individuals at crime scenes. However, the analysis range of the microbiome is wide and there are many variables, such as the distribution and composition of bacteria, environment, and differences among individuals. Considering that the above results showed a concordance rate of approximately 80% or more, the STR analysis is the best test for human-derived samples. However, only human genetic information can be obtained, and additional environmental or circumstantial information cannot be completely obtained [**27**]. As with human DNA, there are no specific marker loci such as STR. Hence, it is difficult to perform specific tests. However, this study suggests that sample-derived tracking and sample-related information tracking are possible using NGS, by preparing a panel list using specific bacteria in a manner similar to the STR analysis. As the bacterial abundance is different for each sample type, it is better to determine the abundance for each cluster rather than use a specific marker for feature tracking. The use of a single strain is risky as the strain may exist extensively in other sites [**16**]. Accordingly, we prepared lists using various microbiomes and presented a proposal to selectively select strains using these lists. Using this approach, additional analyses can be employed for samples of unknown origin. Furthermore, samples that fail STR analysis can be analyzed using a secondary method.

An era has arrived when human DNA alone cannot be used as valid evidence in court. In an actual case, although DNA of the suspect was found and the Y-STR was matched, the evidence used in the case did not contain the suspect's fingerprint. Furthermore, the DNA of the police personnel involved in the case at the time matched the Y-STR of the suspect, thus invalidating the sentence of the perpetrator [**14**]. This indicates that oral descriptions of the victim and the human DNA profile applicable to suspects are not valid in court, and additional evidence must be presented. Therefore, for a clear trial and fair judgment, not only the main body of evidence but also additional evidence supporting and/or linking the suspect or victim to the case is inevitable. For example, supplemental information can be provided by identifying oral bacteria in bite marks that occur in sexual assault cases. In this regard, we would like to emphasize that information of forensic microorganisms that can identify human traces is also necessary for creating secondary profiles and collecting additional evidence. In addition, the method used for DNA extraction is important. Since analyzing traces of microbial genes and human genes via a single sample collection can be applied in a forensic approach, it has the advantage of securing more evidence in case resolution.

In this study, only the V3–V4 region was analyzed using 16S rRNA sequencing. Analysis of other V regions could identify more microbial communities and individual strains. The classification of the V region compared to the actual analyzed region is also different. In addition, if the sample itself is not of acceptable quality, the number of bacteria that can be analyzed is small, and complete information according to individual tracking may not be available. Moreover, high-quality DNA can be obtained using various tools and swabbing solutions in the DNA collection method [**44,45**].

5. AT A GLANCE

In forensic analysis, samples collected are often of varying quality. Non-human evidence is also discovered, and even when explicit human evidence like blood is identified, human DNA can be eradicated or prove challenging to retrieve and isolate. Our research suggests that site tracking can be accomplished by compiling a list of microbiological categories specific to each site. Furthermore, microbiome analysis allows for the implementation of a specific experimental technique, which may be carried out with just one extraction step. Additionally, it enables the secondary examination of samples that were unsuccessful in STR analysis. The outcomes of this study provide a basis for the development of strategies in forensic microbiology to analyze low-quality material.

6. CONCISE SUMMARY

The significance of forensic investigation lies in its ability to scrutinize evidence and expedite the identification of crucial individuals, such as suspects and victims, in a criminal case. The utilization of genomic DNA in forensic applications has grown alongside advancements in DNA sequencing technology, facilitating supplementary analysis in criminal investigations where more legal evidence is necessary. This work employed next-generation sequencing to enhance the acquisition of supplementary data for the examination of human evidence collected via short tandem repeat (STR) analysis. We conducted an investigation into the suitability and capacity of assessing communities of microbial genomes. Confirmation of microbiological supplementation information was obtained for two out of four

STR samples that did not meet the required standards. Furthermore, the gargle sample exhibited a validated accuracy rate of 100% and is highly probable to be categorized as a bodily fluid sample. Our experimental approach validated the acquisition of anthropological and microbiological evidence by the execution of two experiments using a single extraction. We analyze the benefits and drawbacks of employing these methods, investigate potential opportunities in the forensic domain, and emphasize recommendations for future investigation.

REFERENCES

1. Reid, C.A.; Howes, L.M. Communicating forensic scientific expertise: An analysis of expert reports and corresponding testimony in Tasmanian courts. *Sci. Justice* **2020**, *60*, 108–119. **[Google Scholar] [CrossRef] [PubMed]**
2. Dumache, R.; Ciocan, V.; Muresan, C.; Enache, A. Molecular DNA Analysis in Forensic Identification. *Clin. Lab.* **2016**, *62*, 245–248. **[Google Scholar] [CrossRef] [PubMed]**
3. Silva, N.M.; Pereira, L.; Poloni, E.S.; Currat, M. Human neutral genetic variation and forensic STR data. *PLoS ONE* **2012**, *7*, e49666. **[Google Scholar] [CrossRef] [PubMed] [Green Version]**
4. Dziak, R.; Peneder, A.; Buetter, A.; Hageman, C. Trace DNA Sampling Success from Evidence Items Commonly Encountered in Forensic Casework. *J. Forensic Sci.* **2018**, *63*, 835–841. **[Google Scholar] [CrossRef] [PubMed]**
5. Raymond, J.J.; van Oorschot, R.A.; Walsh, S.J.; Roux, C. Trace DNA analysis: Do you know what your neighbour is doing? A multi-jurisdictional survey. *Forensic Sci. Int. Genet.* **2008**, *2*, 19–28. **[Google Scholar] [CrossRef]**
6. Harbison, S.; Fallow, M.; Bushell, D. An analysis of the success rate of 908 trace DNA samples submitted to the Crime Sample Database Unit in New Zealand. *Aust. J. Forensic Sci.* **2008**, *40*, 49–53. **[Google Scholar] [CrossRef]**
7. Alaeddini, R.; Walsh, S.J.; Abbas, A. Forensic implications of genetic analyses from degraded DNA—A review. *Forensic Sci. Int. Genet.* **2010**, *4*, 148–157. **[Google Scholar] [CrossRef]**
8. Kloosterman, A.; Kersbergen, P. Efficacy and limits of genotyping low copy number DNA samples by multiplex PCR of STR loci. *Int. Congr. Ser.* **2003**, *1239*, 795–798. **[Google Scholar] [CrossRef]**
9. Meadow, J.F.; Altrichter, A.E.; Bateman, A.C.; Stenson, J.; Brown, G.Z.; Green, J.L.; Bohannan, B.J. Humans differ in their personal microbial cloud. *PeerJ* **2015**, *3*, e1258. **[Google Scholar] [CrossRef] [Green Version]**

10. Eetemadi, A.; Rai, N.; Pereira, B.M.P.; Kim, M.; Schmitz, H.; Tagkopoulos, I. The Computational Diet: A Review of Computational Methods Across Diet, Microbiome, and Health. *Front. Microbiol.* **2020**, *11*, 393. **[Google Scholar]** **[CrossRef]** **[Green Version]**

11. Pasolli, E.; Asnicar, F.; Manara, S.; Zolfo, M.; Karcher, N.; Armanini, F.; Beghini, F.; Manghi, P.; Tett, A.; Ghensi, P.; et al. Extensive Unexplored Human Microbiome Diversity Revealed by over 150,000 Genomes from Metagenomes Spanning Age, Geography, and Lifestyle. *Cell* **2019**, *176*, 649–662.e20. **[Google Scholar]** **[CrossRef]** **[PubMed]** **[Green Version]**

12. De la Cuesta-Zuluaga, J.; Kelley, S.T.; Chen, Y.; Escobar, J.S.; Mueller, N.T.; Ley, R.E.; McDonald, D.; Huang, S.; Swafford, A.D.; Knight, R.; et al. Age- and Sex-Dependent Patterns of Gut Microbial Diversity in Human Adults. *mSystems* **2019**, *4*, e00261-19. **[Google Scholar]** **[CrossRef]** **[PubMed]** **[Green Version]**

13. Huttenhower, C.; Gevers, D.; Knight, R.; Abubucker, S.; Badger, J.; Chinwalla, A.; Giglio, M.G. Structure, function and diversity of the healthy human microbiome. *Nature* **2012**, *486*, 207. **[Google Scholar]**

14. Metcalf, J.L. Estimating the postmortem interval using microbes: Knowledge gaps and a path to technology adoption. *Forensic Sci. Int. Genet.* **2019**, *38*, 211–218. **[Google Scholar]** **[CrossRef]** **[Green Version]**

15. Schmedes, S.E.; Woerner, A.E.; Novroski, N.M.M.; Wendt, F.R.; King, J.L.; Stephens, K.M.; Budowle, B. Targeted sequencing of clade-specific markers from skin microbiomes for forensic human identification. *Forensic Sci. Int. Genet.* **2018**, *32*, 50–61. **[Google Scholar]** **[CrossRef]**

16. Fierer, N.; Lauber, C.L.; Zhou, N.; McDonald, D.; Costello, E.K.; Knight, R. Forensic identification using skin bacterial communities. *Proc. Natl. Acad. Sci. USA* **2010**, *107*, 6477–6481. **[Google Scholar]** **[CrossRef]** **[Green Version]**

17. Goga, H. Comparison of bacterial DNA profiles of footwear insoles and soles of feet for the forensic discrimination of footwear owners. *Int. J. Legal Med.* **2012**, *126*, 815–823. **[Google Scholar]** **[CrossRef]**

18. Yang, J.; Tsukimi, T.; Yoshikawa, M.; Suzuki, K.; Takeda, T.; Tomita, M.; Fukuda, S. *Cutibacterium acnes* (*Propionibacterium acnes*) 16S rRNA genotyping of microbial samples from possessions contributes to owner identification. *mSystems* **2019**, *4*, e00594-19. **[Google Scholar]** **[CrossRef]** **[Green Version]**

19. Concheri, G.; Bertoldi, D.; Polone, E.; Otto, S.; Larcher, R.; Squartini, A. Chemical elemental distribution and soil DNA fingerprints provide the critical evidence in murder case investigation. *PLoS ONE* **2011**, *6*, e20222. **[Google Scholar]** **[CrossRef]**

20. López, C.D.; González, D.M.; Haas, C.; Vidaki, A.; Kayser, M. Microbiome-based body site of origin classification of forensically relevant blood traces. *Forensic Sci. Int. Genet.* **2020**, *47*, 102280. **[Google Scholar] [CrossRef] [PubMed]**

21. Aroutcheva, A.A.; Simoes, J.A.; Behbakht, K.; Faro, S. *Gardnerella vaginalis* isolated from patients with bacterial vaginosis and from patients with healthy vaginal ecosystems. *Clin. Infect. Dis.* **2001**, *33*, 1022–1027. **[Google Scholar] [CrossRef] [Green Version]**

22. Alcaraz, L.D.; Belda-Ferre, P.; Cabrera-Rubio, R.; Romero, H.; Simón-Soro, A.; Pignatelli, M.; Mira, A. Identifying a healthy oral microbiome through metagenomics. *Clin. Microbiol. Infect.* **2012**, *18* (Suppl. 57–54 ,(4. **[Google Scholar] [CrossRef] [Green Version]**

23. Bisgaard, M.; Nørskov-Lauritsen, N.; De Wit, S.; Hess, C.; Christensen, H. Multilocus sequence phylogenetic analysis of *Avibacterium*. *Microbiology* **2012**, *158*, 993–1004. **[Google Scholar] [CrossRef] [Green Version]**

24. Ho, W. What's the Y-STR Gene? Determined for a 12-Year Sentence of Not Guilty. NEWSIS. 2020. Available online: **https://news.v.daum.net/v/20200529142012559** (accessed on 29 May 2020).

25. Lee, S.H.; You, H.S.; Kang, H.G.; Kang, S.S.; Hyun, S.H. Association between Altered Blood Parameters and Gut Microbiota after Synbiotic Intake in Healthy, Elderly Korean Women. *Nutrients* **2020**, *12*, 3112. **[Google Scholar] [CrossRef] [PubMed]**

26. Dhariwal, A.; Chong, J.; Habib, S.; King, I.L.; Agellon, L.B.; Xia, J. MicrobiomeAnalyst: A web-based tool for comprehensive statistical, visual and meta-analysis of microbiome data. *Nucleic Acids Res.* **2017**, *45*, W180–W188. **[Google Scholar] [CrossRef] [PubMed]**

27. Romsos, E.L.; French, J.L.; Smith, M.; Figarelli, V.; Harran, F.; Vandegrift, G.; Moreno, L.I.; Callaghan, T.F.; Brocato, J.; Vaidyanathan, J. Results of the 2018 rapid DNA maturity assessment. *J. Forensic Sci.* **2020**, *65*, 953–959. **[Google Scholar] [CrossRef]**

28. Abelson, B.; Sun, D.; Que, L.; Nebel, R.A.; Baker, D.; Popiel, P.; Amundsen, C.L.; Chai, T.; Close, C.; DiSanto, M. Sex differences in lower urinary tract biology and physiology. *Biol. Sex Differ.* **2018**, *9*, 45. **[Google Scholar] [CrossRef] [Green Version]**

29. Willis, J.R.; Gabaldón, T. The human oral microbiome in health and disease: From sequences to ecosystems. *Microorganisms* **2020**, *8*, 308. **[Google Scholar] [CrossRef] [PubMed] [Green Version]**

30. Edmonds-Wilson, S.L.; Nurinova, N.I.; Zapka, C.A.; Fierer, N.; Wilson, M. Review of human hand microbiome research. *J. Dermatol. Sci.* **2015**, *80*, 3–12. **[Google Scholar] [CrossRef] [Green Version]**

31. Ohta, J.; Sakurada, K. Oral gram-positive bacterial DNA-based identification of saliva from highly degraded samples. *Forensic Sci. Int. Genet.* **2019**, *42*, 103–112. **[Google Scholar] [CrossRef]**

32. Petrova, M.I.; Lievens, E.; Malik, S.; Imholz, N.; Lebeer, S. *Lactobacillus* species as biomarkers and agents that can promote various aspects of vaginal health. *Front. Physiol.* **2015**, *6*, 81. **[Google Scholar] [CrossRef] [PubMed] [Green Version]**

33. Li, Z.; Xia, J.; Jiang, L.; Tan, Y.; An, Y.; Zhu, X.; Ruan, J.; Chen, Z.; Zhen, H.; Ma, Y. Characterization of the human skin resistome and identification of two microbiota cutotypes. *Microbiome* **2021**, *9*, 47. **[Google Scholar] [CrossRef]**

34. Killer, J.; Kopečný, J.; Mrázek, J.; Havlík, J.; Koppová, I.; Benada, O.; Rada, V.; Kofroňová, O. *Bombiscardoviacoagulans* gen. nov., sp. nov., a new member of the family Bifidobacteriaceae isolated from the digestive tract of bumblebees. *Syst. Appl. Microbiol.* **2010**, *33*, 359–366. **[Google Scholar] [CrossRef] [PubMed]**

35. Weon, H.-Y.; Kim, B.-Y.; Hong, S.-B.; Joa, J.-H.; Nam, S.-S.; Lee, K.H.; Kwon, S.-W. *Skermanellaaerolata* sp. nov., isolated from air, and emended description of the genus *Skermanella*. *Int. J. Syst. Evol. Microbiol.* **2007**, *57*, 1539–1542. **[Google Scholar] [CrossRef]**

36. Mitra, A.; MacIntyre, D.A.; Mahajan, V.; Lee, Y.S.; Smith, A.; Marchesi, J.R.; Lyons, D.; Bennett, P.R.; Kyrgiou, M. Comparison of vaginal microbiota sampling techniques: Cytobrush versus swab. *Sci. Rep.* **2017**, *7*, 9802. **[Google Scholar] [CrossRef] [PubMed] [Green Version]**

37. Tozzo, P.; D'Angiolella, G.; Brun, P.; Castagliuolo, I.; Gino, S.; Caenazzo, L. Skin Microbiome analysis for forensic human identification: What do we know so far? *Microorganisms* **2020**, *8*, 873. **[Google Scholar] [CrossRef]**

38. Thornbury, D.; Goray, M.; van Oorschot, R.A. Indirect DNA transfer without contact from dried biological materials on various surfaces. *Forensic Sci. Int. Genet.* **2021**, *51*, 102457. **[Google Scholar] [CrossRef]**

39. Bondurant, S.W.; Duley, C.M.; Harbell, J.W. Demonstrating the persistent antibacterial efficacy of a hand sanitizer containing benzalkonium chloride on human skin at 1, 2, and 4 hours after application. *Am. J. Infect. Control.* **2019**, *47*, 928–932. **[Google Scholar] [CrossRef] [Green Version]**

40. Tuzil, J.; Filkova, B.; Malina, J.; Kerestes, J.; Doležal, T. Smoking in women with chronic vaginal discomfort is not associated with decreased abundance of Lactobacillus spp. but promotes *Mobiluncus* and *Gardnerella* spp. overgrowth:

Secondary analysis of trial data including microbiome analysis. *Ceska Gynekol.* **2021**, *86*, 22–29. **[Google Scholar] [CrossRef]**

41. Dzidic, M.; Abrahamsson, T.R.; Artacho, A.; Collado, M.C.; Mira, A.; Jenmalm, M. Oral microbiota maturation during the first 7 years of life in relation to allergy development. *Allergy* **2018**, *73*, 2000–2011. **[Google Scholar] [CrossRef] [Green Version]**
42. Du, Y.; Yu, X.; Wang, G. *Massiliatieshanensis* sp. nov., isolated from mining soil. *Int. J. Syst. Evol. Microbiol.* **2012**, *62*, 2356–2362. **[Google Scholar] [CrossRef] [PubMed] [Green Version]**
43. Barka, E.A.; Vatsa, P.; Sanchez, L.; Gaveau-Vaillant, N.; Jacquard, C.; Klenk, H.-P.; Clément, C.; Ouhdouch, Y.; van Wezel, G.P. Taxonomy, physiology, and natural products of *Actinobacteria. Microbiol. Mol. Biol. Rev.* **2016**, *80*, 1–43. **[Google Scholar] [CrossRef] [PubMed] [Green Version]**
44. Hedman, J.; Jansson, L.; Akel, Y.; Wallmark, N.; Liljestrand, R.G.; Forsberg, C.; Ansell, R. The double-swab technique versus single swabs for human DNA recovery from various surfaces. *Forensic Sci. Int. Genet.* **2020**, *46*, 102253. **[Google Scholar] [CrossRef] [PubMed]**
45. You, H.S.; Lee, S.H.; Ok, Y.J.; Kang, H.-G.; Sung, H.J.; Lee, J.Y.; Kang, S.S.; Hyun, S.H. Influence of swabbing solution and swab type on DNA recovery from rig

18 ENHANCED DNA EXTRACTION AND ILLUMINA SEQUENCING OF DNA OBTAINED FROM AGED ROOTLESS HAIR SHAFTS FOUND IN RELICS ASSOCIATED WITH THE ROMANOV FAMILY

1. INTRODUCTION: AN OVERVIEW

The fate of the Romanov dynasty was considered one of the most profound mysteries of the 20th century, still captivating interest and inspiring speculation today. Despite the absence of any written records, rumors indicate that Tsar Nicholas and his family were apprehended and put to death in July 1918 by a Bolshevik firing squad in the Ipatiev House in Ekaterinburg, Russia. As per the accounts, the dead bodies were thereafter buried in a shallow trench that was quickly dug. In 1991, a collective of nine skeleton remains were unearthed in a shallow burial place in close proximity to Ekaterinburg. Preliminarily, Russian authorities ascertained that these remains corresponded to the Tsar, the Tsarina, three of their five children, and four others who were in the service of the family. The confirmation of the identity of the five Romanovs was established through subsequent DNA analysis [3]. The examination involved the analysis of mitochondrial DNA (mtDNA) extracted from the purported Tsarina and three of her daughters. The mtDNA profile of the Tsarina was found to be identical to that of Prince Philip, Duke of Edinburgh, indicating that they shared the same maternal lineage. The testing also included the evaluation of biological sex and examination of autosomal short tandem repeats (STR). The DNA analysis definitively verified the existence of a familial cluster that corresponds to the Romanovs. In addition, the examination of the likely remains of the Tsar through mtDNA testing yielded a sequence that corresponded to the mtDNA profiles of two living maternal relatives of the Tsar, as well as the mtDNA profile of the unearthed remains of the Tsar's brother, Grand Duke George Romanov [3,4]. After a 16-year gap following the initial discovery of the first burial, the location of the missing descendants was finally established when a few little bone fragments were recovered near the site of the previous discoveries. The bone fragments were subjected to anthropological, mitochondrial, and nuclear DNA analyses, which definitively confirmed the presence of the two Romanov children who were not found in the original burial location [5,6].

The FBI laboratory's DNA Support Unit (DSU) has just been notified about two artifacts believed to contain biological material related to the Romanov family. The possessions are owned by Mr. Nikolai Bachmakov, a distinguished expert in the restoration of Fabergé artifacts. Mr. Bachmakov obtained the initial item, a locket believed to be from approximately 1899, from a separate vendor. The pendant, a locket crafted by Karl Fabergé, was embellished with three exceptionally beautiful gemstones (Figure 1A). A historic monochrome portrait of Empress Alexandra Feodorovna Romanov (Figure 1B) was displayed within a glass case. Furthermore, on the other side, also enclosed in glass, there was a cluster of light-colored hair strands.

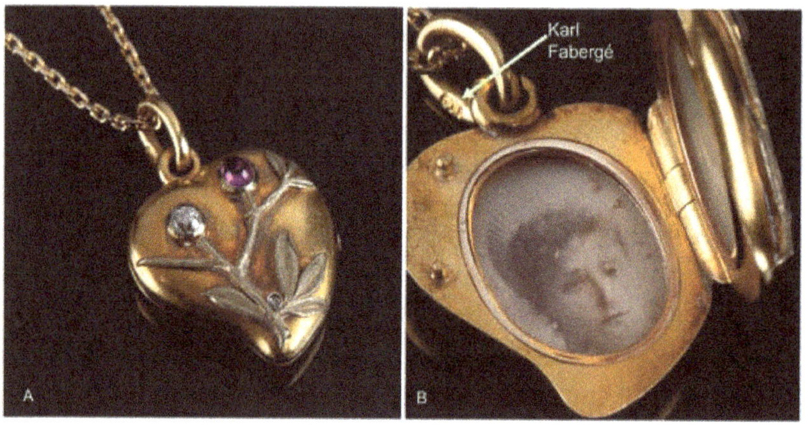

Figure 1. (A) Closed Fabergé locket showing precious stones. (B) The open locket showing a picture of Tsarina Alexandra Feodorovna Romanov. The stamped initials from Karl Fabergé can be seen on the jump ring. Photographed by Richard Walker. Courtesy of the Russian History Museum.

Subsequently, Mr. Bachmakov notified the DSU that he possessed a framed image of Queen Louise of Hesse-Kassel (1817–1898), who was the wife of King Christian IX of Denmark and the maternal grandmother of Tsar Nicholas II (Figure 2). A substantial lock of blonde hair, believed to belong to Queen Louise, was found between the glass of the frame and the image.

Although genetic characterization seemed to be the only way to confirm the provenance of the hairs, it was well understood that DNA recovery from such aged and degraded hairs would be challenging. MtDNA, which is sometimes targeted in forensic casework, is increasingly difficult to recover with increasing sample age. In

2005, Melton et al. [7] showed that the likelihood of obtaining a full hypervariable region I/hypervariable region II (HVI/HVII) mtDNA profile decreased from 90% in single shed hairs 0–5 years old to ~60% in hairs greater than 20 years old. Similarly, in a 2017 study, mtDNA averaged only 61 bp in human hairs that had been collected 50 to 100 years ago [**8**]. For older samples, next generation sequencing (NGS) techniques that do not rely on predefined amplicons have proven necessary for DNA recovery [**9,10,11**]. In these particular cases, it is likely that the quantities of hair that were used (between 20 mg and 2 g), the beneficial conditions of preservation, and/or the presence of some hair roots contributed to successful DNA typing.

Figure 2. Picture frame containing a picture of Queen Louise of Hesse-Kassel after restoration. Courtesy of M. Perekrestov, from the Russian History Foundation.

We have recently utilized Next-Generation Sequencing (NGS) to generate DNA data from common types of hair samples encountered in forensic casework. These samples typically consist of single shed hairs containing less than 1 mg of biological material [12]. In this study, we expanded upon previous research by examining rootless hairs of greater age and degradation, which were thought to be linked to the Russian monarchy. Preliminarily, we evaluated the viability of extracting DNA from hairs presumed to be more than a hundred years old. To enhance the probability of DNA retrieval and optimize DNA recovery techniques, a collective sample of several hairs was utilized for DNA extraction. Subsequently, we utilized these refined procedures to examine the many categories of individual hair strands commonly observed in forensic DNA investigations. In summary, the findings not only support the idea that the hairs may have come from individuals belonging to the Russian Royal family, but also highlight the usefulness of limited sequencing data in answering investigative concerns related to single hairs that are decades old.

2. MATERIAL

2.1. The Fabergé Locket

The Fabergé pendant is shown in **Figure 1**A, B. It is a 14 karat yellow gold heart locket that weighs 8.7 g and measures 24.75 mm × 19.84 mm × 7.15 mm. On the front side, a floral pattern is entwined over the heart and is embellished with three stones: one old-cut round diamond stone of ~0.10 carat (ct), one Cabochon-shaped ruby stone of ~0.12 ct, and a small rose-cut diamond of 0.001 ct (**Figure 1**A).

On the back, the Russian monogram AΘ (Θ being an old Cyrillic equivalent to the letter F) is engraved as well as the great Imperial crown of Russia. The initials of the locket's maker, stamped on the loose jump ring, are KΦ, as in Karl Fabergé (**Figure 1**B). The interior of the locket is fitted with two oval-shaped glazed picture frames and on the right side, the scratched inventory number 15,650 means that the locket was made between late 1897 and early 1898. One side contained a black and white portrait of the Empress Alexandra Feodorovna (1872–1918) and on the other side was a lock of light brown/dark blond hairs.

2.2. The Picture Frame

The frame, made of silver 925 standard, is covered with red enamel, and decorated with silver ribbons. The size of the black and white cabinet card of Queen Louise is 6.7 cm by 4.3 cm (**Figure 2**).

2.3. Hair Samples

The hair samples obtained from both the locket and the frame are documented in Table 1. The FBI trace analysis examiner conducted microscopic examination on three hairs from the locket and one hair from the frame, in addition to DNA testing. The hairs were affixed to slides and permanently archived in the FBI hair reference collection.

Table 1. Description of the hair shaft samples that were used in this study and purification methods that were used during DNA extraction.

Associated Item	Hair Shafts Tested	Experiment	Purification Protocol
	Single hair of ~5 cm (light brown). Lo1a	Microscopic examination	NA
	Single hair of ~2.5 cm (light brown). Lo1b	Microscopic examination	
	Single hair of ~3.8 cm (light brown). Lo1c	Microscopic examination	
Fabergé Locket	5 hairs for a total of ~16 cm. Lo5h	DNA	MinElute column
	Single hair of 4 cm (light brown). Lo2	DNA	Magnetic silica beads
	Single hair of 3.2 cm (light brown). Lo3	DNA	Magnetic silica beads
	Single hair of 6 cm (white). Q3	Microscopic examination	NA
Frame	6 hairs for a total of ~27 cm. Q6h	DNA	MinElute column
	Single hair of 5.5 cm (white). Q1	DNA	Magnetic silica beads
	Single hair of 7 cm (brown). Q2	DNA	Magnetic silica beads

To maximize the chances of DNA recovery, initial experiments were performed with multiple hairs. From the locket, and under the assumption that all the hairs shared the same mtDNA lineage, five hairs totaling ~16 cm were extracted (Locket 5 hairs: Lo5h). Similarly for the frame, six hairs totaling ~27 cm were tested (Queen 6 hairs: Q6h). Following extraction protocol optimization involving the replacement of MinElute columns with silica coated magnetic beads, and successful results when co-extracting multiple hairs at once, DNA extraction from single hair shafts was attempted for both historical items.

3. METHODS

3.1. Microscopic Examination

A total of four hairs were prepared on glass microscope slides using Permount mounting media (**Table 1**; hairs Lo1a, Lo1b, Lo1c, and Q3). The hairs were examined using a high magnification transmitted light microscope (Leica DM4000B with a Leica FS4000 optical bridge) and photographed with a Leica MC190 HD.

3.2. DNA Extraction Protocols

DNA extraction and library preparation were performed in a laboratory dedicated to low-quantity and low-quality DNA samples with limited personnel access. Personal protective equipment to minimize the risk of contamination included Tyvek® IsoClean® single-use garments (frocks, sleeves, masks from DuPont™, Wilmington, DE, USA), goggles, and double gloving. The reagents were aliquoted and irradiated before the samples were introduced and all plastic consumables were irradiated in a cross-linker prior to use.

A total of three DNA extracts were generated from both the locket hairs and the frame hairs, for a total of six extracts. Of these six extracts, two were developed from multiple hairs (Lo5h and Q6h), and four were developed from single hair shafts (Lo2, Lo3, Q1, and Q2).

Two different purification methods were used. Protocol B from Brandhagen et al. [**12**] was applied first for the extractions of multiple hairs samples (Lo5h and Q6h). In brief, DNA was extracted using a lysis buffer optimized for hair digestion and then purified with MinElute silica columns from the Qiagen PCR purification kit (Qiagen, Germantown, MD, USA) using large volumes of binding buffer. To accommodate the large volumes of solution, the MinElute columns were fitted with Qiagen extension columns and placed on a vacuum manifold (VacConnector, Qiagen).

DNA from single strands of hair was extracted using the same optimized digestion buffer and volumes as in protocol B, but instead of using MinElute columns, DNA was purified with magnetic beads coated with silica (cat#: 786–915, G-Bioscience, Maryland Heights, MO, USA) as recommended by Rohland et al. [**13**].

The DNA quantification using qPCR was not conducted on the extracts because earlier investigations have shown that DNA fragments found in hair samples of significant age are typically too tiny to produce a 105 bp amplicon with the mtDNA qPCR assay commonly employed in DSU [12,14]. Fluorometric measurement was conducted using a QubitTM Fluorometer and a dsDNA High Sensitivity Assay Kit (Invitrogen, Waltham, MA, USA), however no dsDNA was detectable. The adapters were set at a final concentration of 300 nM, without any specific criteria.

3.3. Library Preparation

A total of 50 μL from all six DNA extracts (Lo5h, Lo2, Lo3, Q6h, Q1, Q2) and four reagent controls (RB1 to RB4) were converted into double stranded DNA Illumina libraries using the NEBNext® Ultra™ II DNA Library Prep Kit for Illumina® (NEB, Ipswich, MA, USA) and looped adapters from the NEBNext® Multiplex Oligos for Illumina® kit (NEB). Illumina libraries were prepared according to the manufacturer's instructions except for ligation which used highly diluted adapters and was performed overnight at 7 °C. Following ligation, the ligated looped adapters were converted into Y-shaped adapters with the USER® Enzyme kit (NEB). The libraries were then purified with 1.8× AMPure XP beads (Beckman Coulter, Sykesville, MD, USA). All the libraries were dual indexed with primers from the NEBNext® Multiplex Oligos for Illumina® kit and subsequently amplified for 25 cycles with the NEBNext® Q5® Hot Start HiFi PCR Master Mix (NEB). Purification of the amplified libraries was performed using 1.5× AMPure XP beads. The libraries were then shotgun sequenced on an MiSeq Forensic Genomics System (FGx™) instrument (Illumina Inc., San Diego, CA, USA) with a v2 cartridge and 2 × 151 cycles and/or on a NextSeq 500/550 (Illumina) with a High Output kit v2.5 and 2 × 75 cycles in order to obtain additional data.

3.4. Hybridization Capture

Mitochondrial DNA enrichment via hybridization capture was performed on all RBs and on four libraries, one from the locket (Lo5h) and three from the frame (Q6h, Q1, and Q2), using biotinylated RNA baits (myBaits Mito Human, Modern Global, Daicel Arbor Biosciences, Ann Arbor, MI, USA). Each hybridization capture was performed at 62 °C for 24 h using 100 ng of 80 bp long RNA baits. The captured product was then amplified for 16 to 19 cycles and purified with AMPure XP beads. DNA from the purified enriched libraries was successfully quantified using a Qubit™ Fluorometer and the Qubit™ dsDNA Broad Range Assay Kit (Invitrogen), then diluted and analyzed on a 2100 Bioanalyzer with the High Sensitivity DNA Kit (Agilent, Santa Clara, CA, USA). Following quantification, libraries were pooled and sequenced on a MiSeqFGx™ instrument with a v2 cartridge and × 2 151 cycles. The very first capture library that was sequenced, Lo5h-mt, was also sequenced on the NextSeq 500 with 75 × 2 cycles. The comparison of the data from both instruments showed that a MiSeq run was sufficient to accurately sequence the small mtGenome. As a result, all the subsequent capture libraries were only sequenced on the MiSeqFGx™.

3.5. Data Analyses

Raw Fastq reads were first imported as paired-end reads, quality trimmed, and the overlapping pairs were merged in the CLC Genomics Workbench, version 12.0 (CLC Bio/Qiagen, Aarhus, Denmark). Merged reads were then mapped to the human genome by a supercomputer using the Burrows-Wheeler Aligner (BWA aln, v0.7.17–r1188) with seed disabled [**15**] and SAMTools v.1.11 [**16**]. BWA aln was chosen over CLC because this algorithm is more efficient at mapping very short reads (i.e., reads under 50 bp). Reads with mapping quality below 30 were removed and, to circumvent the fact that BWA can't map reads to a circular reference, a chimeric mtDNA reference was created to allow reads that overlap the mtGenome origin to map. The 33,152 bp sequence was developed by combining the sequence of the standard revised Cambridge reference (rCRS) 16,569 bp genome, followed by 14 N and a second mtGenome sequence starting at position 8284 and ending at position 8283. This chimeric mtDNA reference also replaced the mtDNA reference ChrM in the human genome reference hg19 (a.k.a GRCh37). Duplicate reads were removed using Picard v.2.23.8 [**17**]. Mappings were imported back into CLC as bam files and the mtDNA reads were extracted and re-mapped to the standard 16,569 bp circular rCRS reference. Duplicate mtDNA reads were removed and CLC was used to develop variant calls, DNA size distribution, and read depth. The shotgun data from single hair extracts were used to assess biological sex using the R_X method published by Mittnik et al. [**18**]. This method compares the number of sequences originating from the X chromosome to the number of sequences originating from the 22 autosomes. A male assignment is made if the upper bound of the 95% CI is less than 0.60, while a female assignment is made if the lower bound is greater than 0.80.

Next, the shotgun data were used to analyze ~1.3 million nuclear SNPs that were covered by at least one read. From the SNP results, a "forced haploid" approach was used to assess relatedness between the two locket hairs, Lo2 and Lo3 (**Figure 3**).

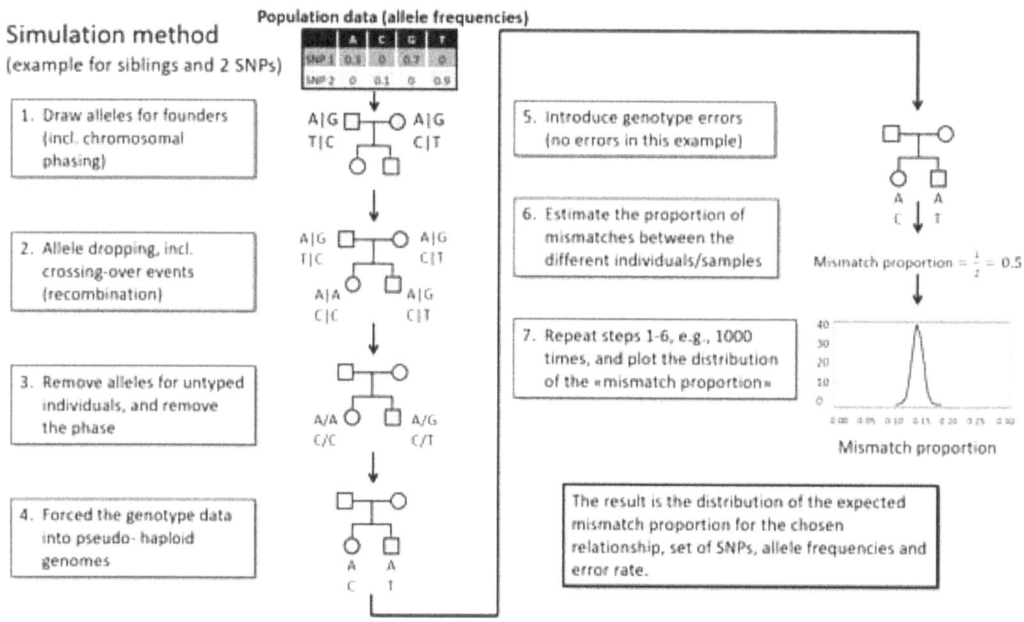

Figure 3. Simulation approach applied to infer the degree of relatedness between low coverage DNA datasets based on allele mismatch proportions.

In brief, relatedness was assessed by comparing allele mismatch proportions between the two locket hairs to expected allele mismatch proportions for unrelated, parent-child, sibling, and same individual relationships. Specifically, the shotgun datasets from Lo2 and Lo3 were first reduced to 1.3 million SNPs that include the

SNPs generally typed by genetic genealogy companies and used in GEDmatch [**19,20**]. The number of SNPs was then further reduced to SNPs that overlapped between the two hair samples. Next, using a custom R script, the datasets were forced into pseudo-haploid genomes for which the allele mismatch proportion was calculated. The resulting SNP marker set (i.e., SNPs with allele calls in both samples) was used as the input in the mismatch proportion simulation method. To obtain expected mismatch proportion distributions, pseudo-haploid datasets for various degrees of relatedness (i.e., same individual, parent/child, full siblings, or unrelated individuals) were simulated using the European allele frequencies of the Genomes Project Consortium [**21**] and SNP-nexus [**22,23**] for the specific set of SNPs that were included in the sample comparison. In addition, a simplistic error model was included to account for the possibility of different genotype errors such as deamination, depurination, PCR error, sequencing error, and mapping error, that may occur for low-quality and low-quantity DNA samples. An error rate of 0 to 2% was used in our simulations. As controls, shotgun data from two 50 year-old-sourced hair samples, completely unconnected to this project and that were known to originate from the same individual (189A and 189B) were analyzed and compared.

Finally, some of the unmapped reads were analyzed using the Metagenomics RAST (MG-RAST) server [**24**] in an effort to identify the source of the non-human reads.

4. RESULTS

4.1. Histology Examination

The four hairs that were examined under a microscope were found to be exceptionally well-preserved. No signs of decomposition, insect damage, or fungal damage were discovered, as depicted in Figure 4A–C and Figure 5. Furthermore, there was no evidence of any artificial manipulation, such as the application of dyes or bleaching agents. The examiner concluded that all four hairs were most likely human head hairs of European ancestry based on factors such as pigment arrangement, pigment density, overall diameter, cross-sectional form of the hair shaft, and cuticle thickness.

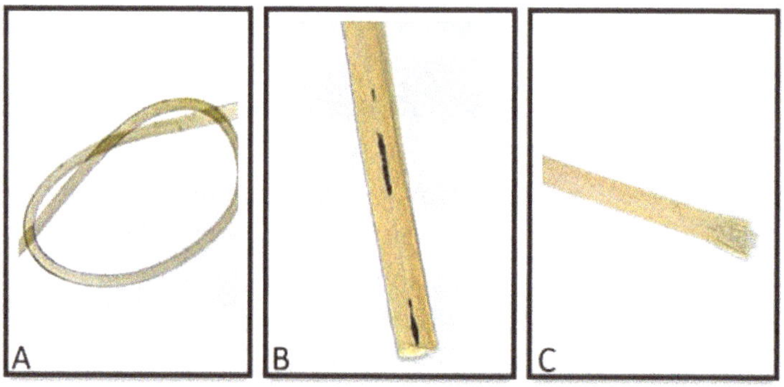

Figure 4. Photographs taken from three different hairs that were found in the Fabergé locket. (A) Magnification 100×. (B) Magnification 200× shows that the proximal end of the hair appears cut. (C) Magnification 200× shows that the proximal end of the hair appears broken.

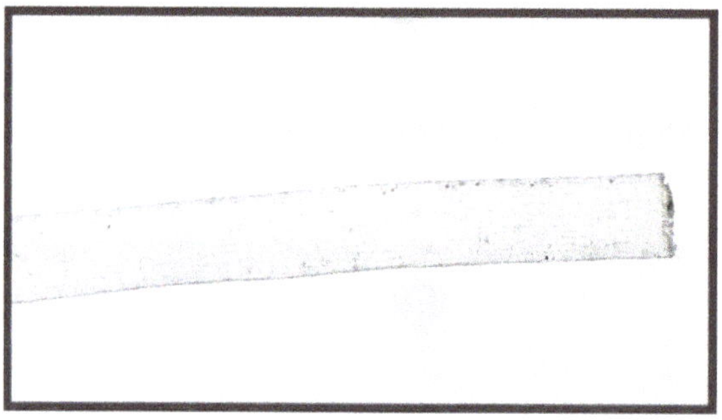

Figure 5. Magnification: 200×. Photograph taken from a white hair that was found in the picture frame. The proximal end of the hair appears cut.

Based on microscopic examination, it was established that no root was present on any of the samples examined.

4.2. Illumina Sequencing Results from the Fabergé Locket Hairs

Three libraries were produced with the locket samples. From sample Lo5h (extract/library from five hairs), a shotgun DNA library (Lo5h-sh, a library not enriched for any marker) as well as a mtDNA-enriched library (Lo5h-mt) were sequenced on a MiSeqFGx™. The mtDNA-enriched library Lo5h-mt was also sequenced on a NextSeq 500 to compare the data that were obtained from both instruments. For single hair extracts/library Lo2 and Lo3, shotgun libraries were only sequenced on a NextSeq 500 and no hybridization capture was performed. Sequencing statistics for the hair samples that were found in the locket are presented in **Table 2**. Shotgun library Lo5h-sh was only sequenced on a MiSeqFGx™ since our focus for the multi-hair extracts was to establish that DNA could be recovered from hairs of this age, and that is best addressed by targeting the mtGenome. In addition, since the locket had a picture of Tsarina Alexandra, it was conceivable that it contained hairs from herself and/or from some of her children. As a result, all the individuals would share the same mtGenome and data analysis would not be compromised by mixtures. Analysis of nuclear DNA, however, would have presented a challenge. Consequently, high throughput sequencing of Lo5h-sh for nuclear DNA analysis was not attempted on the NextSeq 500 platform.

Table 2. Sequencing statistics from the Fabergé locket hairs.

Shotgun Libraries	**Platforms Used**	**Number of Mapped Reads**	**Percentage of Mapped Reads**	**Number of Unique Reads**
Lo5h-sh (5 hairs)	MiSeqFGx™	6,229,985	48%	5,500,712
Lo2-sh (1 hair)	NextSeq 500	199,948,284	66.54%	38,036,020
Lo3-sh (1 hair)	NextSeq 500	151,630,120	78.91%	33,388,550
mtDNA-Enriched Library	Platforms Used	Number of Mapped Reads	Percentage of Mapped Reads	Number of Unique mtDNA Reads
Lo5h-mt (5 hairs)	NextSeq 500	21,109,044	97.50%	35,188

The percentage of endogenous human reads in the Lo5h-sh library was 48%. This percentage was a bit lower than expected considering that all hairs had been thoroughly decontaminated via Terg-a-zyme sonication before digestion. When unmapped reads of 75 bp or longer were analyzed with MG-RAST, close to 63% of the sequences remained unknown and most of the sequences that matched a sequence in GenBank belonged to the Proteobacteria phylum.

The percentage of endogenous human DNA was, however, higher when single hairs were extracted with the optimized protocol (66.54% for Lo2-sh and 78.91% for Lo3-sh). Additionally, as a precaution, purification reagents remained stored at 4°C between use to avoid potential mold growth. In this case, since both Lo2 and Lo3 were single source samples, high throughput sequencing of the libraries was carried out on the NextSeq 500 platform to collect as many DNA sequences as possible.

4.2.1. Mitochondrial DNA Results

MtDNA-enriched library Lo5h-mt produced 35,188 unique mtDNA reads representing >3.2 million bases. The entire mtGenome was successfully sequenced at an average depth of 198× (range: 159–273×; standard deviation (SD): 11.38). Shotgun sequencing of the Lo5h-sh produced 29,879 mtDNA reads representing >1.9 million bases. Once again, the complete mtGenome was sequenced at an average depth of 119× (range: 5–173×; SD: 17.77). The DNA length distributions of Lo5h-sh and Lo5h-mt are presented in **Figure 6**A.

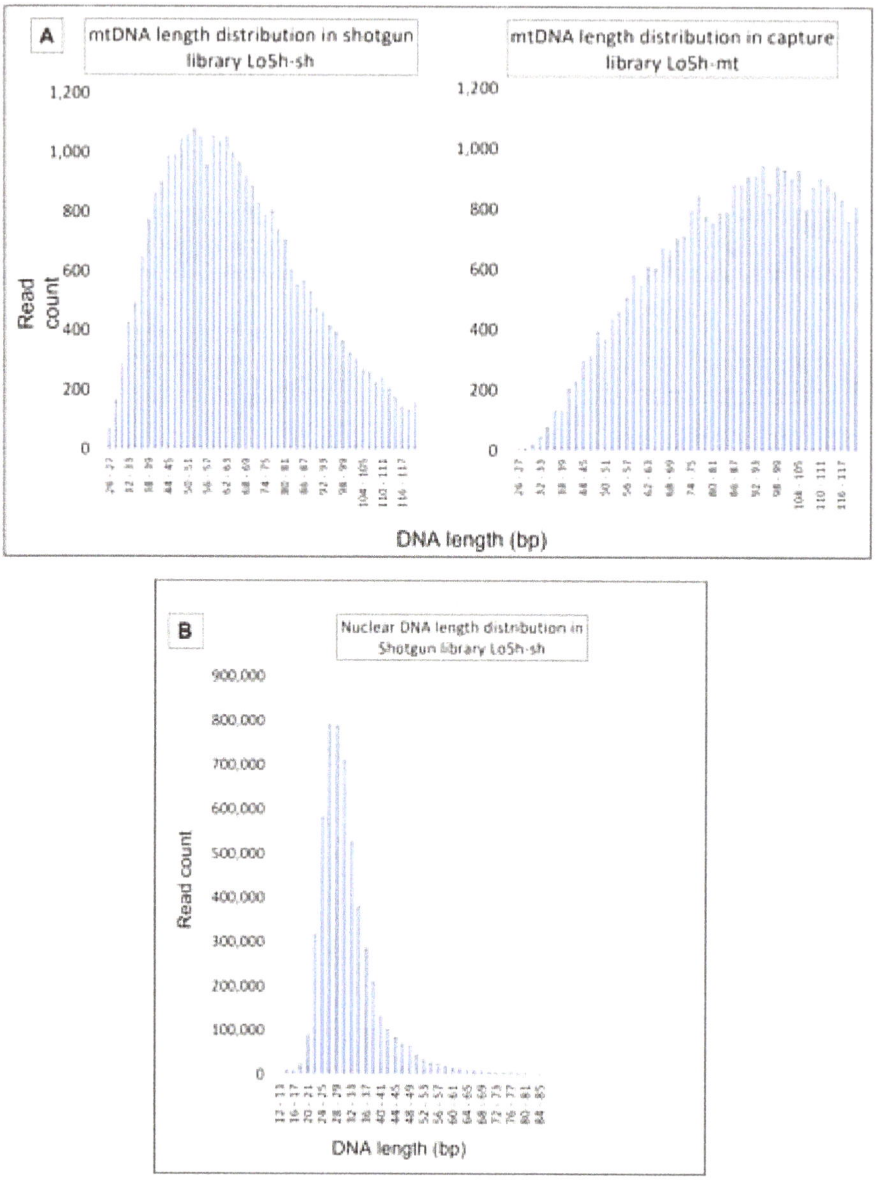

Figure 6. Length distribution of (A) mtDNA reads in Lo5h-sh and Lo5h-mt, and (B) nuclear DNA reads in Lo5h-sh.

Shotgun libraries Lo2-sh and Lo3-sh yielded 27,058 and 32,099 unique mtDNA sequences with an average depth of 111× (range: 22–148×; SD:13.08) and 164× (range: 26–219×; SD:12.06) respectively. MtDNA enrichment of the Lo2 and Lo3 libraries via hybridization capture was deemed unnecessary since there was sufficient coverage of the mtGenome in the shotgun data that were produced by the NextSeq.

The average mtDNA fragment length varied between 66.04 bp and 93.46 bp, depending on the library and the reference used for mapping (**Table 3**). Interestingly, the hybridization capture library that was developed from Lo5h (Lo5h-mt) yielded an average mtDNA fragment size that was much larger than the one that was observed in the Lo5h shotgun library (Lo5h-sh).

Table 3. Average length of the mtDNA molecules in the locket libraries when mapped to hg19 (left) or to rCRS (right).

Library	Mapped mtDNA Reads Average Length (bp)
Lo2-sh	67.82/67.89
Lo3-sh	84.99/85.17
Lo5h-sh	66.04/69.16
Lo5h-mt	93.46

At 93.46 bp, the mean length of the mtDNA reads in the enriched Lo5h-mt library was substantially higher than the mean mtDNA length that were observed in the shotgun libraries (**Table 3**). This pattern between the shotgun and enriched libraries has been observed in several other studies involving ancient DNA [**25,26,27**]. Our hypothesis that the noticeable disparity in average length could be caused by a preferential binding of larger mtDNA molecules during hybridization capture was confirmed by Krasnenko et al. [**28**] who stated that "[...] insert size crucially impact(s) enrichment results" and that "the maximum efficiency of enrichment (>90%) is achieved with 250–330 bp insertion length". Of note, average DNA length estimates should be viewed with the following caveats in mind: they can be greatly influenced by factors such as DNA extraction protocol, library preparation protocol, number of sequencing cycles, and bioinformatics pipelines (trimming in particular).

Shotgun libraries Lo2-sh and Lo3-sh also showed substantial variation in mean mtDNA fragment length (67.82 versus 84.99 bp). These differences could be the result of a number of factors, including sample variability between individual hairs or variability resulting from the distance between the tested section and the root [**12**].

All the locket hair libraries produced the same mtGenome profile: 263G, 315.1C, 524.1A, 524.2C, 750G, 1438G, 3010A, 4137T, 4769G, 8860G, 15326G, 16111T, 16357C, and 16519C. The sequence belongs to haplogroup H1af2 and is identical to the mtGenome of Empress Alexandra's maternal lineage published as "Queen Victoria's profile" by Rogaev et al. [**6**], (GenBank # FJ656214).

4.2.2. Nuclear DNA Results

Sequencing of the shotgun libraries produced 38,008,962 (Lo2-sh), 33,328,275 (Lo3-sh), and 5,470,833 (Lo5h-sh) unique human nuclear sequences representing 1,256,322,853 (Lo2-sh), 1,104,095,334 (Lo3-sh), and 174,843,798 (Lo5h-sh) nucleotides, respectively. The average read depth was 0.40× (Lo2-sh), 0.35× (Lo3-sh) and 0.06× (Lo5h-sh), and the average nuclear DNA fragment length was 33.05 (Lo2-sh), 33.13 (Lo3-sh) and 31.96 bp (Lo5h-sh) (**Table 4**). In every library, at least 98.9% of the human DNA that was sequenced belonged to the nuclear genome.

Table 4. Average length of the nuclear DNA molecules in the locket libraries.

Library	Mapped Nuclear DNA Reads Average Length (bp)
Lo2-sh	33.05
Lo3-sh	33.13
Lo5h-sh	31.96

As has been previously observed in Brandhagen et al. [**12**], the average length of nuclear DNA fragments tends to be smaller than the average length of mtDNA fragments.

4.2.3. Biological Sex

The two single hairs that were shotgun sequenced (Lo2-sh and Lo3-sh) were assessed for biological sex. In both cases, the lower bound of the Rx confidence interval was >0.8, resulting in a biological sex of female (**Table 5**). This result excludes the Romanov's son, Tsarevich Alexis, as a possible source of these two hairs.

Table 5. Determination of the biological sex of the donor of the hairs that were found in the Fabergé locket using Rx as described in [18]. SD: standard deviation. CI: Confidence interval. The values indicating a biological sex of female are in bold.

Library	Rx	SD	CI	CI Low	CI Sup
Lo2-sh	0.965	0.231	0.097	**0.869**	1.062
Lo3-sh	0.935	0.229	0.097	**0.838**	1.031

4.2.4. Relatedness Estimates

In addition to being assessed for biological sex, hairs Lo2 and Lo3 were tested for relatedness. Only reads >35 bp were used for this analysis to avoid spurious mapping of bacterial sequences, as recommended by Meyer et al. [**29**]. Out of 1.3 million SNPs, 15,331 and 33,591 were covered by at least one read for Lo2-sh and Lo3-sh, respectively, and 1196 of these SNPs were covered in both samples. The proportion of allele mismatches was calculated to 0.11 after forcing the Lo2-sh and Lo3-sh datasets into pseudo-haploid genomes. Assuming a non-zero genotype error rate, such a proportion is expected among samples originating from the same individual and is, at the same time, unexpected for parent/child, siblings, and unrelated individuals (**Figure 7**). As controls for the forced haploid simulation approach, the mismatch proportions for the two known samples originating from the same individual as well as two samples originating from two known unrelated individuals were determined. The observed mismatch proportions for those comparisons fit well with the expected mismatch distributions for same individual and unrelated individuals, respectively. Based on the observed forced haploid mismatch proportion as evaluated against the simulated data and considering the error rate that is most appropriate for the given comparison, the data indicate that Lo2 and Lo3 originate from the same individual.

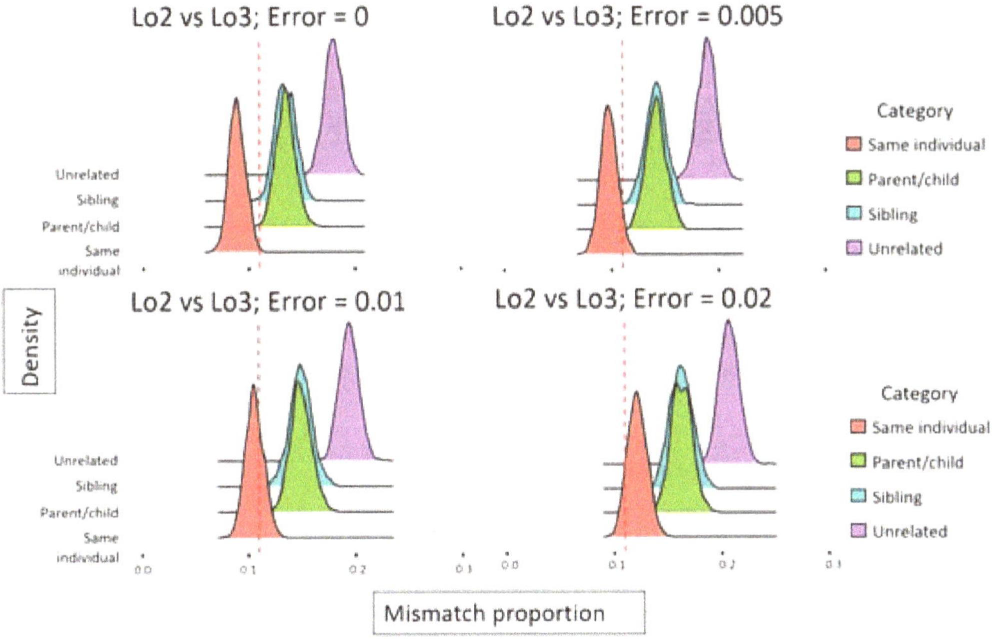

Figure 7. Observed mismatch proportion for the Lo2 vs. Lo3 comparison (red dashed line) in relation to the expected mismatch proportions for various degrees of relatedness and different genotype error rates.

4.3. Illumina Sequencing Results from the Picture Frame Hairs

Sequencing statistics for the hair samples that were found in the frame are shown in **Table 6**. The low percentage of endogenous human DNA in the library derived from six hairs (Q6h-sh) (21.17%) is even lower than the 48% that was observed with the library from five locket hairs (Lo5h-sh). However, as with the locket hairs, the percentage of human DNA improved substantially in the libraries that were derived from single hair shafts. Sample Q1-sh produced 70.12% human DNA, while Q2-sh produced 89%. The three hybridization capture libraries, not surprisingly, yielded the highest percentages of target DNA. In all three cases, ≥88% of the human reads were mtDNA sequences.

Table 6. Sequencing data from libraries made with DNA extracted from hair shafts found in the picture frame.

Shotgun Libraries	Platforms Used	Number of Mapped Raw Reads	Percentage of Raw Mapped Reads	Number of Unique Human Reads
Q6h-sh (6 hairs)	MiSeqFGx	2,072,278	21.17%	1,817,968
Q1-sh (1 hair)	MiSeqFGx	267,392	70.12%	239,692
Q2-sh (1 hair)	NextSeq 500	278,531,240	89%	56,390,706
mtDNA-Enriched Libraries	Platforms Used	Number of Mapped Raw Reads	Percentage of Raw Mapped mtDNA Reads	Number of Unique mtDNA Reads
Q6h-mt (6 hairs)	MiSeqFGx	7,059,775	90.68%	26,038
Q1-mt (1 hair)	MiSeqFGx	3,586,989	96.69%	14,514
Q2-mt (1 hair)	MiSeqFGx	8,104,414	88%	24,636

Among the shotgun libraries, the number of endogenous molecules in Q1-sh was noticeably lower than in the other samples (**Table 6**). The high quantity of damage in this sample, exacerbated by an over- amplification of the DNA, resulted in a high percentage of noise in the data. As a result, the Q1-sh library was not sequenced on the NextSeq 500 instrument and fewer unique human reads were, therefore, recovered.

4.3.1. Mitochondrial DNA Results

Since the frame hairs seemed more degraded than the locket hairs, every Q library was enriched for mtDNA to produce the highest possible data quality.

For Q6h-mt, 26,038 unique mtDNA reads were obtained (>1.2 million bp) and the entire mtGenome was sequenced at an average depth of 74× (range: 27–117×; SD:11.45). At 47.26 bp, the average length of the DNA fragments was significantly shorter than the 93.08 bp average mtDNA size that was observed with Lo5h-mt.

For Q1-mt, a total of 14,514 unique mtDNA reads, spanning 563,192 bp were sequenced. In this case, the average read depth was 34× (range: 1–100×; SD: 14.98) and the average mtDNA length was 38.80 bp.

Finally, for Q2-mt, 24,636 unique reads totaling >1.1 million bp were recovered. Reads spanned the entire mtGenome, with an average read depth of 67× (range: 13–125×; SD: 15.49) and an average fragment length of 45.33 bp.Mitochondrial DNA was also analyzed from the shotgun sequence data of Q6h-sh and Q2-sh (**Table 7**).

Table 7. Number of unique mitochondrial DNA reads and mean mtDNA lengths in the shotgun and capture libraries.

Sequencing Platform	Library	Number of Unique mtDNA Reads	Average mtDNA Size (bp). hg19/rCRS
MiSeqFGx™	Q6h-sh	11,052	45.78/45.56
MiSeqFGx™	Q6h-mt	26,038	47.26
MiSeqFGx™	Q1-sh	1281	NA
MiSeqFGx™	Q1-mt	14,514	38.80
NextSeq 500	Q2-sh	15,267	42.32/43.9
MiSeqFGx™	Q2-mt	24,636	45.33

The 1281 mtDNA reads obtained with Q1-sh were insufficient to produce reliable variant calls. Similar to what was observed with the locket hairs, the average mtDNA fragment size in the capture libraries was larger, although the magnitude of the difference was relatively small (Figure 8). This is likely due to the fact that the range of the average lengths for the mtDNA fragments in the frame hairs was smaller than the range that was observed with the locket hairs.

Capture library Q6h-mt and shotgun library Q6h-sh both produced the following mtDNA profile: 73G, 263G, 315.1C, 709A, 750G, 1438G, 1842G, 1888A, 2706G, 2850C, 4216C, 4769G, 4917G, 6257A, 7022C, 7028T, 8697A, 8860G, 10463C, 11251G, 11719A, 11812G, 13368A, 13965C, 14233G, 14687G, 14766T, 14905A, 15326G, 15452A, 15607G, 15928A, 16126C, 16169Y, 16294T, 16296T, 16519C. The profile belongs to haplogroup T2a1a and is identical to the published sequence of Tsar Nicholas Romanov, including position 16169 which is characterized by a point heteroplasmy in the Tsar and his brother's mtGenome [**3,4,5,6**], (GenBank # FJ656215). Position 16169 had a read depth of 75× with 57 Ts and 18 Cs (76% T). In the shotgun data, the percentage of T versus C varies between 80% and 81% depending on the reference used for mapping. The reference can cause bias as a result of pseudogenes in the human genome. If data are mapped against hg19, which includes both the nuclear and the mitochondrial genome, unspecific sequences will be randomly mapped to one genome or the other, and true mtDNA sequences may be erroneously mapped to nuclear pseudogenes. On the other hand, if the data are mapped only to the rCRS, pseudogene sequences (although rare in these types of degraded samples) could erroneously map to the mtGenome. For this reason, the shotgun data were mapped to both references.

Since the Q6h library was prepared using six hairs, it is not possible to determine whether individual hairs were heteroplasmic or homoplasmic. That is, the mixed position at 16169 in both the shotgun and capture libraries could be the result of individual hairs being homoplasmic for the two variants.

The mtGenome profile that was produced by Q1-mt is similar to the Q6h-mt profile with one exception: position 16169 appeared homoplasmic for thymine, with 27 out of 27 reads exhibiting a T.

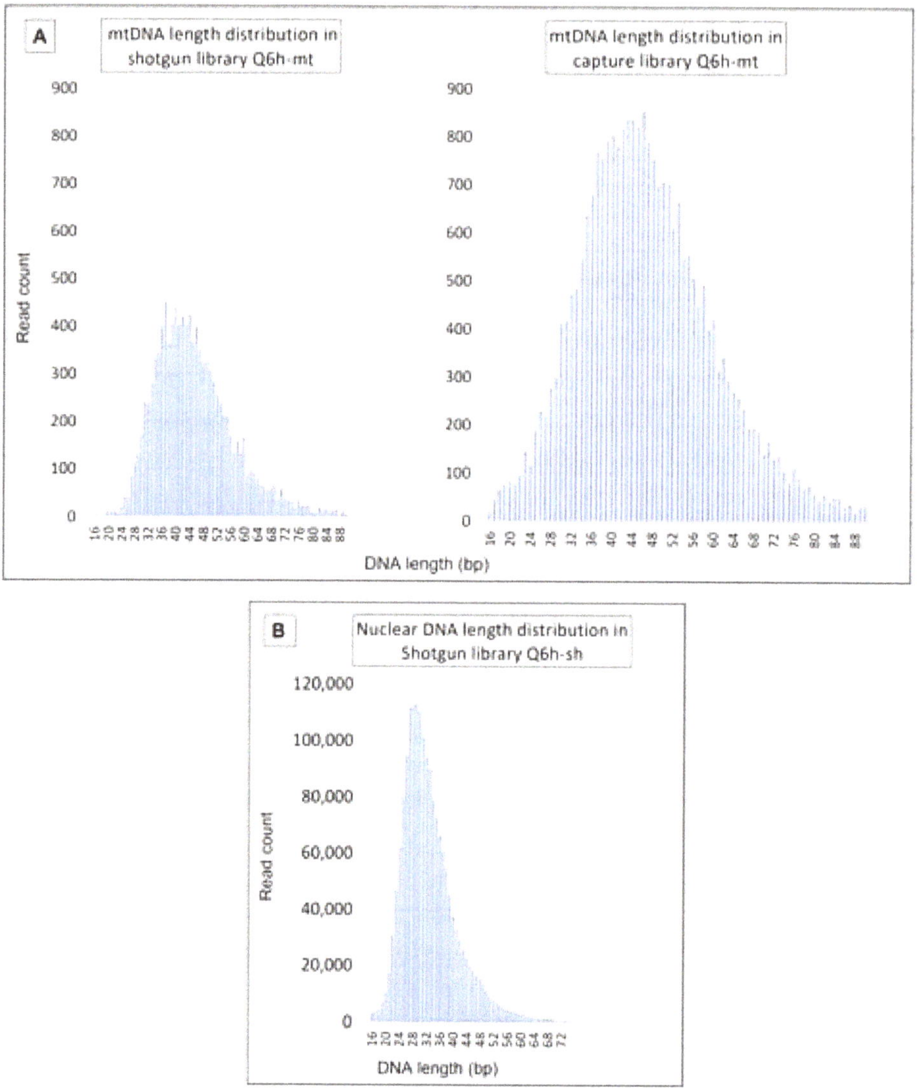

Figure 8. Length distribution of (A) mtDNA reads in Q6h-sh and Q6h-mt, and (B) nuclear DNA reads in Q6h-sh.

In contrast to Q1-mt, position 16169 in library Q2-mt reflected primarily cytosine bases, with 53 C and 1 T. Since hydrolytic deamination is one of the most abundant forms of damage in ancient molecules, it is possible that the thymine base that was detected in the data was a deaminated cytosine [**30,31**]. In fact, in the shotgun library, all 25 reads have a C which supports 16169 being homoplasmic for cytosine.

4.3.2. Nuclear DNA Results

Sequencing of the library Q6h-sh produced 1,806,916 nuclear reads representing close to 60 million bases, an average read depth of 0.02× (range: 1–109×; SD: 1.04), and a mean nuclear DNA length of 33.18 bp. Sequencing of library Q1-sh on a MiSeqFGx™ produced 238,411 unique human nuclear sequences for a total of 8.4 million bases. The average read depth was <0.01× and the mean DNA length was 35.51 bp.

Finally, sequencing of library Q2-sh on a NextSeq 500 platform produced 56,375,439 unique human nuclear reads representing a total of 1.8 billion nucleotides. The average read depth was 0.58× (range: 1–184; SD: 4.65) and the median nuclear read length was 32.06 bp.

Unfortunately, the number of quality reads that were recovered from Q1-sh was insufficient to perform a relatedness analysis between Q1 and Q2. However, the biological sex could be assessed.

For both hairs, the lower bound of the confidence interval was >0.8 (**Table 8**), resulting in a biological sex of female, the expected result if the hairs belonged to Queen Louise.

Table 8. Determination of the biological sex of the donor of two hairs that were found in the frame using Rx as described in [18]. SD: standard deviation. CI: Confidence interval. The values indicating a biological sex of female are in bold.

Library	Rx	SD	CI	CI Low	CI Sup
Q1-sh	0.903	0.201	0.084	**0.819**	0.987
Q2-sh	0.964	0.228	0.095	**0.868**	1.059

5. DISCUSSION

The primary goal of this study was to explore the origin of the hairs in the locket and the frame using techniques and, more importantly, sample sizes that are relevant to forensic casework. By using these historical samples as a proxy for evidentiary material, protocols for extremely limited and compromised samples could be further optimized for routine casework. Overall, the results demonstrate that this method can successfully recover both full mitochondrial genomes and nuclear DNA data from the types of single, aged, and degraded hair samples that are routinely encountered in forensic casework. Furthermore, the results illustrate that limited genetic information can be useful for answering questions that are relevant to DNA identification. Specifically, the comparison between the single strands of hair that were tested from the locket indicate that despite partial SNP information for any given hair sample and limited overlapping data between any two hair samples, valuable information can be acquired. Typical forensic samples rarely, if ever, meet or exceed the suspected age and degradation states of the hairs tested here, yet the integrity of DNA in older cases with hairs recovered from less optimal storage conditions or that were originally collected from harsher environmental settings than those that were encountered here may make these studies directly applicable to some casework scenarios. Hence, the development of protocols for samples that challenge or exceed the lower detection limits of the currently employed methods demonstrates potential for the utility of the protocols in even the worst, or at least the vast majority of casework scenarios.

5.1. DNA Extraction Modifications

To maximize DNA recovery from the hair shafts, modifications were made to the purification step of the DNA extraction. Previous work has shown that both mtDNA and nuclear DNA can be obtained from single shed hairs up to 60 years old with a DNA extraction protocol that employed the MinElute PCR purification kit [**12**]. This particular protocol/purification method was used as a starting point to assess the feasibility of recovering DNA from hair this old, and extractions were performed with five (Lo5h) and six (Q6h) hairs to maximize the chances of detecting DNA.

When these two libraries were evaluated with the 2100 Bioanalyzer, the electropherograms of the reagent blanks showed a large curve that was similar to the ones that were observed with the hair libraries (**Figure 9**A).

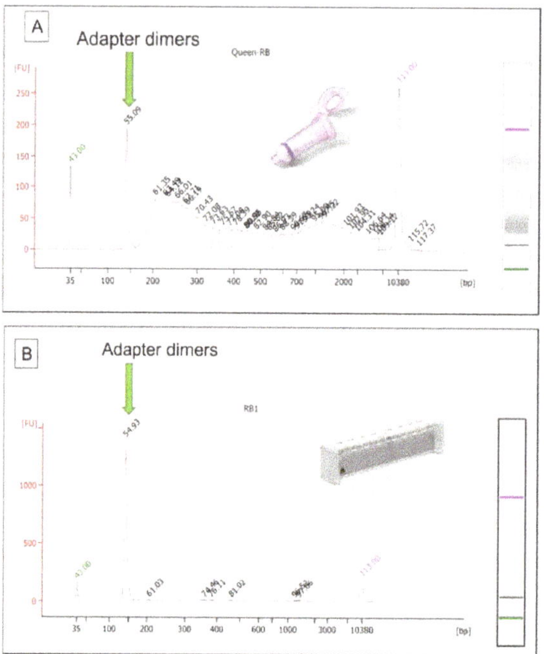

Figure 9. Bioanalyzer electropherograms of reagent blank libraries when DNA purification was performed using (A) MinElute columns, or (B) magnetic beads from G-Biosciences.

The sequencing of these reagent blank libraries produced data but the reads did not map to the human genome reference. These results suggest that the MinElute columns are contaminated with trace amounts of non-human DNA. For most human DNA identification applications, this type of very low-level contamination would have little impact on the final sequence data. Given that the contaminant is non-human and present only in trace quantities, the ratio of target human DNA to contaminant DNA would be high enough for the contaminant to go undetected. However, when the number of PCR cycles is high and when only a few target molecules are present in the extract, such as in the aged and degraded single hair shaft extracts that were tested here, the ratio of endogenous DNA to non-human contaminant DNA drops substantially. The MinElute columns were, therefore, replaced first with the Ultraclean Production (UCP) MinElute columns (cat # 56204, Qiagen). Though Bioanalyzer profiles of hair libraries that were purified with the UCP columns did not reflect the presence of any DNA, the columns were not used on the samples that were presented in this study for two reasons: (1) the use of a vacuum and extender columns was cumbersome and contamination-prone and (2) use of the columns resulted in a near doubling of library purification costs from ~$6 USD per sample to over $11 USD per sample. The columns were thus replaced with silica beads from G-Biosciences. These beads produced the same library purity as the UCP columns, but at a cost of ~$5 USD per sample. In addition, experiments that were performed on other aged hairs showed that the average size of DNA molecules that were collected by silica beads was slightly smaller than the DNA that was purified with MinElute and UCP MinElute columns. As a result, the columns were replaced by the more practical and cost-efficient DNA-free silica magnetic beads for all subsequent experiments.

The sequencing results of the bead-purified hair extracts demonstrated a substantial increase in the percentage of human DNA that was recovered (**Table 2** and **Table 6**). For the locket hairs, the recovery of human DNA improved from 48% in the MinElute-purified library (Lo5h-sh) to 66.54% (Lo2-sh) and 78.91% (Lo3-sh) in the two bead-purified shotgun libraries. Similarly, for the frame hairs, human DNA recovery increased from 21.17% (Q6h-sh) to 70.12% and 89% in Q1-sh and Q2-sh, respectively. Finally, unlike the Bioanalyzer electropherograms from the MinElute-purified libraries, the bead-purified RB library electropherograms exhibited only a single 140 bp peak of adapter dimers (**Figure 9**B).

In Brandhagen et al. [**12**], the previous version of our protocol had been tested as a way to save ultrashort DNA fragments that were lost during a wash step from the PrepFiler Kit protocol (Applied Biosystems, Waltham, MA, USA). We confirmed that the optimized silica beads protocol works equally well, or better, than the previous one in small fragment retention.

5.2. Authenticity of the Items

DNA testing is often employed to authenticate items of historical significance [32]. However, successful DNA testing is dictated by the type of item, its age and condition, how it has been stored, whether it has been handled/contaminated, etc. Items that can be decontaminated, such as hairs and calcified tissues, generally offer the best chance of success. Yet, even in cases for which the target DNA can often be recovered, authentication can still prove difficult, particularly when the direct or familial references that are required for identification are not straightforward to establish or acquire [33]. Fortunately, for historical items that are believed to be associated with the Romanov family, sufficient DNA reference information exists. Thus, on top of the historical evidence tying the locket and frame artifacts to the Romanov family, the DNA data that were recovered from the hairs may also prove useful.

5.2.1. Fabergé Locket

Historical Evidence of Authenticity

The photograph of Princess Alexandra displayed in the locket comes from a cabinet card, a type of photography mounted on a piece of cardboard that was very popular in the 1860s. It could be confidently identified as a copy of a photograph that was made in St Petersburg in 1895 by Alexander Pasetti, the photographer of the Imperial Russian court

The pendant was crafted by the famous Russian jeweler and goldsmith, Karl Gustavovich Fabergé (also known as Peter Carl Fabergé; 1846–1920). Karl was the son of Gustav Fabergé, a German jeweler who moved to St. Petersburg in 1842 and opened the House of Fabergé. The association between Karl Fabergé and the Russian Imperial family is well documented. On Orthodox Easter of 1885, Tsar Alexander III Romanov offered his wife, Empress Maria Feodorovna, a jeweled Eastern Hen egg that was made by Fabergé [34]. The Empress liked it so much that Alexander named Fabergé goldsmith by special appointment to the Imperial crown and commissioned an Easter egg every following year. After Alexander's death in 1894, his son, Tsar Nicholas II, took over the tradition and asked Fabergé to create two unique eggs each year, one for his mother, the Dowager Empress Maria, and one for his wife Alexandra Feodorovna [35]. While Easter eggs made the House of Fabergé famous, they only represent a fraction of the Fabergé creations that were made for the Tsars and their families [34].

Genetic Evidence

The Tsarina's mtGenome was characterized in 2009 [6]. Hence, it was possible to directly compare the mtGenome haplotype that was developed from the locket hairs to the reference haplotype for the Tsarina; the two sequences are identical.

The haplotype belongs to haplogroup H1af2 [36], which is characterized by a C4137T transition. To understand the rarity of the haplotype, three different mtDNAdatasets were searched: the public FamilyTreeDNA© mitochondrial DNA Haplotree database [37], GenBank, and the European DNA Profiling Group mtDNA Population database (EMPOP, [38]). Sequence level data could not be compared in the FamilyTreeDNA© dataset. However, only ten of 170,000 complete mtGenome sequences were reported to belong to the H1af2 haplogroup. When the complete mtGenome sequence is compared to the GenBank database, no match is returned besides the profiles of the Tsarina, Alexei, and Maria, that were published by Rogaev et al. [6]. In fact, of more than 52,000 mtGenome profiles present in GenBank, only 38 sequences have a C4137T, and only five of those (including the three profiles mentioned above) belongs to haplogroup H [39]. H1af2 sequences are not currently present in EMPOP. While it is not possible to establish the identity based on mtDNA, the rarity of this profile increases the odds that the hairs that were found in the locket belonged to the Tsarina or a member of her maternal lineage.

To gain further insight into the source of the hairs and, in more general terms, explore the potential probative value of the nuclear DNA data that were recovered, autosomal SNPs were also analyzed. Autosomal SNP reference data was not available for the Tsarina or her relatives, so our goal with the SNP analysis was not source identity. Instead, the SNP results were used to assess if the strands of hair in the locket originated from the same individual or from multiple individuals from the same maternal lineage. Assuming the hairs belong to the Tsarina or one of her immediate maternal relatives (i.e., her children), the hairs would be over 100 years old. If this is indeed the case, the presence of mtDNA fragments that were larger than 90 bp in both the shotgun and the enriched library of Lo5h suggests that at least a portion of the molecules was well preserved.

In previous studies, hairs of somewhat similar age have yielded average mtDNA fragment sizes of approximately 60 bp [8], while younger hairs (between 40 and 60 years old) were shown to harbor mtDNA fragments averaging 70–80 bp [12]. There could be at least two explanations for the great preservation of the locket hair DNA. First, according to the locket owner, Mr. Bachmakov, the locket was extremely difficult to open and required specialized tools to do so. In Mr. Bachmakov's expert opinion, the locket had not been opened for years, perhaps even decades, and it is likely that the hairs remained safely sealed and protected from oxygen, light, and humidity. Second, saw dust that was found inside the locket may have acted as a desiccant. In the 19th and 20th centuries, jewels that were made of gold were treated with acid to enrich the proportion of gold on the surface and to change the color from a pink to a yellow gold [40]. After being rinsed with water, the objects were dried and saw dust was used to absorb moisture. In some cases, as in the case of this Fabergé locket, small amounts of saw dust remained inside the jewels.

Unbeknown to the maker of the pendant, the desiccant properties of the saw dust may have played a key role in preserving the hairs and their genetic material.

5.2.2. Picture Frame

Historical Evidence

The origin of the framed picture of Queen Louise of Hesse Kassel is supported by documents that were provided to Mr. Bachmakov when the object was purchased. The paperwork indicates that the framed photograph had originally belonged to Maria Feodorovna (known as Princess Dagmar of Denmark prior to her marriage), one of Louise's daughters. It was passed down to one of Maria Feodorovna's daughters, Grand Duchess Xenia (1875–1960), and then again through various family members to a living descendant who kept it in his/her possession until 2017 when it was purchased by Mr. Bachmakov. The picture also comes from a cabinet card and the initials on the back show that the photograph was taken by Mary Steen, a famous Danish photographer who, in 1896, was appointed court photographer to Alexandra, Princess of Wales (1844–1925), later Queen Alexandra.

Given the picture of Queen Louise and the fact that the strands of hair were white, it is presumed that they came from the Queen and were collected from her in her senior years. If these assumptions are correct, the hairs would be at least 120 years old.

Genetic Evidence and mtDNAHeteroplasmy

Mitochondrial DNA sequencing was performed to test the hypothesis that the frame hairs are from Tsar Nicholas' maternal grandmother, Queen Louise of Hesse-Kassel. Fortunately, the mtDNA sequence of the Tsar's maternal lineage is well documented. His control region (CR) haplotype was first reported in a study on the DNA analysis of skeletal remains that were discovered in Russia in 1991 and attributed to Nicholas II. The profile was compared to the CR of two living relatives who served as maternal references: James George Alexander Bannerman Carnegie, 3rd Duke of Fife (1929–2015), the Tsar's first cousin twice removed and Countess Xenia CheremetevaSfiris (b. 1942), his great grandniece [3] (**Figure 10**). The profile of Tsar Nicholas revealed a C/T point heteroplasmy at position 16169 that was not present in the two references CR profiles that only had the 16169T variant. Since little was known about the occurrence of mitochondrial DNA heteroplasmy variants in the early 1990s, the mixed C/T position was originally viewed by skeptics as a "mismatch" between the tsar's remains and his living relatives. However, two studies, both published in 1996, settled the debate. Ivanov et al. [4] analyzed the skeletal remains of the Tsar's brother, Grand Duke George (1871–1899) and the study revealed once more, the presence of a point heteroplasmy at position 16169, albeit with opposite ratios. While C was the predominant base in Nicholas' DNA, T was the dominant allele that was observed in his brother's profile. Concurrently, the CR from Tikhon Nikolaevich Kulikovsky-Romanov (1917–1993), one of Tsar Nicholas nephews, was also Sanger sequenced and, in this case, only cytosines were observed at position 16169, adding credence to the presence of an heteroplasmic variant in the Tsar's DNA [**41**].

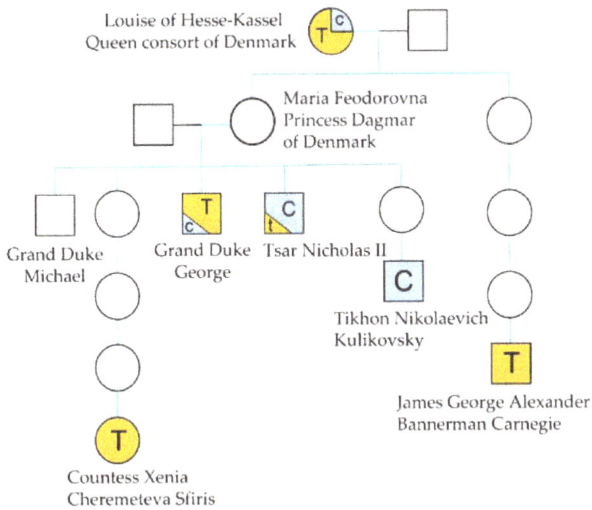

Figure 10. Family tree of the Romanov family showing the major allele (T or C) at heteroplasmic position 16169.

The point heteroplasmy in the Tsar's lineage was again confirmed in 2009, following the discovery of skeletal remains that were believed to be the two missing Romanov children. Additional genetic analyses of all the skeletal remains that were found in 1991 not only confirmed that the two sets of newly discovered skeletal remains were the two missing Romanov children [**5**], but also re-established the presence of point heteroplasmy in the Tsar's remains [**6**].

In this study, the 16169 pointheteroplasmy was not observed in skeletal remains, but in hair shafts. The occurrence of mtDNAheteroplasmy in human hairs has been extensively characterized [**42,43**]. Studies show that different levels of heteroplasmy can be observed in hairs from a single individual, as was seen in [**44,45,46,47**] and even along the length of a single hair shaft [**48**]. Our study followed the transmission of heteroplasmy through six generations. Interestingly, while it endured between the two generations that separate the Tsar from his grandmother, it seems to have rapidly evolved towards homoplasmy in the three different branches of the family that were studied. The perfect match between the sequence of the Tsar's mtGenome and the sequence from the frame hairs, as well as the presence of a point heteroplasmy at position 16169 are consistent with the hairs coming from Louise of Hesse-Kassel, Queen of Denmark or another individual that is maternally related to Tsar Nicholas II.

6. AT A GLANCE

Our prior research has demonstrated the successful retrieval of nuclear DNA from individual hair shafts lacking roots [12]. In this study, we expand upon previous research by showing that it is possible to retrieve both mitochondrial and nuclear DNA from individual rootless hairs of considerable age. This achievement is made possible by utilizing an extraction process specifically designed for extremely fragmented DNA. In addition, by the optimization of the extraction technique to eliminate or diminish the presence of low-level contaminating DNA in the reagents and supplies, we were able to enhance both the quantity of DNA obtained and the quality of the sequencing results. Since routine nuclear and mitochondrial DNA type assays typically focus on DNA fragments bigger than 100–150 bp, extraction procedures that prioritize the preservation of smaller pieces are often unneeded. For the most severely contaminated specimens, utilizing this specific extraction method in combination with shotgun sequencing may be the most effective and possibly exclusive opportunity to retrieve conclusive DNA data. The hair samples analyzed here, with an average mtDNA and nuclear DNA fragment sizes of around 40–85 bp and 33 bp respectively, would not have been able to provide DNA results even with the smallest mtDNA amplicons commonly employed in operational casework. Moreover, conventional extraction techniques would probably have eliminated a significant portion, if not the majority, of the native target DNA. Through the utilization of an enhanced extraction methodology and advanced next generation shotgun sequencing, it became feasible to successfully retrieve intact mtGenomes from hair samples that are very probable to be older than any typical hair sample submitted to a forensic laboratory for analysis. These ancient samples yielded a significantly higher amount of mtDNA sequencing data compared to the control region. Additionally, they provided sufficient nuclear DNA to determine the biological sex and, in one instance, establish relatedness between the samples. Our findings provide additional evidence that valuable nuclear DNA can potentially be extracted from the specific type of hair shafts without roots that are commonly found in forensic investigations and typically analyzed only for mitochondrial DNA. Furthermore, our findings imply that aged hair obtained from keepsake lockets, hairbrushes, and other aged sources could potentially be used as an alternate source of direct or family reference material in missing persons situations that necessitate nuclear DNA analysis and when hair samples are the sole available material. Ultimately, it is probable that this or comparable extraction procedure(s) can be advantageous for any form of severely deteriorated sample. Furthermore, we have applied this approach to samples other than hairs and can affirm that it can yield highly favorable results with any sample containing extensively deteriorated DNA molecules.

7. CONCISE SUMMARY

This study presents a refined method for extracting DNA from extremely small DNA molecules found in individual hair strands without roots. We applied it to the most ancient samples at our disposal: strands of hair discovered in artifacts linked to the Romanov dynasty. The mitochondrial DNA genome sequences of Tsar Nicholas II and his wife, Tsarina Alexandra, were used to test the effectiveness of this DNA extraction process and determine the range of genetic information that may be obtained by sequencing very short fragments. The mtGenome of the Tsarina's ancestry was detected in hairs that were hidden within a pendant crafted by Karl Fabergé for Alexandra Feodorovna Romanov, using this technique. Furthermore, in order to ascertain whether the lock was handled by many individuals, two strands of hair were separately removed from the locket and transformed into Illumina libraries for shotgun sequencing using a NextSeq 500 platform. The data was used to analyze autosomal SNPs in order to evaluate relatedness. The findings suggested that the two hairs originated from a solitary individual. Analysis of the hairs discovered in the second artifact, which was a framed photograph of Louise of Hesse-Kassel, Queen of Denmark and the maternal grandmother of Tsar Nicholas II, determined that the hair came from a woman who shared the same maternal lineage as Tsar Nicholas. This included the notable point heteroplasmy at position 16169.

REFERENCES

1. Massie, R.K. *The Romanovs: The Final Chapter*; Random House: New York, NY, USA, 1995. **[Google Scholar]**
2. Byard, R.W. The execution of the Romanov family at Yekatarinberg. *Forensic Sci. Med. Pathol.* **2020**, *16*, 552–556. **[Google Scholar] [CrossRef] [PubMed]**
3. Gill, P.; Ivanov, P.L.; Kimpton, C.P.; Piercy, R.; Benson, N.; Tully, G.; Evett, I.; Hagelberg, E.; Sullivan, K. Identification of the remains of the Romanov family by DNA analysis. *Nat. Genet.* **1994**, *6*, 130–135. **[Google Scholar] [CrossRef] [PubMed]**
4. Ivanov, P.L.; Wadhams, M.J.; Roby, R.K.; Holland, M.M.; Weedn, V.W.; Parsons, T.J. Mitochondrial DNA sequence heteroplasmy in the Grand Duke of Russia Georgij Romanov establishes the authenticity of the remains of Tsar Nicholas II. *Nat. Genet.* **1996**, *12*, 417–420. **[Google Scholar] [CrossRef] [PubMed]**
5. Coble, M.; Loreille, O.; Wadhams, M.J.; Edson, S.; Maynard, K.; Meyer, C.E.; Niederstätter, H.; Berger, C.; Berger, B.; Falsetti, A.B.; et al. Mystery Solved: The Identification of the Two Missing Romanov Children Using DNA Analysis. *PLoS ONE* **2009**, *4*, e4838. **[Google Scholar] [CrossRef] [Green Version]**

6. Rogaev, E.I.; Grigorenko, A.P.; Moliaka, Y.K.; Faskhutdinova, G.; Goltsov, A.; Lahti, A.; Hildebrandt, C.; Kittler, E.L.W.; Morozova, I. Genomic identification in the historical case of the Nicholas II royal family. *Proc. Natl. Acad. Sci. USA* **2009**, *106*, 5258–5263. **[Google Scholar] [CrossRef] [Green Version]**

7. Melton, T.; Dimick, G.; Higgins, B.; Lindstrom, L.; Nelson, K. Forensic Mitochondrial DNA Analysis of 691 Casework Hairs. *J. Forensic Sci.* **2005**, *50*, 73–80. **[Google Scholar] [CrossRef]**

8. Tobler, R.; Rohrlach, A.; Soubrier, J.; Bover, P.; Llamas, B.; Tuke, J.; Bean, N.; Abdullah-Highfold, A.; Agius, S.; O'Donoghue, A.; et al. Aboriginal mitogenomes reveal 50,000 years of regionalism in Australia. *Nature* **2017**, *544*, 180–184. **[Google Scholar] [CrossRef]**

9. Rasmussen, M.; Li, Y.; Lindgreen, S.; Pedersen, J.S.; Albrechtsen, A.; Moltke, I.; Metspalu, M.; Metspalu, E.; Kivisild, T.; Gupta, R.; et al. Ancient human genome sequence of an extinct Palaeo-Eskimo. *Nature* **2010**, *463*, 757–762. **[Google Scholar] [CrossRef]**

10. Rasmussen, M.; Guo, X.; Wang, Y.; Lohmueller, K.E.; Rasmussen, S.; Albrechtsen, A.; Skotte, L.; Lindgreen, S.; Metspalu, M.; Jombart, T.; et al. An Aboriginal Australian Genome Reveals Separate Human Dispersals into Asia. *Science* **2011**, *334*, 94–98. **[Google Scholar] [CrossRef] [Green Version]**

11. Moltke, I.; Korneliussen, T.S.; Seguin-Orlando, A.; Moreno-Mayar, J.V.; LaPointe, E.; Billeck, W.; Willerslev, E. Identifying a living great-grandson of the Lakota Sioux leader Tatanka Iyotake (Sitting Bull). *Sci. Adv.* **2021**, *7*, 2013. **[Google Scholar] [CrossRef]**

12. Brandhagen, M.D.; Loreille, O.; Irwin, J.A. Fragmented Nuclear DNA Is the Predominant Genetic Material in Human Hair Shafts. *Genes* **2018**, *9*, 640. **[Google Scholar] [CrossRef] [PubMed] [Green Version]**

13. Rohland, N.; Glocke, I.; Aximu-Petri, A.; Meyer, M. Extraction of highly degraded DNA from ancient bones, teeth and sediments for high-throughput sequencing. *Nat. Protoc.* **2018**, *13*, 2447–2461. **[Google Scholar] [CrossRef] [PubMed]**

14. Kavlick, M.F. Development of a triplex mtDNA qPCR assay to assess quantification, degradation, inhibition, and amplification target copy numbers. *Mitochondrion* **2019**, *46*, 41–50. **[Google Scholar] [CrossRef] [PubMed]**

15. Li, H.; Durbin, R. Fast and accurate short read alignment with Burrows–Wheeler transform. *Bioinformatics* **2009**, *25*, 1754–1760. **[Google Scholar] [CrossRef] [PubMed] [Green Version]**

16. Li, H.; Handsaker, B.; Wysoker, A.; Fennell, T.; Ruan, J.; Homer, N.; Marth, G.; Abecasis, G.; Durbin, R.; 1000 Genome Project Data Processing Subgroup. The Sequence Alignment/Map format and SAMtools. *Bioinformatics* **2009**, *25*, 2078–2079. **[Google Scholar] [CrossRef] [PubMed] [Green Version]**

17. Picard. Available online: **http://broadinstitute.github.io/picard** (accessed on 15 September 2021).

18. Mittnik, A.; Wang, C.C.; Svoboda, J.; Krause, J. A Molecular Approach to the Sexing of the Triple Burial at the Upper Paleolithic Site of DolniVestonice. *PLoS ONE* **2016**, *11*, e0163019. **[Google Scholar] [CrossRef] [PubMed] [Green Version]**

19. Tillmar, A.; Sjölund, P.; Lundqvist, B.; Klippmark, T.; Älgenäs, C.; Green, H. Whole-genome sequencing of human remains to enable genealogy DNA database searches–A case report. *Forensic Sci. Int. Genet.* **2020**, *46*, 102233. **[Google Scholar] [CrossRef] [PubMed]**

20. Tillmar, A.; Fagerholm, S.A.; Staaf, J.; Sjölund, P.; Ansell, R. Getting the conclusive lead with investigative genetic genealogy –A successful case study of a 16 year old double murder in Sweden. *Forensic Sci. Int. Genet.* **2021**, *53*, 102525. **[Google Scholar] [CrossRef]**

21. 1000 Genomes Project Consortium; Auton, A.; Brooks, L.D.; Durbin, R.M.; Garrison, E.P.; Kang, H.M.; Korbel, J.O.; Marchini, J.L.; McCarthy, S.; McVean, G.A.; et al. A global reference for human genetic variation. *Nature* **2015**, *526*, 68–74. **[Google Scholar]**

22. Nexus. Available online: **https://www.snp-nexus.org** (accessed on 15 December 2020).

23. Oscanoa, J.; Sivapalan, L.; Gadaleta, E.; Ullah, A.Z.D.; Lemoine, N.R.; Chelala, C. SNPnexus: A web server for functional annotation of human genome sequence variation (2020 update). *Nucleic Acids Res.* **2020**, *48*, W185–W192. **[Google Scholar] [CrossRef]**

24. Meyer, F.; Paarmann, D.; Souza, M.D.; Olson, R.; Glass, E.M.; Kubal, M.; Paczian, T.; Rodriguez, A.; Stevens, R.; Wilke, A.; et al. The metagenomics RAST server—A public resource for the automatic phylogenetic and functional analysis of metagenomes. *BMC Bioinform.* **2008**, *9*, 386. **[Google Scholar] [CrossRef] [PubMed] [Green Version]**

25. Carpenter, M.; Buenrostro, J.D.; Valdiosera, C.; Schroeder, H.; Allentoft, M.; Sikora, M.; Rasmussen, M.; Gravel, S.; Guillén, S.; Nekhrizov, G.; et al. Pulling out the 1%: Whole-Genome Capture for the Targeted Enrichment of Ancient DNA Sequencing Libraries. *Am. J. Hum. Genet.* **2013**, *93*, 852–864. **[Google Scholar] [CrossRef] [PubMed] [Green Version]**

26. Enk, J.M.; Devault, A.M.; Kuch, M.; Murgha, Y.E.; Rouillard, J.M.; Poinar, H.N. Ancient whole genome enrichment using baits built from modern DNA. *Mol. Biol. Evol.* **2014**, *31*, 1292–1294. **[Google Scholar] [CrossRef] [PubMed]**

27. Ávila-Arcos, M.C.; Sandoval-Velasco, M.; Schroeder, H.; Carpenter, M.; Malaspinas, A.-S.; Wales, N.; Peñaloza, F.; Bustamante, C.D.; Gilbert, M.T.P. Comparative performance of two whole-genome capture methodologies on ancient DNA Illumina libraries. *Methods Ecol. Evol.* **2015**, *6*, 725–734. **[Google Scholar] [CrossRef]**

28. Krasnenko, A.; Tsukanov, K.; Stetsenko, I.; Klimchuk, O.; Plotnikov, N.; Surkova, E.; Ilinsky, V. Effect of DNA insert length on whole-exome sequencing enrichment efficiency: An observational study. *Adv. Genom. Genet.* **2018**, *8*, 13–15. **[Google Scholar] [CrossRef] [Green Version]**

29. Meyer, M.; Arsuaga, J.L.; de Filippo, C.; Nagel, S.; Aximu-Petri, A.; Nickel, B.; Martinez, I.; Gracia, A.; de Castro, J.M.B.; Carbonell, E.; et al. Nuclear DNA sequences from the Middle Pleistocene Sima de losHuesos hominins. *Nature* **2016**, *531*, 504–507. **[Google Scholar] [CrossRef]**

30. Hofreiter, M.; Jaenicke, V.; Serre, D.; von Haeseler, A.; Pääbo, S. DNA sequences from multiple amplifications reveal artifacts induced by cytosine deamination in ancient DNA. *Nucleic Acids Res.* **2001**, *29*, 4793–4799. **[Google Scholar] [CrossRef]**

31. Rathbun, M.M.; McElhoe, J.A.; Parson, W.; Holland, M.M. Considering DNA damage when interpreting mtDNAheteroplasmy in deep sequencing data. *Forensic Sci. Int. Genet.* **2017**, *26*, 1–11. **[Google Scholar] [CrossRef]**

32. King, T.E.; Fortes, G.G.; Balaresque, P.; Thomas, M.G.; Balding, D.; Delser, P.M.; Neumann, R.; Parson, W.; Knapp, M.; Walsh, S.; et al. Identification of the remains of King Richard III. *Nat. Commun.* **2014**, *5*, 5631. **[Google Scholar] [CrossRef]**

33. Parson, W.; Loreille, O.; Niederstätter, H.; Parsons, T.J.; Scheithauer, R. DNA Investigations of the putative Mozart cranium. In Proceedings of the 17th International Symposium on Human Identification, Nashville, TN, USA, 10 October 2006. **[Google Scholar]**

34. Bainbridge, C.; Peter, C.F. *Goldsmith and Jeweller to the Russian Imperial Court and the Principal Crowned Heads of Europe*; Batsford: London, UK, 1 January 1949. **[Google Scholar]**

35. Available online: **https://www.christies.com/features/Faberge-15-things-a-collector-needs-to-know-8353-1.aspx** (accessed on 15 September 2021).

36. van Oven, M.; Kayser, M. Updated comprehensive phylogenetic tree of global human mitochondrial DNA variation. *Hum. Mutat.* **2009**, *30*, e386–e394. **[Google Scholar] [CrossRef]**

37. FamilyTree DNA. Available online: **https://www.familytreedna.com/public/mt-dna-haplotree/H** (accessed on 20 December 2021).

38. EMPOP. Available online: **https://www.empop.org** (accessed on 20 December 2021).

39. MITOMASTER. Available online: **https://www.mitomap.org/foswiki/bin/view/MITOMASTER/WebHome** (accessed on 20 December 2021).

40. Bachmakov, N.; Expert in Antique Restoration in New York, New York, NY, USA. Personal communication, 2020.

41. Rogaev, I.E.; Ovchinnikov, I.V.; Dzhorzh-Khislop, P.; Rogaeva, A.E. Comparison of mitochondrial DNA sequences of T.N. Kulikovski? -Romanov, the nephew of Tsar Nikola? II Romanov, with DNA from the putative remains of the Tsar. *Генетика* **1996**, *32*, 1690–1692. **[Google Scholar]**

42. Comas, D.; Pääbo, S.; Bertranpetit, J. Heteroplasmy in the control region of human mitochondrial DNA. *Genome Res.* **1995**, *5*, 89–90. **[Google Scholar] [CrossRef] [Green Version]**

43. Bendall, K.E.; Macaulay, V.A.; Sykes, B.C. Variable Levels of a Heteroplasmic Point Mutation in Individual Hair Roots. *Am. J. Hum. Genet.* **1997**, *61*, 1303–1308. **[Google Scholar] [CrossRef] [Green Version]**

44. Alonso, A.; Salas, A.; Albarrán, C.; Arroyo-Pardo, E.; Castro, A.; Crespillo, M.; di Lonardo, A.M.; Lareu, M.V.; Cubría, C.L.; Soto, M.L.; et al. Results of the 1999–2000 collaborative exercise and proficiency testing program on mitochondrial DNA of the GEP-ISFG: An inter-laboratory study of the observed variability in the heteroplasmy level of hair from the same donor. *Forensic Sci. Int.* **2002**, *125*, 1–7. **[Google Scholar] [CrossRef]**

45. Sekiguchi, K.; Sato, H.; Kasai, K. Mitochondrial DNA heteroplasmy among hairs from single individuals. *J. Forensic Sci.* **2004**, *49*, 986–991. **[Google Scholar] [CrossRef] [Green Version]**

46. Tully, G.; Barritt, S.; Bender, K.; Brignon, E.; Capelli, C.; Dimo-Simonin, N.; Eichmann, C.; Ernst, C.; Lambert, C.; Lareu, M.; et al. Results of a collaborative study of the EDNAP group regarding mitochondrial DNA heteroplasmy and segregation in hair shafts. *Forensic Sci. Int.* **2004**, *140*, 1–11. **[Google Scholar] [CrossRef]**

47. Van Der Gaag, K.J.; Desmyter, S.; Smit, S.; Prieto, L.; Sijen, T. Reducing the Number of Mismatches between Hairs and Buccal References When AnalysingmtDNAHeteroplasmic Variation by Massively Parallel Sequencing. *Genes* **2020**, *11*, 1355. **[Google Scholar] [CrossRef]**

48. Desmyter, S.; Bodner, M.; Huber, G.; Dognaux, S.; Berger, C.; Noël, F.; Parson, W. Hairy matters: MtDNA quantity and sequence variation along and among human head hairs. *Forensic Sci. Int. Genet.* **2016**, *25*, 1–9. **[Google Scholar]** **[CrossRef]**

19 SCANDINAVIAN EYE COLOUR PREDICTION UTILIZING THE EYE COLOUR 11 (EC11) SINGLE NUCLEOTIDE POLYMORPHISM (SNP) SET

1. INTRODUCTION: AN OVERVIEW

Predictions of phenotypic traits (observable characteristics) derived from DNA can be vital in criminal investigations where there are no apparent suspects. Multiple assays have been devised for the purpose of forecasting eye color, hair color, skin color, and biogeographic heritage in forensic investigation. These assays vary in terms of their accuracy levels [1,2,3,4,5]. Biogeographic ancestry prediction can serve as an indirect indicator of an individual's physical characteristics. However, there is a notable variation in eye color among individuals of European ancestry. Offering a specific description of the offender's eye color could aid law enforcement in focusing their inquiry to a smaller group of individuals. The IrisPlex assay is widely recognized as the most prominent technique for forecasting eye color. The method depends on six distinct genetic markers, referred to as single nucleotide polymorphisms (SNPs), which are present in six separate genes (HERC2, OCA2, SLC24A4, SLC45A2, TYR, and IRF4). These genes have been associated with genetic differences that result in different eye colors. The IrisPlex algorithm precisely categorizes eye color into three separate groups: blue, intermediate, and brown [3,6]. The occurrence of blue or brown eye color is strongly correlated with a single SNP in the HERC2 gene, particularly rs12913832, as widely acknowledged [7,8]. The single nucleotide polymorphism (SNP) is located in the promoter region of the OCA2 gene and shows a significant association with the expression of OCA2 [9]. In 2012, Visser and colleagues demonstrated that the presence of the rs12913832: A allele led to increased expression of OCA2 in melanocytes, while the rs12913832: G allele resulted in lower production of OCA2 in melanocytes. Hence, those possessing the rs12913832: A allele (genotypes rs12913832: AA and rs12913832:AG) are expected to exhibit brown eye color, whereas those with the genotype rs12913832:GG are expected to display blue eye color. The accuracy of predicting eye color using the

IrisPlex approach relies heavily on the rs12913832 gene variant [10]. As to the IrisPlex forecast, the accuracy is high for blue and brown eye colors, but low for intermediate eye color [3]. In some cases, the genotype-phenotype patterns of rs12913832 can result in inappropriate estimates of eye color, resulting to incorrect IrisPlex eye color predictions [10,11]. In our previous investigation, we performed comprehensive sequencing investigations on people whose IrisPlex eye color forecasts were found to be incorrect. Through this research, we identified genetic differences in the OCA2, TYRP1, TYR, and SLC24A4 genes that may explain the inaccurate predictions of eye color [11,12]. In this study, we examined various types of genetic variants and the particular IrisPlex single nucleotide polymorphisms (SNPs) in a group of 757 individuals of European ancestry. Through this approach, we have identified 11 Single Nucleotide Polymorphisms (SNPs) that are specifically significant in predicting eye color. Consequently, we created a new and innovative model known as Eye Colour 11 (EC11), which was built using the chosen SNPs. We conducted a study in which we examined and confirmed the accuracy of the EC11 by analyzing a distinct dataset consisting of 523 individuals from Norway. Subsequently, we conducted a comparison between the predictions generated by the EC11 model and those produced by a model solely based on the rs12913832 SNP, as well as a model utilizing the six IrisPlex SNPs (referred to as IrisPlex Norway). We employed multiple reporting methodologies to assess the efficacy of these models. Ultimately, we contrasted the results with those obtained through the use of the IrisPlex online prediction tool (IrisPlex Web) [3].

2. MATERIALS AND METHODS

2.1. Samples and DNA Extraction

The variant discovery data set consisted of 757 healthy individuals residing in Denmark, Sweden, or Italy. The dataset consisted of 523 individuals who were in good health and resided in Tromsø and Bodø, Norway. Blood samples were obtained from university personnel and students, residents and employees at healthcare centers, hospital employees, and through the Danish Blood Donor Study, as previously described [10,13,14]. The DNA was isolated using the QIAamp DNA Blood Mini Kit (Qiagen, Hilden, Germany), the QIAsymphony DNA Midi Kit (Qiagen, Hilden, Germany), or the PrepFiler™ Express DNA Extraction kit (Thermo Fisher Scientific, Waltham, MA, USA) according to the manufacturers' instructions. The utilization of the material was sanctioned by the Danish Ethical Committee (M-20090237, H-4-2009-125, and H-3-2012-023), the Ethical Committee of AziendaOspedalieraOspedal Sant'Anna di Como (U.0026484.23-11-2012), the Ethical Committee of the University of Milan-Bicocca (P.U. 0033373/12), and the Faculty of Health Sciences, UiT–The Arctic University of Norway (2021/2034). Each subject provided signed and informed consent, and all samples were anonymized.

2.2. Quantitative measurements of eye color and categorization of eye color

A digital photograph of one of the eyes of each participant was taken as described previously [13]. The pixel index of the eye score (PIE-score) was calculated from jpeg images 639 × 426 pixels as: (number of blue pixels–number of brown pixels)/(number of blue pixels + number of brown pixels) using the Digital Iris Analysis Tool (DIAT) [13]. The PIE-scores ranged from -1 to 1; eye colour photos with a PIE-score of 1 had only blue pixels, and eye colour photos with a PIE-score of -1 had only brown pixels. Eye colour was also categorised in a two-category (blue and brown) and a three-category (blue, intermediate, and brown) system. The categorisation was based on the PIE-score and our recent study on the perception of blue and brown eye colours [15]. In the two-category system, eye colours with PIE-score > 0.2 were categorised as blue, and eye colours with PIE-score ≤ 0.2 were categorised as brown. In the three-category system, eye colours with PIE-score > 0.8 were categorised as blue, eye colours with PIE-score ≤ 0.8 and ≥-0.5 were categorised as intermediate, and eye colours with PIE-score < -0.5 were categorised as brown. The number of individuals in each eye colour category and the mean PIE-score in the discovery data set and the model data set are shown in **Table 1**.

Table 1. Eye colour variation in the variant discovery and model data sets.

	Quantitative System [1]	Two-Category System [2]		Three-Category System [3]		
	Mean PIE-Score	Blue	Brown	Blue	Intermediate	Brown
Discovery data set (n = 757)	0.24	447 (59%)	310 (41%)	368 (49%)	148 (20%)	241 (31%)
Model data set (n = 523)	0.44	376 (72%)	147 (28%)	293 (56%)	123 (24%)	107 (20%)

[1.] Statistically significant difference in mean PIE-scores between the two data sets ($p < 0.05$). [2.] PIE-score > 0.2: blue, and PIE-score ≤ 0.2: brown. [3.] PIE-score > 0.8: blue, PIE-score ≤ 0.8 to ≥0.5-: intermediate, and PIE-score < 0.5-: brown.

2.3. Variant Typing (Discovery Data Set)

A total of 44 pigmentary variants were investigated. The variants included all six IrisPlex SNPs [3], five variants in *OCA2* found to be associated with blue eye colour in individuals with the rs12913832:GA genotype [11], as well as 33 variants in *IRF4*, *TYRP1*, *SLC24A4*, and *TYR*, associated with brown eye colour in individuals with the rs12913832: GG genotype [12]. We included three variants in *TYRP1* (rs201447946, rs74606098, and rs79586719, $r^2 \geq 0.91$) to tag one large haploblock associated with eye colour [12]. Two variants in *TYR* were also in strong LD (rs1126809 and rs1393350,

r^2 = 0.92). rs1393350 was part of the IrisPlex [**3**], and rs1126809 was found to be of importance in individuals with brown eye colour [**12**]. The 757 individuals in the variant discovery data set were previously typed for the IrisPlex SNPs and the five *OCA2* variants [**10,11**]. The 33 variants in *IRF4*, *TYRP1*, *SLC24A4*, and *TYR* were typed using two multiplexes, a 24 plex and an 11 plex, respectively. The 24 plex included rs12913832 and 23 variants in *SLC24A4* and *TYRP1* (rs10131374, rs12590749, rs12880508, rs12894551, rs17128288, rs17128324, rs34755843, rs35617057, rs4904887, rs4904891, rs4904897, rs4904927, rs59977926, rs7144273, rs7152962, rs7401792, rs10491745, rs1408799, rs201447946, rs62538950, rs62538956, rs74606098, and rs79586719). The 11 plex comprised rs1393350 from the IrisPlex and 10 variants in *IRF4* and *TYR* (rs1393350, rs1050976, rs10530949, rs12211228, rs9378807, rs11018509, rs1126809, rs2047512, rs34749698, rs7120151, and rs9919559). Samples were typed with the iPLEX™ Assay (Agena Bioscience, Hamburg, Germany) and analysed with matrix-assisted laser desorption-ionization time of flight mass spectrometry (MALDI-TOF MS) using the MassARRAY Analyzer 4 System (Agena Bioscience, Hamburg, Germany). The PCRs contained 0.5 μL 10X PCR buffer, 0.4 μL MgCl25) 2 mM), 0.1 μL dNTP mix (25 mM), 0.2 μL PCR Enzyme, 1 μL primer mix (1 μM of each primer), and 1 μL DNA (≥1 ng). Thermal cycling comprised initial denaturation for 2 min at 94 °C followed by 45 cycles of denaturation at 94 °C for 20 s, annealing at 56–58 °C for 30 s (56 °C for the 11 plex and 58 °C for the 24 plex), and extension at 72 °C for 1 min. A final extension step at 72 °C for 3 min was included. Shrimp alkaline phosphate (SAP) treatment, single-base extension (SBE) reactions, and preparation for mass analysis was carried out following the manufacturer's recommendations. In the MassARRAYNanodispenser RS1000, 5–15 nl sample was robotically spotted onto a SpectroCHIP® II (Agena Bioscience, Hamburg, Germany). Mass analysis was carried out using the MassARRAY Analyzer 4, and the mass spectra were analysed with the Typer 4.0 software (Agena Bioscience, Hamburg, Germany). The Hardy–Weinberg equilibrium (HWE) and pairwise r^2 values for linkage disequilibrium (LD) were calculated with HaploView version 4.2 [**16**].

2.4. Selection of Variants for Eye Colour Prediction Model (Discovery Data Set)

Eye colour was considered in three different reporting systems, the quantitative system (PIE-score), the two-category system, and the three-category system. For numerical reasons (statistical modelling and stability), the PIE-score (values from -1 to 1) in the quantitative system was transformed to resemble unbounded values. The transformation had an inverse, such that any real number could be transformed into a PIE-score. The PIE-score, r, was transformed by:

$$y = f(r) = \text{logit}\,(0.5 + r^{*}0.499)$$

Genotypes were coded as 0, 1, or 2 according to the number of minor alleles, where the minor alleles were determined based on the allele frequencies in all samples. Due to low allele frequency, the two variants rs121918166 and rs74653330 were combined [**11**]. rs12913832 was considered dominant with AA and AG as 0 and GG as 1. For stability, the discovery data set (757 individuals) was randomly divided into a training set (2/3) and a test set (1/3). This was repeated 100 times. If a training set resulted in fixed variants (i.e., only one genotype observed at a base position), a new set was randomly selected. In the quantitative system (transformed PIE-score) and in the two-category system, data were analysed with three different mathematical models: (i) LASSO model with main effects [**17,18**]; (ii) LASSO model (with the distribution family depending on the system; Gaussian for the quantitative system and binomial for the two-category system) with main effects and all pairwise interactions between rs12913832 and all other variants (still obeying the hierarchical principle such that the main effects must also be included); and (iii) a regression tree [**19**]. In the three-category system, data were analysed with a classification tree [**19**]. All seven mathematical models were fitted on the training set, which was also used to compute variable importance of the top 20 variables (variants). The test data were used for testing and estimating the test error. For LASSO regressions, the variables were standardised (such that all had a standard deviation of (1)), and the tuning parameter was chosen by cross-validation with 10 folds to get the most regularised model with an error within one standard error of the minimum error amongst the folds. For LASSO regressions, the absolute value of the estimated effect was used as variable importance. For regression and classification trees, the variable importances provided by the R package *rpart* version 15–4.1 were used [**19**]. Variable importances were standardised by dividing the importance of each variant with the sum of importances within each model. The mean variable importance (across all models) was used to rank variables (i.e., the variants). The top 12 performing variants were selected for an eye colour prediction model.

2.5. Variant Typing (Model Data Set)

The 12 selected variants and two non-selected SNPs from the IrisPlex [**3**] were typed with SNaPshot (rs12913832, rs12896399, rs16891982, rs1800407, rs10131374, rs1126809, rs121918166, rs1408799, rs1800401, rs4904927, rs7120151, rs74653330, rs12203592, and rs1393350) in the model data set comprising 523 Norwegian individuals. DNA was amplified in one multiplex reaction (14-plex) using the Qiagen Multiplex PCR kit (Qiagen, Hilden, Germany) in a final reaction volume of 10 µL. The PCR comprised denaturation for 15 min at 95 °C followed by 35 cycles of 94 °C for 30 s, 58 °C for 30 s, 72 °C for 30 s, and a final extension step of 72 °C for 10 min. A total of 5 µL amplified DNA/PCR product was treated with 2 µL

ExoSAP-IT™ (Thermo Fisher Scientific, Waltham, MA, USA) for 60 min at 37 °C and 15 min at 75 °C. Single-base extension was performed with 1 μL cleaned PCR product, 2 μL SNaPshot™ Multiplex Ready Reaction Mix, 1 μL nuclease-free water (G-Biosciences), and 1 μL SBE primer mix. Thermal cycling comprised 30 cycles of 96 °C for 10 s, 55 °C for 5 s, and 60 °C for 30 s. SBE products were treated with 1 μL SAP (Thermo Fisher Scientific, Waltham, MA, USA) for 60 min at 37 °C and 15 min at 75 °C. Separation and detection of SBE products were carried out on an ABI 3500 Genetic Analyser (Thermo Fisher Scientific, Waltham, MA, USA) using the FragmentAnalysis36_POPxl run module (POP-4™ polymer, 36 cm capillary, Dye Set E5). Capillary electrophoresis was performed with 1 μL SAP-treated SBE products and 20 μL Hi-Di formamide mixed with GeneScan™-120 LIZ® Size Standard (200:1). The results were analysed with GeneMapper® ID-X v.1.5 (Thermo Fisher Scientific, Waltham, MA, USA). One of the selected variants, rs7120151, could not be typed and was excluded. Hence, we used 11 selected variants for the eye colour prediction model (EC11).

2.6. Eye Colour Prediction Modelling (Model Data Set)

Eye color prediction models were created by the application of leave-one-out cross-validation (LOOCV) utilizing a dataset containing 523 individuals from Norway and the selected genetic variants. During the ith observation, all observations except the ith one were used for training the model. The model was used to predict the eye colour of observation *i*, and the predicted and observed values were compared. Three different reporting systems were used (quantitative system, two-category system, and three-category system), and thereby three different ways of measuring the prediction error were employed. For the quantitative system, a linear regression model was used where the prediction error was the mean squared error. For the two-category system, a logistic regression model was used where blue was chosen as 1 and brown as 0 (without loss of generality). The prediction error for a predicted probability (p) was log(p) if the true eye colour was blue, and log(1-p) if the true eye colour was brown. For the three-category system, a multinomial logistic regression model was used [**20**]. The prediction error was the Kullback–Leibler divergence between the observed distribution (the observed eye colour has probability 1, and the other two categories probability 0) and the estimated probabilities. All prediction errors were out-of-sample prediction errors. For each reporting system, the modelling was performed with three different variant (SNP) sets: the six IrisPlex SNPs, the 11 SNPs selected in this study (EC11), and rs12913832 alone. This resulted in a total of nine models. Lastly, we used the IrisPlex online tool (model called IrisPlex Web) (**https://hirisplex.erasmusmc.nl/**, accessed 1 July 2020) to predict eye colour in three categories.

3. RESULTS

3.1. Allele Frequencies of 44 Variants in the Discovery Data Set

In our discovery data set of 757 people, we examined 44 different pigmentary variants. In this study, we utilized single-base extension to type 33 variations [12] in two multiplexes. Previous studies have documented the typing of eleven variations [11,13]. The allele frequencies of the 44 variants are shown in **Table 2.** The allele frequencies were similar in the Danish, Swedish, and Italian populations with small discrepancies between the Scandinavian (Danish and Swedish) and Italian populations, especially in rs12913832 and rs1800407. Moreover, two variants, rs12913832 and rs16891982, deviated significantly from the Hardy–Weinberg equilibrium (HWE) (p-value < 0.001) in the discovery data set. This could be explained by positive selection in the European population or a lack of random mating between the Scandinavian (Danish and Swedish) and the Italian populations.

Table 2. Allele frequencies of 44 variants typed in the discovery data set and 13 variants typed in the model data set.

Gene	Variant [1]	Reference Allele	Variant Allele	Variant Allele Frequency	
				Discovery Data Set (n = 757)	Model Data Set (n = 523)
HERC2	**rs12913832 ***	A	G	0.74	0.83
IRF4	**rs1050976**	C	T	0.44	
IRF4	rs10530949	TCT	-	0.43	
IRF4	rs12211228	G	C	0.14	
IRF4	rs9378807	C	T	0.49	
TYR	rs11018509	T	A	0.29	
TYR	**rs1126809**	G	A	0.24	0.24
TYR	**rs1393350 ***	G	A	0.23	0.23
TYR	**rs2047512**	T	C	0.35	
TYR	rs34749698	T	C	0.23	
TYR	rs7120151 [2]	A	G	0.74	NA
TYR	rs9919559	T	C	0.33	
SLC24A4	**rs10131374**	G	A	0.14	0.15
TYRP1	**rs10491745**	T	C	0.82	
SLC24A4	rs12590749	C	A	0.37	
SLC24A4	rs12880508	C	T	0.74	
SLC24A4	rs12894551	T	C	0.65	
TYRP1	**rs1408799**	T	C	0.68	0.69

Gene	Variant [1]	Reference Allele	Variant Allele	Variant Allele Frequency	
				Discovery Data Set (n = 757)	Model Data Set (n = 523)
SLC24A4	**rs17128288**	A	G	0.30	
SLC24A4	rs17128324	C	T	0.17	
TYRP1	rs201447946	T	TA	0.06	
SLC24A4	rs34755843	CGACTCT	-	0.16	
SLC24A4	rs35617057	G	T	0.41	
SLC24A4	rs4904887	C	G	0.36	
SLC24A4	rs4904891	G	C	0.35	
SLC24A4	rs4904897	C	T	0.22	
SLC24A4	**rs4904927**	A	G	0.87	0.89
SLC24A4	**rs59977926**	T	C	0.18	
TYRP1	rs62538950	A	T	0.10	
TYRP1	rs62538956	T	C	0.11	
SLC24A4	rs7144273	C	T	0.49	
SLC24A4	rs7152962	G	A	0.23	
SLC24A4	rs7401792	G	A	0.62	
TYRP1	rs74606098	C	T	0.06	
TYRP1	rs79586719	G	A	0.06	
OCA2	rs1800407 *	C	T	0.07	0.03
IRF4	**rs12203592** *	C	T	0.08	0.09
SLC24A4	**rs12896399** *	G	T	0.49	0.52
SLC45A2	**rs16891982** *	C	G	0.93	0.95
OCA2	**rs1800401**	G	A	0.04	0.05
OCA2	**rs1800414**	T	C	<0.01	
OCA2	rs62008729	C	T	0.09	
OCA2	**rs121918166** [2]	C	T	<0.01	0.01
OCA2	**rs74653330** [2]	C	T	<0.01	0.02

[1] Variants in bold were part of the EC11 SNPs and typed in the model data set. rs7120151 was not included in the final prediction modelling. [2] The combined frequency of rs121918166 and rs74653330 was 0.01 in the discovery data set. * Part of the IrisPlex prediction model [3]. NA: not analysed.

3.2. Selection of Variants for Eye Colour Prediction (Discovery Data Set)

The discovery data set with information on eye colour and the 44 pigmentary variants was analysed with seven different mathematical models. Variants were ranked according to the mean variable importance across the mathematical models (**Table 3**). rs12913832 was the top-performing SNP in all seven mathematical models and ranked number one (**Table 3**). The mean variable importance for rs12913832 was 74.6% (**Table 3**). rs121918166 and rs74653330 were combined as one variable and ranked second (**Table 3**). We saw a drop in mean variable importance after rank 11. Hence, we selected the top 11 performing variables (comprising 12 variants) for a new eye colour SNP set (**Table 3**). These variants had mean variable importances of at least 0.3% (**Table 3**). Four of six variants in the IrisPlex assay [**3**] were among the selected variants (rs12913832, rs16891982, rs1800407, and rs12896399), whereas the *TYR* variant rs1393350 (ranked 18) and the *IRF4* variant rs12203592 (not in top 20) from the IrisPlex were not selected.

Table 3. Twelve variants selected for the EC11 SNP set.

Rank	Gene	Variant	Mean Variable Importance
1	*HERC2*	rs12913832 *	74.63%
2	*OCA2*	rs121918166 + rs74653330	8.54%
3	*SLC45A2*	rs16891982 *	6.23%
4	*OCA2*	rs1800407 *	5.26%
5	*TYRP1*	rs1408799	1.54%
6	*SLC24A4*	rs4904927	1.19%
7	*SLC24A4*	rs12896399 *	0.68%
8	*TYR*	rs1126809	0.47%
9	*TYR*	rs7120151 [1]	0.46%
10	*SLC24A4*	rs10131374	0.32%
11	*OCA2*	rs1800401	0.32%

[1] Variant not included in the final prediction modelling. * Part of the IrisPlex prediction model [**3**].

3.3. Typing of Selected Variants and IrisPlex SNPs (Model Data Set)

The 12 selected variants (SNPs) were typed with single-base extension in an independent data set, the model data set of 523 Norwegians (bold in **Table 2** and **Table 3**). To enable comparison with the IrisPlex, we included two SNPs from the IrisPlex assay, rs12203592 and rs1393350, in a multiplex comprising 14 SNPs. One

SNP, rs7120151, that was ranked as number 9 (**Table 3**), was excluded due to poor amplification of the A allele. We obtained 523 complete profiles, including the 11 selected SNPs (EC11) (**Table 2**) and the two additional IrisPlex SNPs.

3.4. Eye Colour Prediction Models with EC11, IrisPlex SNPs, and rs12913832

We developed nine distinct models for predicting eye color using a dataset of 523 individuals from Norway. This was achieved by employing three different reporting systems and three different sets of single nucleotide polymorphisms (SNPs), as shown in Table 2 and Table 3. We also used the IrisPlex Web for eye colour prediction using the three-category system. Prediction errors for each eye colour prediction model are presented in **Table 4**. Since we used three different reporting systems, the models within each system have their own measure of prediction error. Therefore, the performance of SNP sets (including prediction errors) can only be directly compared within the same but not across different reporting systems. The EC11 models had the smallest error under all reporting systems, followed by IrisPlex Norway and rs12913832. Under the three-category system, the IrisPlex Web resulted in the highest prediction error (**Table 4**).

Table 4. Prediction errors for the nine eye colour prediction models (three reporting systems modelled with three SNP sets) and the IrisPlex online tool (IrisPlex Web).

Eye Colour Prediction Model	Quantitative System [1]	Two-Category System [2]	Three-Category System [3]
EC11	5.07	0.26	0.59
IrisPlex Norway	5.90	0.30	0.66
rs12913832	6.96	0.32	0.69
IrisPlex Web *	NA	NA	0.80

[1] Prediction error is the mean squared error. [2] Prediction error for a predicted probability, p, is log(p) if the true eye colour was blue, and log(-1p) if the true eye colour was brown. [3] Prediction error is the Kullback–Leibler divergence. * The IrisPlex Web predicts eye colour according to a three-category system. NA: not analysed

The sensitivity and specificity of eye colour prediction models in the two-category and the three-category reporting systems were determined without applying a probability threshold (pmax) (**Table 5** and **Table 6**). Hence, the predicted eye colour was the eye colour with the highest probability value. The rs12913832 and IrisPlex Norway models showed the same sensitivity and specificity (0.92 and 0.84, respectively) in the two-category system. The EC11 was slightly more sensitive (0.96), and in turn, slightly less specific (0.82) (**Table 5** and **Table S5**).

Table 5. Sensitivity and specificity of eye colour prediction models in the two-category reporting system modelled with three SNP sets. No probability threshold was applied (pmax).

Two-Category System	Sensitivity [1]	Specificity [1]
rs12913832	0.92	0.84
IrisPlex Norway	0.92	0.84
EC11	0.96	0.82
[1] Reference is blue eye colour.		

Table 6. Sensitivity and specificity of eye colour prediction models in the three-category reporting system modelled with three SNP sets and the IrisPlex Web model. No probability threshold was applied (pmax).

Three-Category System		Sensitivity [1]	Specificity [1]
rs12913832	Blue	0.95	0.61
	Intermediate	0.00	1.00
	Brown	0.95	0.87
IrisPlex Norway Blue 0.94			0.61
	Intermediate	0.10	0.97
	Brown	0.86	0.90
EC11	Blue	0.95	0.59
	Intermediate	0.15	0.95
	Brown	0.88	0.96
IrisPlex Web	Blue	0.95	0.60
	Intermediate	0.00	1.00
	Brown	0.95	0.88
[1] Reference is blue eye colour.			

In the three-category system, the sensitivity was highest for blue and brown eye colours and lowest for intermediate eye colour with all three SNP sets and the IrisPlex Web (**Table 6**). The rs12913832 and IrisPlex Web predictions resulted in the highest sensitivity for brown eye colour: 0.95, whereas the IrisPlex Norway and EC11 models had slightly lower sensitivities (**Table 6**). For blue eye colour, the sensitivities were similar for all models. No individuals were predicted to have intermediate eye colour with either rs12913832 or the IrisPlex Web. Hence, the sensitivity was 0, and the specificity was 1 (**Table 6**). Of the individuals with intermediate eye colours, 69% and 72% were incorrectly predicted to have blue eye colours with the two models, respectively. In contrast, intermediate eye colour predictions were obtained with IrisPlex Norway and EC11. However, only 48% and 46% of the predictions were correct. Thus, the sensitivity was low (0.10 and 0.15, respectively), and the specificity was high (0.97 and 0.95, respectively) (**Table 6**).

Figure 1 shows the percentages of correct, incorrect, and inconclusive predictions for prediction models in the two-category reporting system. Using pmax, %89 of the predictions with both rs12913832 and IrisPlex Norway were correct (**Figure 1**). The EC11 model resulted in 92% correct predictions (**Figure 1**). Here, 93% of the blue eye colour predictions and 90% of the brown eye colour predictions were correct. We also evaluated the prediction with a probability threshold of 0.7 in the two-category reporting system (**Figure 1**). If the highest prediction value was below 0.7, the prediction was defined as inconclusive. No eye colours were inconclusive with rs12913832 (**Figure 1**). The IrisPlex Norway and EC11 resulted in 9% and 5% inconclusive predictions, respectively (**Figure 1**). With the IrisPlex Norway, 62% of the inconclusive eye colours were brown according to the PIE score. With EC11, it was only 46% .

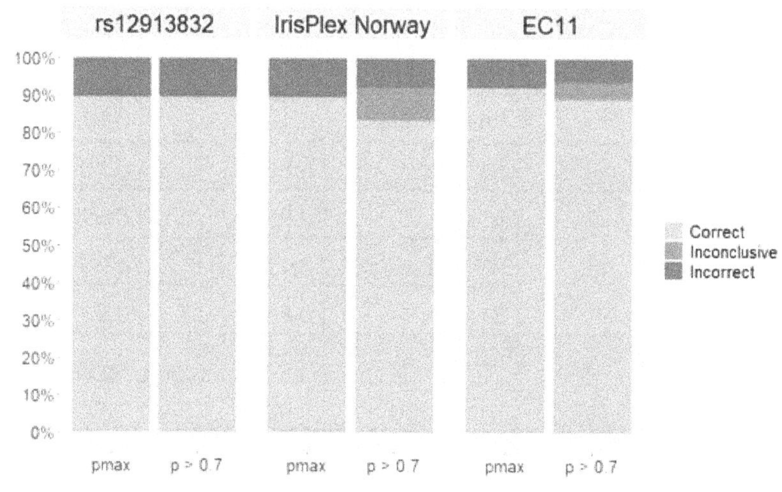

Figure 1. Performance of eye colour prediction models in the two-category reporting system modelled with three SNP sets: rs12913832, IrisPlex Norway, and EC11. Bars represent the percentage of correct, incorrect, and inconclusive predictions with no probability threshold (pmax) and a probability threshold of 0.7 (p > 0.7).

In the three-category reporting system, we tested the prediction with pmax, a probability threshold of 0.5, and a probability threshold of 0.7 (**Figure 2**). With no probability threshold, the rs12913832, IrisPlex Norway, and IrisPlex Web models all resulted in 72% correct predictions (**Figure 2**). The EC11 model resulted in a slightly higher number of correct predictions (75%) (**Figure 2**). When applying a probability threshold of 0.5, predictions with rs12913832 were unchanged compared with predictions without probability threshold (**Figure 2**). Predictions with the IrisPlex Web tool were also similar though 3% of the total predictions were inconclusive. There was a slight decrease in the number of correct and incorrect predictions with EC11 and IrisPlex Norway as both models resulted in 2% inconclusive predictions (**Figure 2**). When applying a probability threshold of 0.7, blue eye colour was correctly predicted in 95% of the blue-eyed individuals with rs12913832, but no individuals were predicted to have brown eye colours. Thus, the total number of correct predictions with rs12913832 was only 53%. The number of correct predictions using the 0.7 probability threshold was highest with IrisPlex Web (**Figure 2**). However, the IrisPlex Web also resulted in the highest percentage of incorrect predictions (20%) and resulted in 15% inconclusive predictions (**Figure 2**). The IrisPlex Norway model resulted in only 8% incorrect predictions but a high number of inconclusive predictions (51%) (**Figure 2**). Prediction with EC11 resulted in 12% incorrect predictions, 32% inconclusive predictions, and only 54% correct predictions (**Figure 2**). Of the 32% inconclusive eye colour predictions, 49% had blue eye colour, 39% had intermediate eye colour, and only 12% had brown eye colour based on the PIE-score. Especially the percentage of brown-eyed individuals with inconclusive predictions was much lower than compared with rs12913832, IrisPlex Norway, and IrisPlex Web. For these models, the percentages were 32–65%.

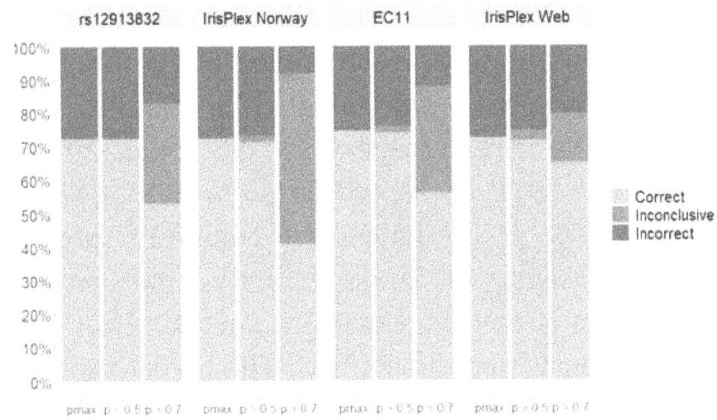

Figure 2. Performance of eye colour prediction models in the three-category reporting system modelled with three SNP sets: rs12913832, IrisPlex Norway and EC11, as well as performance of the IrisPlex Web prediction model. Bars represent the percentage of correct, incorrect, and inconclusive predictions with no probability threshold (pmax), probability threshold of 0.5 (p > 0.5), and probability threshold of 0.7 (p > 0.7).

4. DISCUSSION

For this research, we specifically chose 11 SNPs (EC11) to predict eye color. We then created new models to predict eye color using a two-category and a three-category method. These models outperformed the IrisPlex and rs12913832 prediction models that were previously used. The 11 SNPs in EC11 were selected from a group of 44 pigmentary variants that were originally identified in eye colour association studies [**3**,**6**] and from detailed sequence analyses of individuals with eye colours that deviated from the expected eye colour based on the genotype of rs12913832 [**11**,**12**]. The 44 variants were typed in 757 Europeans whose eye colours were quantitatively determined. Seven different mathematical models were used to rank the variants according to informativeness, and all variants with more than 0.3% mean variable importance were selected. Four of the six SNPs in the IrisPlex assay [**3**] were included in EC11. However, the *TYR* SNP, rs1393350 (ranked 18), was replaced by another *TYR* SNP, rs1126809 (ranked 8), and the *IRF4* gene represented by the SNP rs12203592 (not in top 20) from the IrisPlex, was not included in EC11. For the selection of SNPs, we combined individuals of Scandinavian (Danish and Swedish) and Italian descent and treated these as one population. We are aware that we may had selected different SNPs if the selection was performed solely on either the Scandinavian or the South European population. We typed the EC11 SNPs and the two additional IrisPlex SNPs in an independent data set of 523 Norwegians whose eye colours were determined with the same quantitative method as the 757 individuals in the discovery data set. We modelled nine different eye colour prediction models on the Norwegian population using the LOOCV method. Each eye colour prediction model consisted of a combination of one of three reporting systems: the quantitative system (prediction of PIE score), the two-category system (blue and brown), and the three-category system (blue, intermediate, brown), and one of three SNP sets: EC11, IrisPlex SNPs, and rs12913832. We also evaluated the IrisPlex Web model for prediction of eye colour in three categories. After examining the error rates, sensitivity, and specificity of various eye color prediction models, three key factors should be taken into account before introducing eye color prediction in a forensic genetic context: The three components are: (1) the population used as a reference, (2) the list of single nucleotide polymorphisms (SNPs), and (3) the approach employed for reporting.

4.1. The Reference Population

The rs12913832, IrisPlex Norway, and IrisPlex Web models showed almost identical results in the three-category system (**Figure 2**, **Table 6**). Nevertheless, a detailed comparison between the three-category IrisPlex Norway and the IrisPlex Web models highlights the importance of the reference population. The two models were based on the exact same SNPs. However, the IrisPlex Norway model was developed on the Norwegian population, which was the intended target population, whereas the IrisPlex Web model was developed on 9466 individuals of primarily European descent [**6,21,22**]. The two models resulted in the same number of correct predictions (72%) when no probability threshold was applied. With probability thresholds, inconclusive results were possible, and the IrisPlex Web model resulted in a higher number of correct predictions than the IrisPlex Norway model (**Figure 2**). However, when applying the recommended threshold for the IrisPlex Web model ($p > 0.7$) [**21**], the number of incorrect predictions was also higher (**Figure 2**). The two models showed similar sensitivities and specificities for blue and brown eye colour but differed for the intermediate eye colour category (**Table 6**). A total of 123 Norwegian individuals had intermediate eye colours according to the PIE-score (**Table 1**). Intermediate eye colour predictions were obtained with the IrisPlex Norway model using both pmax and $p > 0.5$, whereas no individuals were predicted to have intermediate eye colours with the IrisPlex Web model (**Table S6**). Hence, the IrisPlex Norway model showed an overall lower prediction error than the IrisPlex Web model (**Table 4**), and this emphasises the importance of modelling a prediction model on the appropriate reference population. This is in agreement with previous evaluations of the IrisPlex model which showed that prediction models based on the intended target population (the reference population) performed better than the IrisPlex Web tool [**23,24**]. However, it is important to note that the differences between the IrisPlex Web and the IrisPlex Norway models may not only be due to the different reference populations. Different strategies on phenotyping and categorisation of eye colour may also have contributed here. In our study, we determined the eye colour quantitatively, and the eye colour prediction models were modelled accordingly. For the IrisPlex Web model, the eye colour was evaluated by a medical researcher who categorised eye colours in blue, brown, and non-blue/non-brown (called intermediate) categories [**3,6,21,22**].

4.2. The SNP Set

The eye color prediction models utilizing EC11 demonstrated the most accurate prediction error rates across all three reporting systems. These models consistently outperformed the rs12913832, IrisPlex Norway, and IrisPlex Web models in the Norwegian population, as shown in Table 4. In the three-category reporting system, only EC11 and IrisPlex Norway were able to predict intermediate eye colours (**Table 6**). For the EC11 model, intermediate eye colour predictions were even obtained with a probability threshold of 0.7. We did expect the prediction errors to decrease when the number of loci increased. However, the prediction errors obtained with EC11 (11 SNPs) were only slightly smaller than with IrisPlex Norway (six SNPs), closely followed by rs12913832 (one SNP) and IrisPlex Web (six SNPs) (**Table 4**). This shows that a single SNP, rs12913238, may be sufficient for prediction of eye colour.

This SNP was ranked as number one across all mathematical models with a mean variable importance of 74.6% (**Table 3**). We modelled rs12913832 in a dominant matter and acknowledge that it is unreasonable to predict eye colour in three categories with a predictor variable containing only two levels (AA/AG and GG). However, even in the three-category reporting system, prediction with rs12913832 showed a lower prediction error than prediction with the IrisPlex Web (**Table 4**). In both the three-category reporting system and the two-category reporting system, prediction with rs12913832 and IrisPlex Norway performed almost identically (**Figure 1**, **Figure 2**). This stresses the importance of rs12913832 for prediction of eye colour and shows that the remaining five SNPs in the IrisPlex SNP set have very small effects on the outcome of the eye colour prediction in the studied population.

4.3. The Reporting Strategy

Within this study, we conducted experiments to evaluate three distinct methods for categorizing eye color: the quantitative system (which predicts the PIE score), the two-category system (classifying eyes as either blue or brown), and the three-category system (classifying eyes as blue, intermediate, or brown). While it is feasible to express the projected eye color as a PIE-score, it is probable that either the end-user or the reporting laboratory would need to convert the PIE-score into a specific eye color category. Therefore, the use of a quantitative system for reporting is not relevant in a forensic genetic setting. The difference between the two- and three-category system is the definition of the intermediate eye colour, which is very difficult to perceive. When multiple individuals were asked to evaluate eye colourscategorised as intermediate, they often disagreed, whereas they agreed much more frequently when eye colours were blue or brown [**15**]. Intermediate eye colour is predicted as the most likely eye colour (without any probability threshold) for only 8% (60 out of 729) and 14% (7.508 out of 52,488) of the possible genotype combinations in the IrisPlex Web and EC11 models, respectively [**10**] . For the IrisPlex Web, the maximum probability value for prediction of intermediate eye colour is 0.62 [**10**]. With EC11, it is 0.94. In this study, we did observe intermediate eye colour predictions with EC11 with high probability values (maximum: 0.82). However, the intermediate eye colour predictions were incorrect more than half of the time. Overall, eye colour prediction in two categories resulted in more correct predictions than eye colour prediction in three categories (92% vs. 75% for EC11; 89% vs. 72% for IrisPlex Norway and 89% vs. 72% for rs12913832) (**Figure 1** and **Figure 2**). Hence, the definition of an intermediate eye colour category is counterintuitive, as it is both difficult to identify and predict. A recent study discusses the need for standardised methods for reporting forensic DNA phenotyping predictions to the police [**25**]. Reducing the complexity of eye colour predictions to only two categories results in only two hypotheses (H_1: The person has brown eyes and H_2:

The person has blue eyes). Hence, it is possible to report the weight of the evidence with a single likelihood ratio, which resembles standard STR-profiling reports. The likelihood ratio could be supplemented with picture examples of eye colours represented by each category. This may overcome any misunderstandings or subjective opinions of eye colour interpretation, especially for eye colours that may appear non-blue and non-brown.

4.4. DNA Phenotyping in Forensic Genetics

The Section of Forensic Genetics in Denmark has just been providing eye color prediction to the police using a two-category method that relies on the genotype of rs12913832. The anticipation of EVCs can give rise to ethical considerations, as detailed in reference [26]. This is especially apparent if the genetic markers used for prediction a certain trait are also linked to diseases [**26**]. That is not the case for rs12913832. The SNP is included in the Precision ID Ancestry Panel (Thermo Fisher Scientific, Waltham, MA, USA), which has already been validated for case work [**5,27,28**]. The weight of the evidence for both ancestry and eye colour predictions are reported as likelihood ratios.

For eye colour predictions in the Danish population, LR = (rs12913832:GG|H_1/ rs12913832:GG|H_2) = 0.1, LR = (rs12913832:AG|H_1/rs12913832:AG|H_2) = 19, and LR = (rs12913832:AA|H_1/rs12913832:AA|H_2) = 54 [**15**], where H_1: The person has brown eyes and H_2: The person has blue eyes. The EC11 model may be implemented at the Section of Forensic Genetics in Denmark in the future once the EC11 markers are included in a validated massively parallel sequencing (MPS) assay. The most important shortcoming of the rs12913832 two-category prediction model is the lack of information gained when including the two *OCA2* variants, rs121918166 and rs74653330. These variants were previously shown to be of importance for blue eye colour in Scandinavians with the rs12913832:AG genotype [**11**]. The variants had low frequencies in Danes and Swedes and were completely absent in Italians [**11**]. These variants combined were ranked as second most important, with a mean variable importance of 8.5% (**Table 3**). In the Norwegian data set, 30 individuals with the rs12913832:AG genotype had blue eye colours according to the PIE-score. Seventeen of the 30 individuals had at least one of the *OCA2* variants and were correctly predicted to have blue eyes using the EC11 two-category model.

By contrast, only one of the 30 individuals was correctly predicted with the IrisPlex Norway model, and none were correctly predicted with rs12913832. Moreover, two Norwegians with the rs12913832:AA genotype had blue eye colours according to the PIE-score. One of the two individuals was correctly predicted to have blue eye colour with the EC11 two-category model. This individual was homozygous for the rs74653330 variant. We hypothesise that the second individual has other variants in or around the *OCA2* gene that could explain the formation of blue eye colour in the rs12913832:AA genotype background. Both individuals were incorrectly predicted to have brown eye colours with the IrisPlex Norway and rs12913832 models. Lastly, 10 Norwegians with the rs12913832:GG genotype had brown eyes according to the PIE-score. Only one of them was correctly predicted to have brown eyes with the two-category EC11 and IrisPlex Norway models. This shows that we do not fully understand the formation of brown eye colour in rs12913832:GG individuals and that the EC11 model may have to be expanded further.

5. CONCISE SUMMARY

Providing a detailed account of the eye color of a suspect can be a crucial lead in forensic investigations if there are no obvious individuals under suspicion. In this study, we introduce 11 single nucleotide polymorphisms (SNPs) known as Eye Colour 11-EC11, which play a crucial role in predicting eye colour. We also develop eye colour prediction models for both a two-category reporting system (blue and brown) and a three-category system (blue, intermediate, and brown). The EC11 single nucleotide polymorphisms (SNPs) were meticulously chosen from a pool of 44 variations in pigment-related genes.

These genes were previously identified as being linked to eye color in a sample of 757 individuals of European descent, specifically from Denmark, Sweden, and Italy. The variants were ranked using three distinct reporting systems: a quantitative technique known as PIE-score, a two-category system with blue and brown categories, and a three-category system with blue, intermediate, and brown categories. SNPs that had a mean variable importance above 0.3% were chosen for EC11. Eye color prediction models were created using the EC11 SNPs and evaluated using leave-one-out cross-validation (LOOCV) on a separate dataset of 523 Norwegian people. The performance of the EC11 models was compared to models utilizing the IrisPlex SNPs and the key eye color locus, rs12913832, for both the two-category and three-category systems. We additionally assessed the model's performances by utilizing the IrisPlex online tool, known as IrisPlex Web. The EC11 eye colour prediction models outperformed the IrisPlex and rs12913832 models in all reporting systems and were superior to the IrisPlex Web in the three-category scheme. The text discusses three crucial factors that should be taken into account when introducing eye color prediction in a forensic genetic context. The three components are: (1) the population used as a reference, (2) the list of single nucleotide polymorphisms (SNPs), and (3) the approach used for reporting the results.

REFERENCES

1. Walsh, S.; Liu, F.; Wollstein, A.; Kovatsi, L.; Ralf, A.; Kosiniak-Kamysz, A.; Branicki, W.; Kayser, M. The HIrisPlex system for simultaneous prediction of hair and eye colour from DNA. *Forensic Sci. Int. Genet.* **2013**, *7*, 98–115. **[Google Scholar] [CrossRef] [PubMed] [Green Version]**
2. Chaitanya, L.; Breslin, K.; Zuñiga, S.; Wirken, L.; Pośpiech, E.; Kukla-Bartoszek, M.; Sijen, T.; de Knijff, P.; Liu, F.; Branicki, W.; et al. The HIrisPlex-S system for eye, hair and skin colour prediction from DNA: Introduction and forensic developmental validation. *Forensic Sci. Int. Genet.* **2018**, *35*, 123–135. **[Google Scholar] [CrossRef] [PubMed] [Green Version]**
3. Walsh, S.; Liu, F.; Ballantyne, K.N.; van Oven, M.; Lao, O.; Kayser, M. IrisPlex: A sensitive DNA tool for accurate prediction of blue and brown eye colour in the absence of ancestry information. *Forensic Sci. Int. Genet.* **2011**, *5*, 170–180. **[Google Scholar] [CrossRef] [PubMed]**
4. Ruiz, Y.; Phillips, C.; Gomez-Tato, A.; Alvarez-Dios, J.; Casares de Cal, M.; Cruz, R.; Maroñas, O.; Söchtig, J.; Fondevila, M.; Rodriguez-Cid, M.; et al. Further development of forensic eye color predictive tests. *Forensic Sci. Int. Genet.* **2013**, *7*, 28–40. **[Google Scholar] [CrossRef] [PubMed]**
5. Mogensen, H.S.; Tvedebrink, T.; Børsting, C.; Pereira, V.; Morling, N. Ancestry prediction efficiency of the software GenoGeographer using a z-score method and the ancestry informative markers in the Precision ID Ancestry Panel. *Forensic Sci. Int. Genet.* **2020**, *44*, 102154. **[Google Scholar] [CrossRef]**
6. Liu, F.; van Duijn, K.; Vingerling, J.R.; Hofman, A.; Uitterlinden, A.G.; Janssens, A.C.J.; Kayser, M. Eye color and the prediction of complex phenotypes from genotypes. *Curr. Biol.* **2009**, *19*, R192–R193. **[Google Scholar] [CrossRef] [PubMed] [Green Version]**
7. Sturm, R.A.; Duffy, D.L.; Zhao, Z.Z.; Leite, F.P.; Stark, M.; Hayward, N.K.; Martin, N.; Montgomery, G.W. A single SNP in an evolutionary conserved region within Intron 86 of the HERC2 gene determines human blue-brown eye color. *Am. J. Hum. Genet.* **2008**, *82*, 424–431. **[Google Scholar] [CrossRef] [Green Version]**
8. Eiberg, H.; Troelsen, J.T.; Nielsen, M.; Mikkelsen, A.; Mengel-From, J.; Kjaer, K.W.; Hansen, L. Blue eye color in humans may be caused by a perfectly associated founder mutation in a regulatory element located within the HERC2 gene inhibiting OCA2 expression. *Hum. Genet.* **2008**, *123*, 177–187. **[Google Scholar] [CrossRef]**
9. Visser, M.; Kayser, M.; Palstra, R.-J. HERC2 rs12913832 modulates human pigmentation by attenuating chromatin-loop formation between a long-range enhancer and the OCA2 promoter. *Genome Res.* **2012**, *22*, 446–455. **[Google Scholar] [CrossRef] [Green Version]**

10. Pietroni, C.; Andersen, J.D.; Johansen, P.; Andersen, M.M.; Harder, S.; Paulsen, R.R.; Børsting, C.; Morling, N. The effect of gender on eye colour variation in European populations and an evaluation of the IrisPlex prediction model. *Forensic Sci. Int. Genet.* **2014**, *11*, 1–6. **[Google Scholar] [CrossRef] [Green Version]**

11. Andersen, J.D.; Pietroni, C.; Johansen, P.; Andersen, M.M.; Pereira, V.; Børsting, C.; Morling, N. Importance of nonsynonymous OCA 2 variants in human eye color prediction. *Mol. Genet. Genom. Med.* **2016**, *4*, 420–430. **[Google Scholar] [CrossRef] [Green Version]**

12. Meyer, O.S.; Lunn, M.M.B.; Garcia, S.L.; Kjærbye, A.B.; Morling, N.; Børsting, C.; Andersen, J.D. Association between brown eye colour in rs12913832:GG individuals and SNPs in TYR, TYRP1, and SLC24A4. *PLoS ONE* **2020**, *15*, e0239131. **[Google Scholar] [CrossRef] [PubMed]**

13. Andersen, J.D.; Johansen, P.; Harder, S.; Christoffersen, S.R.; Delgado, M.C.; Henriksen, S.T.; Nielsen, M.M.; Sørensen, E.; Ullum, H.; Hansen, T.; et al. Genetic analyses of the human eye colours using a novel objective method for eye colour classification. *Forensic Sci. Int. Genet.* **2013**, *7*, 508–515. **[Google Scholar] [CrossRef]**

14. Salvo, N.M.; Janssen, K.; Kirsebom, M.K.; Meyer, O.S.; Berg, T.; Olsen, G.-H. Forensic DNA phenotyping—Predicting eye and hair colour in a Norwegian population using Verogen'sForenSeqTM DNA Signature Prep Kit. *Forensic Sci. Int. Genet.* **2021**. submitted. **[Google Scholar]**

15. Meyer, O.S.; Børsting, C.; Andersen, J.D. Perception of blue and brown eye colours for forensic DNA phenotyping. *Forensic Sci. Int. Genet. Suppl. Ser.* **2019**, *7*, 476–477. **[Google Scholar] [CrossRef]**

16. Barrett, J.C.; Fry, B.; Maller, J.; Daly, M.J. Haploview: Analysis and visualization of LD and haplotype maps. *Bioinformatics* **2005**, *21*, 263–265. **[Google Scholar] [CrossRef] [PubMed] [Green Version]**

17. Santosa, F.; Symes, W.W. Linear inversion of band-limited reflection seismograms. *SIAM J. Sci. Stat. Comput.* **1986**, *7*, 1307–1330. **[Google Scholar] [CrossRef]**

18. Tibshirani, R. Regression shrinkage and selection via the Lasso. *J. R. Stat. Soc. Ser. B* **1996**, *58*, 267–288. **[Google Scholar] [CrossRef]**

19. Breiman, L.; Friedman, J.H.; Olshen, R.A.; Stone, C.J. *Classification and Regression Trees*; Taylor & Francis: Abingdon, UK, 1984; ISBN 9781351460491. **[Google Scholar]**

20. Venables, W.N.; Ripley, B.D. *Modern Applied Statistics with S*, 4th ed.; Springer: New York, NY, USA, 2002; pp. 435–446. **[Google Scholar]**

21. Walsh, S.; Wollstein, A.; Liu, F.; Chakravarthy, U.; Rahu, M.; Seland, J.H.; Soubrane, G.; Tomazzoli, L.; Topouzis, F.; Vingerling, J.R.; et al. DNA-based eye colour prediction across Europe with the IrisPlex system. *Forensic Sci. Int. Genet.* **2012**, *6*, 330–340. **[Google Scholar] [CrossRef] [PubMed]**

22. Walsh, S.; Chaitanya, L.; Breslin, K.; Muralidharan, C.; Bronikowska, A.; Pospiech, E.; Koller, J.; Kovatsi, L.; Wollstein, A.; Branicki, W.; et al. Global skin colour prediction from DNA. *Qual. Life Res.* **2017**, *136*, 847–863. **[Google Scholar] [CrossRef] [Green Version]**

23. Dembinski, G.M.; Picard, C.J. Evaluation of the IrisPlex DNA-based eye color prediction assay in a United States population. *Forensic Sci. Int. Genet.* **2014**, *9*, 111–117. **[Google Scholar] [CrossRef]**

24. Salvoro, C.; Faccinetto, C.; Zucchelli, L.; Porto, M.; Marino, A.; Occhi, G.; de los Campos, G.; Vazza, G. Performance of four models for eye color prediction in an Italian population sample. *Forensic Sci. Int. Genet.* **2019**, *40*, 192–200. **[Google Scholar] [CrossRef]**

25. Atwood, L.; Raymond, J.; Sears, A.; Bell, M.; Daniel, R. From identification to intelligence: An assessment of the suitability of forensic DNA phenotyping service providers for use in Australian law enforcement casework. *Front. Genet.* **2021**, *11*, 11. **[Google Scholar] [CrossRef]**

26. Kayser, M. Forensic DNA phenotyping: Predicting human appearance from crime scene material for investigative purposes. *Forensic Sci. Int. Genet.* **2015**, *18*, 33–48. **[Google Scholar] [CrossRef] [PubMed]**

27. Pereira, V.; Mogensen, H.S.; Børsting, C.; Morling, N. Evaluation of the precision ID Ancestry Panel for crime case work: A SNP typing assay developed for typing of 165 ancestral informative markers. *Forensic Sci. Int. Genet.* **2017**, *28*, 138–145. **[Google Scholar] [CrossRef] [PubMed]**

28. Tvedebrink, T.; Eriksen, P.S.; Mogensen, H.S.; Morling, N. Weight of the evidence of genetic investigations of ancestry informative markers. *Theor. Popul. Biol.* **2018**, *120*, 1–10. **[Google Scholar] [CrossRef] [PubMed]**

20 ASSESSMENT OF VARIOUS CLEANING METHODS FOR ELIMINATION OF CONTAMINATING DNA MOLECULES

1. INTRODUCTION: AN OVERVIEW

DNA analysis of trace evidence can present many challenges in a forensic investigation, such as samples that contain low quantities of biological material and degraded DNA. However, improved typing kits for forensic analysis may allow STR profiling on DNA extracted from only a few cells [**1**,**2**]. In addition, using mitochondrial DNA (mtDNA) instead of nuclear DNA (nDNA) markers allows testing of even lower DNA input amounts, due to the high copy number of mtDNA in each cell [**3**,**4**]. Another challenge, especially when analyzing mtDNA, is that exogenous DNA can be present in the samples and cause ambiguous or erroneous results. Contaminating DNA may be introduced to a sample by the first responding officers, rescue workers, or crime scene investigators present at the crime scene, archaeologists during excavations or later by laboratory personnel during DNA analysis [**5**,**6**,**7**]. Contamination may also originate from other samples processed at the same laboratory (cross-contamination) [**6**,**8**] or even from disposables and consumables contaminated during manufacturing [**9**,**10**].

Several preventive measures should be taken to reduce the risk of contamination in order to assure a high quality of DNA profiling results. Contamination is monitored by establishing elimination databases with DNA profiles from staff [**11**], periodic wipe tests, and the use of no template control (NTC) reactions [**8**]. Moreover, personal protective clothing, disposable consumables, aerosol resistant tips, and dedicated instruments are used to prevent the introduction of contaminating DNA into the laboratory environment and casework samples. A laboratory construction with designated pre- and post-PCR rooms equipped with dead-air cabinets and a unidirectional workflow is also important. In addition, several decontamination strategies (e.g., chemical cleaning agents and ultraviolet radiation [UV]) can be used for the decontamination of laboratory surfaces and equipment. While decontamination agents have various DNA damaging mechanisms, UV radiation is destructive because it oxidizes bases to introduce single- and double-strand breaks in the DNA molecules [**12,13,14,15,16,17,18**].

This study evaluates the decontamination efficiency of ten different cleaning strategies: ethanol, UV radiation, ethanol in combination with UV radiation, fresh and stored household bleach, DAX Ytdesinfektion Plus, Rely+On™ Virkon®, Trigene®, DNA Remover®, and sodium hypochlorite. Most forensic materials have a majority of the DNA present within cells, but it also exists in free form, so-called cell-free DNA. As cells deteriorate with time and in challenging environments, the cell walls become disrupted, resulting in a release of cell content, including the DNA [**19,20**]. Therefore, both cell-free (extracted DNA) and cell-contained DNA (blood) were placed on three different surfaces (plastic, metal, and wood) in the experiments. The artificially contaminated surfaces were cleaned with the different treatments and then swabbed for residual DNA. Real-time PCR quantification of mtDNA extracted from the collected samples was thereafter used to compare the efficiency of the different cleaning strategies.

2. MATERIALS AND METHODS

Human male DNA in a 10 μL volume (6 ng/μL, PE Biosystems, Foster City, CA, USA) was deposited on different surfaces. The applied DNA solution consisted of 60 ng DNA, with approximately 18 million mtDNA copies, assuming 2000 mtDNA copies per cell. The mtDNA and nuclear DNA (nDNA) ratio for the DNA sample used in this study was estimated previously (by comparison to a plasmid clone of mtDNA) [**21**]. As cell-contained DNA, 10 μL of whole blood (unknown DNA concentration) from one anonymous donor was deposited on several surfaces. The sample solutions were deposited in marked 25 mm-wide circles on plastic, metal, or wood surfaces. The wood consisted of a skirting board with one layer of water-based transparent white base paint, the metal of aluminum foil and the plastic of plastic document folders. These surfaces were chosen to mimic common types of material in surfaces of workbenches and equipment in the laboratory environment. The liquid was spread with a pipette tip within the circle and left to dry for two hours.

The different decontaminating strategies to be evaluated were chosen following an informal survey of the most commonly used treatments among European forensic laboratories. Briefly, the strategies consisted of 70% aqueous ethanol, UV radiation (20 min at a distance of 60 to 70 cm, 254 nm), 70% aqueous ethanol in combination with UV radiation, 15% freshly diluted Klorin Original household bleach (Colgate-Palmolive AB, New York, NY, USA), (3.6% sodium hypochlorite diluted to 0.54%), 15% stored Klorin household bleach (2.7% sodium hypochlorite diluted to 0.4%), DAX Ytdesinfektion Plus, 1% Rely+On™ Virkon®, 10% Trigene® Disinfectant Cleaner, DNA Remover®, and 0.4% freshly diluted sodium hypochlorite.

The commercial cleaning agents were prepared according to the manufacturers' instructions. The stability of diluted sodium hypochlorite for longer periods has been discussed as the concentration of available chlorine decreases with time [22]. Therefore, both diluted household bleach stored in a dark refrigerator at 8 °C for 80 days and freshly made dilutions were evaluated.

The liquid cleaning agents were administered to the artificially contaminated surfaces by one spray from a calibrated spray bottle. The calibration was performed by weighing five replicates of sprayed volumes followed by adjustments when needed. The cleaned areas were wiped in three circular movements in a similar manner and were performed by the same person using dust-free paper. The cleaned areas were thereafter left to dry for 120 min, with the exception of areas cleaned with Trigene®, which was sprayed with a single spray of water before wiping, and then left for 10 min. The entire marked area for each sample was swabbed with a cotton swab (Selefa, OneMed, Danderyd, Sweden) moistened in 0.9% sodium chloride. Five replicates were collected for each test parameter (surface and treatment). No-treatment controls, consisting of the addition of DNA or blood without any decontamination treatment, were collected in three replicates for each surface. The no-treatment controls revealing a maximum DNA yield were used to estimate the ability of the cleaning strategies to remove DNA contamination by comparing the percentage of recovered DNA for the different cleaning strategies. Negative control swabs (background controls) were obtained by sampling three replicates from surfaces without addition of DNA or blood. Thus, at least three replicates were available and analyzed for all different tests.

DNA extraction was performed using the DNeasy Blood and Tissue kit (Qiagen, Hilden, Germany), with the final extracts being eluted in a total volume of 100 μL. Mitochondrial DNA was quantified using a real-time PCR based assay described by Andréasson et al. [21]. This assay was chosen in order to achieve a highly sensitive DNA quantification allowing determination of trace residues of DNA left after cleaning. The PCR reaction consisted of 2X SsoAdvanced™ Universal SYBR Green Supermix (Bio-Rad Laboratories Inc., Hercules, CA, USA), 400 nM each primer (mt-8294F/mt-8436R), and 5 μL DNA extract in a total reaction volume of 25 μL [21]. Real-time PCR was performed using the Bio-Rad CFX96™ Real-Time System (C1000 Touch Thermal Cycler, Bio-Rad Laboratories, Hercules, CA, USA) and Bio-Rad CFX Manager 2.1 software. The PCR conditions consisted of 98 °C for 2 min followed by 40 cycles at 95 °C for 5 s, 60 °C for 20 s, and 95 °C for 5 s. Negative controls were included in all extraction (no swabs added) and quantification (water instead of DNA added) steps.

Data analysis was performed using Microsoft® Excel (version 2019). The 1.5 × interquartile rule was applied to the five biological replicates for each condition to identify potential outliers. The efficiencies of the cleaning strategies were calculated as a percentage of the positive controls. RStudio® (version 1.3.1093) was used to produce **Figure 1** and **Figure 2** using the following packages: stats [**23**], tidyverse, dplyr, ggpubr, ggplot2, and bibtex.

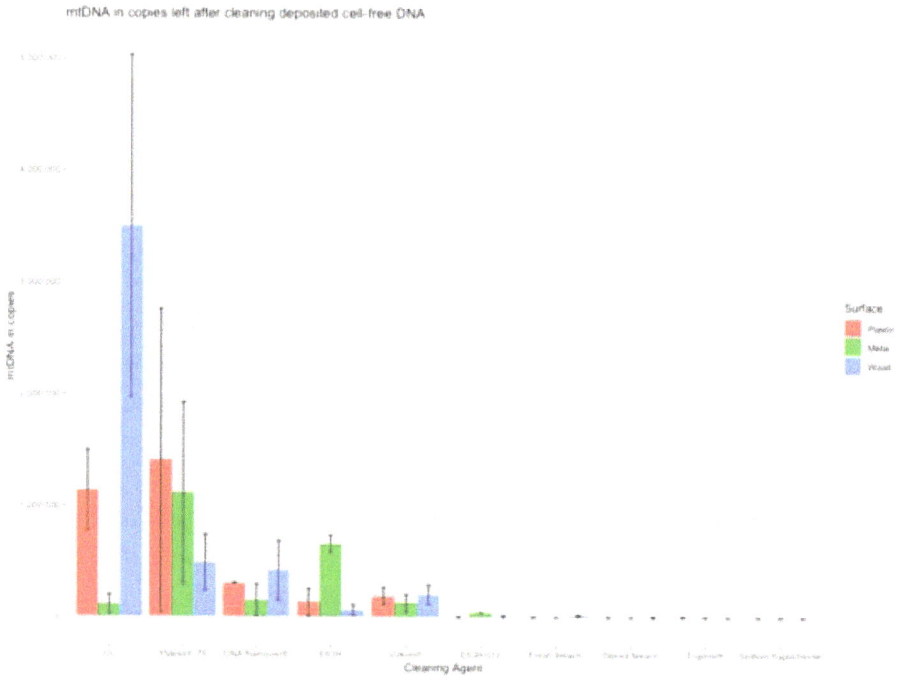

Figure 1. The amount of mtDNA recovered from cell-free DNA deposited on the different surfaces following cleaning with the different strategies. The results are presented as the mean of mtDNA copies (and standard deviations) for a minimum of three replicates recovered from each surface after using the cleaning treatments. The agents are displayed according to the mean recovery for the three surfaces from the highest to the lowest recoveries.

3. RESULTS

An evaluation of the decontamination efficiency of ten cleaning strategies was performed by quantifying the amount of mitochondrial DNA (from both cell-free DNA and cell-contained DNA) recovered after the cleaning of contaminated plastic, metal, and wood surfaces. The mtDNA quantification results for each combination of cleaning treatment and surface are illustrated in **Figure 1** and **Figure 2.** In general, there were large differences in the efficiency of DNA removal between the tested cleaning strategies.

For the cell-free DNA, recoveries of approximately 52, 32, and 27% of the total 60 ng DNA deposited were observed for the no-treatment controls on the plastic, metal, and wood surfaces, respectively. These quantities were used to compare the recovery after the different cleaning strategies. The background controls generated less than 400 mtDNA copies (out of 18 million copies deposited) and were, therefore, ignored in the further evaluation. The recoveries expressed in mtDNA copies and the percent of recovery relative to the no-treatment controls are shown in **Table 1.**

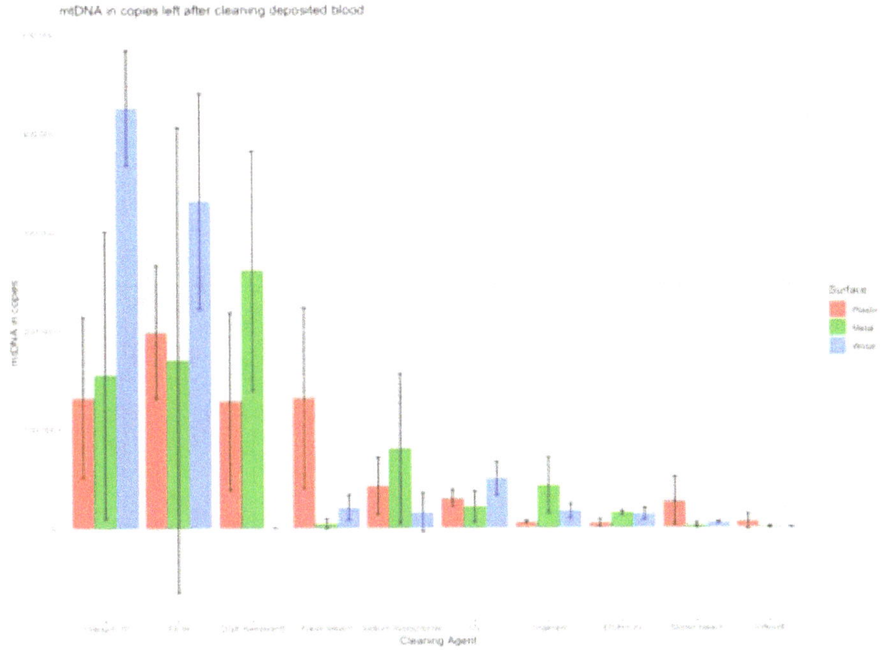

Figure 2. The amount of mtDNA recovered from blood deposited on the different surfaces following cleaning with the different strategies. The results are presented as the mean of mtDNA copies (and standard deviation) for a minimum of three replicates recovered from each surface after using the cleaning treatments. The agents are displayed according to the mean recovery for the three surfaces from the highest to the lowest recoveries.

The cleaning procedures of ethanol and UV were quite inefficient when used alone, with DNA recoveries of up to 11% and 73%, respectively, from all surfaces. However, the combination of both treatments was much more efficient, with recoveries between 0.1 to 0.7% from the three surfaces. Moreover, fresh bleach, stored bleach, Trigene®, and sodium hypochlorite were very efficient in removing DNA, with recoveries from 0.0 to 0.3% from all surfaces. These five decontamination strategies (EtOH+UV, fresh bleach, stored bleach, Trigene®, and sodium hypochlorite) resulted in less than 1% of the DNA recovered from the surface. In addition, if accepting up to 5% recovered DNA, Virkon® is also efficient. The least efficient decontamination strategies were UV and DAX Ytdesinfektion Plus, with up to %73 and %19 DNA recovered, respectively.

Table 1. The efficiency of cell-free DNA removal after cleaning with different treatments on plastic, metal, and wood surfaces. The table is summarizing the mean in mtDNA copies (and standard deviation) per category (3 to 5 biological replicates). The efficiency (Yield) is expressed as the percentage of recovered mtDNA copies of the no-treatment controls.

Cell-Free DNA/ Surface	PLASTIC			METAL			WOOD		
Cleaning Agent	Mean	SD	Yield	Mean	SD	Yield	Mean	SD	Yield
No-treatment controls	9,396,667	4,074,633		5,701,333	4,395,753		4,792,667	3,528,892	
EtOH	131,640	120,771	1.4%	655,333	78,520	11.4%	63,460	45,714	1.3%
UV	1,136,000	361,494	12.1%	115,507	84,110	2.0%	3,500,000	1,530,768	73.0%
EtOH+UV	4724	3020	0.1%	42,567	1570	0.7%	11,480	8569	0.2%
Fresh bleach	1991	2900	0.0%	23	30	0.0%	15,313	8443	0.3%
Stored bleach	918	1238	0.0%	85	77	0.0%	4337	5040	0.1%
DAX Ytdesinfektion Plus	1,399,800	1,357,424	14.9%	1,108,198	812,671	19.2%	486,000	252,391	10.1%
Virkon®	188,720	75,615	2.0%	123,350	77,649	2.1%	201,400	84,417	4.2%
DNA Remover®	304,333	3055	3.2%	149,910	142,169	2.6%	416,000	265,640	8.7%
Trigene®	4223	4026	0.0%	970	1657	0.0%	0	0	0.0%
Sodium hypochlorite	199	151	0.0%	0	0	0.0%	8	16	0.0%
Background controls	15	26		37	65		354	352	

Table 2. The efficiency of cell-contained DNA (blood) removal after cleaning with different treatments on plastic, metal, and wood surfaces. The table is summarizing the mean in mtDNA copies (and standard deviation) per category (3 to 5 biological replicates). The efficiency (Yield) is expressed as the percentage of recovered mtDNA copies of the no-treatment controls.

Blood/Surface	PLASTIC			METAL			WOOD		
Cleaning Agent	Mean	SD	Yield	Mean	SD	Yield	Mean	SD	Yield
No-treatment controls	720,000	66,144		607,667	345,048		469,333	150,111	
EtOH	198,200	67,277	27.5%	170,078	235,148	28.0%	330,600	108,933	70.4%
UV	29,640	8644	4.1%	21,308	15,775	3.5%	49,660	16,808	10.6%
EtOH+UV	4530	3678	0.6%	14,475	1987	2.4%	13,624	6104	2.9%
Fresh bleach	131,380	91,357	18.3%	4028	5124	0.7%	20,465	13,006	4.4%
Stored bleach	26,898	24,457	3.7%	2118	2527	0.4%	5003	1221	1.1%
DAX Ytdesinfektion Plus	131,880	81,201	18.3%	154,488	145,119	25.4%	425,000	58,151	90.6%
Virkon®	5824	7503	0.8%	395	790	0.1%	160	321	0.0%
DNA Remover®	128,060	89,646	17.8%	260,000	121,290	42.8%	0	0	0.0%
Trigene®	5598	1517	0.8%	42,775	28,180	7.0%	16,960	7688	3.6%
Sodium hypochlorite	42,310	28,842	5.9%	79,700	75,524	13.1%	15,618	19,392	3.3%
Background controls	15	26		37	65		354	352	

For cell-contained DNA in the form of blood, the recoveries expressed in mtDNA copies and the percent of recovery relative to the no-treatment controls are shown in **Table 2**. Also, for blood, the cleaning procedures of UV and ethanol were relatively inefficient when used alone, with recoveries of up to 11% and 70%, respectively, from all surfaces. Again, the combination of both treatments (EtOH+UV) was much more efficient, with recoveries between 0.6 to 2.9% from the three surfaces, representing the second most efficient strategy for blood. The most efficient strategy was Virkon®, with recoveries between 0.0 and 0.8% for the three different surfaces. If accepting up to 5% recovered DNA, also stored bleach fulfils the criterion. The least efficient decontamination strategy was DAX Ytdesinfektion Plus, with up to 91% DNA recovered after cleaning.

Large differences were observed between the DNA removal of both DNA sources (cell-free DNA and blood) from the plastic, metal, and wood surfaces, indicating differences in surface absorbance and sampling efficiency. On the wood surface, more than 70% of the added DNA was recovered, and, thus, not efficiently removed by the cleaning with ethanol (for blood), UV (for cell-free DNA,) and DAX Ytdesinfektion Plus (for blood). For these three decontamination strategies, the recoveries differed at least 40% between wood and the other surfaces, indicating that the porous wood presents a challenge for decontamination with large amounts of DNA remaining on the surface (**Table 1** and **Table 2**). For cell-free DNA, UV radiation left the highest amount of mtDNA copies (73% on wood), followed by the second highest, which was DNA in blood left by DAX Ytdesinfektion Plus (27% on metal). DAX Ytdesinfektion Plus left the highest DNA amount from blood (91% on wood) of all agents, followed by aqueous ethanol (70% of blood DNA on wood) and DNA Remover® (%43 of blood DNA on metal). It was observed that decontamination of DNA is less efficient in general for DNA from blood compared to cell-free DNA, which may be a result of the protection by the nucleus and the cell wall (**Table 1** and **Table 2**). This is illustrated by the fact that more of the decontamination agents exceed 5% of recovered DNA for blood than for cell-free DNA. Moreover, recovery of more than 5% of DNA from blood was observed on all three surfaces for DAX Ytdesinfektion Plus (ranging from 18 to 91%) and EtOH (ranging from 28 to 70%).

The most successful decontamination strategies identified in this study for cell-free DNA were fresh and stored bleach, Trigene® and sodium hypochlorite. These cleaning strategies resulted in %0.3 or less of cell-free DNA being recovered from all surfaces, which is equivalent to 0.05 ng DNA or approximately 15,000 mtDNA copies (approximately eight cells) out of the 18 million mtDNA copies initially deposited on the surface. In addition, the combination of EtOH and UV was highly efficient on all three surfaces (0.1 to %0.7). For blood, the best decontamination efficiency was obtained by Virkon® (0.0 to %0.8). Thus, some of the decontamination treatments were demonstrated as highly efficient in this study, while others were less efficient in removing contamination.

4. DISCUSSION

Our results demonstrate that using different cleaning strategies on three different surfaces can remove between 27 and 100% of cell-free DNA and between 9 and 100% of the DNA in the blood. While most treatments showed moderate differences between the recoveries from the three surfaces, the DNA recoveries following UV treatment ranged from 2 to 73% for cell-free DNA and from 4 to 11% for DNA in the blood (**Table 1** and **Table 2**). The differences in the DNA recoveries from the three surfaces may be attributed to the different structures, absorbances, and porousness of the materials as well as the efficiencies of the different treatments. In addition, the agents may result in a chemical reaction with components of the surface, causing for example, corrosion.

In addition to the type of cleaning agents, several other factors can influence the efficiency of removing contaminating DNA during the cleaning process. For example, the mechanical energy from the wiping action, combined with the chemicals used, may enhance the cleaning effect [**24**]. Moreover, substantial DNA loss is expected in the sampling by swabbing and the extraction process [**25**]. In this study, the sampling and extraction efficiency was demonstrated by the no-treatment positive controls, resulting in a loss of 48, 68, and 73% of cell-free DNA from the plastic, metal and wood surfaces, respectively. It is important to note, however, that most extraction kits are developed for the purification of cell-contained DNA and not cell-free DNA, which may result in larger losses in this case than for other sample types. The amount of DNA to simulate contamination to be removed or destroyed was very high in this study, with a total of 60 ng of DNA or 10 μL of blood added to the surface. The high amounts of DNA were used to compensate for expected loss in the initial sample preparation process and to challenge the different cleaning strategies efficiently. The added DNA amount is excessively more than expected from most contamination events by aerosols or touched surfaces (without gloves) and may only be possible to transfer to laboratory surfaces by contamination from a concentrated DNA extract or items with relatively fresh blood stains. Despite all factors that may cause variation, this study could demonstrate decontamination efficiency, by measurements of DNA quantities made before and after cleaning.

Apart from efficiency, several parameters are important when choosing a decontamination strategy, such as simplicity of use, safety, and cost. Ethylene oxide is a highly efficient but poisonous agent and, therefore, unsuitable for decontamination in a laboratory environment [**26**]. Moreover, certain combinations of agents and strategies may be hazardous, such as 1% sodium hypochlorite and 70% ethanol, which form gaseous chlorine and chloroform when combined [**17**]. It is also essential to allow practical and straightforward daily routines. A commonly used decontaminating agent in laboratories is 1–10% sodium hypochlorite solutions, and a low-cost solution can easily be prepared by diluting household bleach. [17]. A low-cost hypochlorite solution can easily be prepared by diluting household bleach. In this study, the in-house sodium hypochlorite dilutions were made at 15% for easy preparations. Due to discontinuation of the stock solution used for the stored bleach, the bleach solutions were prepared from two different variants of the same brand, Klorin Original and Klorin, with sodium hypochlorite concentrations of 3.6 and 2.7%, respectively. Thus, the 15% dilutions resulted in final Cl_2 ion concentrations of 0.54 and %0.4 for fresh and stored bleach, respectively. It is, therefore, important to note that the ratio between the fresh and the stored bleach is 1:0.74 rather than 1:1 when comparing the decontamination efficiencies. To further simplify the routines, the lasting properties of diluted household bleach are important. The available chlorine concentration may be affected in hypochlorite solutions when exposed to light, air, and heat, among other factors. This may reduce the stability, but it has been shown that solutions stored up to 200 days at 4 °C are stable [22]. Our study demonstrated that a hypochlorite solution stored at 8 °C for 80 days is not less efficient than the freshly made solution. Therefore, preparing a new solution is not needed more often than every second month if stored properly.

Many cleaning agent solutions contain hazardous chemical compounds listed in their safety data sheets. The treatments used in this study have a variety of health risks. Some are safe (e.g., DNA Remover®) with no risks listed, while others present several risks. Health risks are very important, but also, the cost is a relevant parameter. For example, DNA Remover® costs 100 times more than household bleach. In contrast, Trigene® has a low cost but is associated with multiple health and environmental hazards. Another important aspect is that the cleaning agent might inhibit the PCR reaction and affect the downstream profiling process. An inhibitory effect has been observed for minute amounts of Trigene® introduced into the PCR reaction, whereas Virkon® or ethanol did not affect STR profiling [14]. All this may influence laboratories final choice of decontamination agent.

Although sodium hypochlorite is highly efficient for the removal of DNA contamination, there are several drawbacks, such as corrosion of metal instruments (e.g., scissors and tweezers). For this reason, alternatives to bleach may be preferred for repeated decontamination of metal items. We performed a limited experiment with metal scissors (stainless steel) submerged in 15% household bleach, DNA Remover®, Virkon®, and water. Preliminary observations showed that corrosion was apparent after only one day in the bleach solution and after five days for water, while no corrosion was observed after 14 days in DNA Remover® or Virkon®.Thus, as Virkon® appears to have a high DNA removal ability, this agent may be a good option when decontaminating metal instruments, despite a somewhat lower ability to remove cell-free contaminating DNA compared to household bleach. However, disposable instruments for evidence sampling and recovery (e.g., single-use scalpels or plastic tweezers) may eliminate the need for a non-corrosive decontamination treatment either completely or at least to some extent. Alternatively, if using sodium hypochlorite for decontamination, rinsing or wiping the surface with distilled water should be performed to avoid corrosion [17].

To conclude, this study showed large differences in the decontamination efficiency between commonly used cleaning strategies on different surfaces and types of traces, cell-free or cell-contained. Of all cleaning agents, six left around 1% or less of either cell-free DNA or DNA in blood on all surfaces. For cell-free DNA, EtOH+UV, fresh bleach, stored bleach, Trigene®, and sodium hypochlorite demonstrated the highest efficiency. Out of these, the highest remaining DNA amounts of 0.7% (on metal and with removal of outliers) correspond to 0.14 ng or around 43,000 copies of mtDNA. For cell-contained DNA as blood, Virkon® left less than %1 DNA and the highest remaining amount corresponds to 0.025 ng or around 7500 mtDNA copies. In addition to efficiency, parameters such as health hazards, simplicity, and cost should also be considered. Therefore, different agents or a combination of treatments may be optimal for the decontamination in laboratories performing forensic DNA analysis. Considering our own requirements and priorities, the updated routines for our laboratory will be to perform decontamination with a %15 dilution of stored household bleach for laboratory work surfaces and with Virkon® for metal instrumentation to avoid corrosion.

5. CONCISE SUMMARY

The selection of decontamination agents is influenced by various aspects and is of utmost importance in routine forensic investigation. This study assessed the efficacy of ten distinct cleaning techniques in eliminating contaminated DNA molecules. Cell-free DNA or blood was applied onto three different surfaces (plastic, metal, and wood) and then subjected to decontamination using different procedures. The quantities of DNA that were recovered, collected through swabbing the surfaces after cleaning using various techniques, were evaluated using real-time PCR. Significant variations in the efficacy of DNA removal were discovered among various cleaning methods, as well as across different surfaces. The most effective methods for cleaning cell-free DNA involved the use of various sodium hypochlorite solutions and Trigene®. These methods resulted in a maximum recovery of 0.3% DNA on all three surfaces. When Virkon® was used for decontamination, only up to 0.8% of the DNA that was deposited on the surface could be retrieved from the blood sample. The recoveries following the implementation of these cleaning techniques are limited to DNA derived from a small number of cells, representing just a fraction of the total 60 ng of cell-free DNA or the numerous deposited blood cells.

REFERENCES

1. Ludeman, M.J.; Zhong, C.; Mulero, J.J.; Lagacé, R.E.; Hennessy, L.K.; Short, M.L.; Wang, D.Y. Developmental Validation of GlobalFiler™ PCR Amplification Kit: A -6Dye Multiplex Assay Designed for Amplification of Casework Samples. *Int. J. Leg. Med.* **2018**, *132*, 1555–1573. **[Google Scholar] [CrossRef] [PubMed] [Green Version]**

2. McLaren, R.S.; Bourdeau-Heller, J.; Patel, J.; Thompson, J.M.; Pagram, J.; Loake, T.; Beesley, D.; Pirttimaa, M.; Hill, C.R.; Duewer, D.L.; et al. Developmental Validation of the PowerPlex® ESI 17/16 Fast and PowerPlex® ESX 17/16 Fast Systems. *Forensic Sci. Int. Genet.* **2014**, *13*, 195–205. **[Google Scholar] [CrossRef] [Green Version]**

3. Bogenhagen, D.; Clayton, D.A. The Number of Mitochondrial Deoxyribonucleic Acid Genomes in Mouse L and Human HeLa Cells. Quantitative Isolation of Mitochondrial Deoxyribonucleic Acid. *J. Biol. Chem.* **1974**, *249*, 7991–7995. **[Google Scholar] [CrossRef]**

4. Allen, M.; Engström, A.-S.; Meyers, S.; Handt, O.; Saldeen, T.; von Haeseler, A.; Pääbo, S.; Gyllensten, U. Mitochondrial DNA Sequencing of Shed Hairs and Saliva on Robbery Caps: Sensitivity and Matching Probabilities. *J. Forensic Sci.* **1998**, *43*, 453–464. **[Google Scholar] [CrossRef]**

5. Kloosterman, A.; Sjerps, M.; Quak, A. Error Rates in Forensic DNA Analysis: Definition, Numbers, Impact and Communication. *Forensic Sci. Int. Genet.* **2014**, *12*, 77–85. **[Google Scholar] [CrossRef]**

6. Codes of Practice and Conduct. Protocol: DNA contamination detection—The management and use of staff elimination DNA databases. In *Forensic Science Regulator, FSR-P-302, Issue 1*; 2014; p. 49. Available online: **https://assets.publishing.service.gov.uk/government/uploads/system/uploads/attachment_data/file/915377/302-Elimination_database_protocol_v2.pdf** (accessed on 15 November 2021).

7. Fonneløp, A.E.; Johannessen, H.; Egeland, T.; Gill, P. Contamination during Criminal Investigation: Detecting Police Contamination and Secondary DNA Transfer from Evidence Bags. *Forensic Sci. Int. Genet.* **2016**, *23*, 121–129. **[Google Scholar] [CrossRef] [PubMed]**

8. Gill, P. Role of Short Tandem Repeat DNA in Forensic Casework in the UK-Past, Present, and Future Perspectives. *BioTechniques* **2002**, *32*, 366–385. **[Google Scholar] [CrossRef] [PubMed]**

9. Neuhuber, F.; Dunkelmann, B.; Höckner, G.; Kiesslich, J.; Klausriegler, E.; Radacher, M. Female Criminals—It's Not Always the Offender! *Forensic Sci. Int. Genet. Suppl. Ser.* **2009**, *2*, 145–146. **[Google Scholar] [CrossRef]**

10. Daniel, R.; van Oorschot, R.A.H. An Investigation of the Presence of DNA on Unused Laboratory Gloves. *Forensic Sci. Int. Genet. Suppl. Ser.* **2011**, *3*, e45–e46. **[Google Scholar] [CrossRef]**

11. Lapointe, M.; Rogic, A.; Bourgoin, S.; Jolicoeur, C.; Séguin, D. Leading-Edge Forensic DNA Analyses and the Necessity of Including Crime Scene Investigators, Police Officers and Technicians in a DNA Elimination Database. *Forensic Sci. Int. Genet.* **2015**, *19*, 50–55. **[Google Scholar] [CrossRef]**

12. Gefrides, L.A.; Powell, M.C.; Donley, M.A.; Kahn, R. UV Irradiation and Autoclave Treatment for Elimination of Contaminating DNA from Laboratory Consumables. *Forensic Sci. Int. Genet.* **2010**, *4*, 89–94. **[Google Scholar] [CrossRef]**

13. Preuße-Prange, A.; Renneberg, R.; Schwark, T.; Poetsch, M.; Simeoni, E.; von Wurmb-Schwark, N. The Problem of DNA Contamination in Forensic Case Work-How to Get Rid of Unwanted DNA? *Forensic Sci. Int. Genet. Suppl. Ser.* **2009**, *2*, 185–186. **[Google Scholar] [CrossRef]**

14. Bright, J.A.; Cockerton, S.; Harbison, S.A.; Russell, A.; Samson, O.; Stevenson, K. The Effect of Cleaning Agents on the Ability to Obtain DNA Profiles Using the Identifiler™ and PowerPlex® Y Multiplex Kits. *J. Forensic Sci.* **2011**, *56*, 181–185. **[Google Scholar] [CrossRef] [PubMed]**

15. Hoofstat, V.; Groote, D.; Van Thuyne, S.; Haerinck, N. Sources of DNA Contamination and Decontamination Procedures in the Forensic Laboratory. *J. Forensic Res.* **2011**, *2*, 1. **[Google Scholar] [CrossRef] [Green Version]**

16. Champlot, S.; Berthelot, C.; Pruvost, M.; Andrew Bennett, E.; Grange, T.; Geigl, E.M. An Efficient Multistrategy DNA Decontamination Procedure of PCR Reagents for Hypersensitive PCR Applications. *PLoS ONE* **2010**, *5*, e13042. **[Google Scholar] [CrossRef] [PubMed] [Green Version]**

17. Ballantyne, K.N.; Salemi, R.; Guarino, F.; Pearson, J.R.; Garlepp, D.; Fowler, S.; Van Oorschot, R.A.H. DNA Contamination Minimisation-Finding an Effective Cleaning Method. *Aust. J. Forensic Sci.* **2015**, *47*, 428–439. **[Google Scholar] [CrossRef]**

18. Shaw, K.; Sesardić, I.; Bristol, N.; Ames, C.; Dagnall, K.; Ellis, C.; Whittaker, F.; Daniel, B. Comparison of the Effects of Sterilisation Techniques on Subsequent DNA Profiling. *Int. J. Leg. Med.* **2008**, *122*, 29–33. **[Google Scholar] [CrossRef] [PubMed]**

19. Vandewoestyne, M.; Van Hoofstat, D.; Franssen, A.; Van Nieuwerburgh, F.; Deforce, D. Presence and Potential of Cell Free DNA in Different Types of Forensic Samples. *Forensic Sci. Int. Genet.* **2013**, *7*, 316–320. **[Google Scholar] [CrossRef] [PubMed]**

20. Quinones, I.; Daniel, B. Cell Free DNA as a Component of Forensic Evidence Recovered from Touched Surfaces. *Forensic Sci. Int. Genet.* **2012**, *6*, 26–30. **[Google Scholar] [CrossRef]**

21. Andréasson, H.; Gyllensten, U.; Allen, M. Real-Time DNA Quantification of Nuclear and Mitochondrial DNA in Forensic Analysis. *BioTechniques* **2002**, *33*, 402–411. **[Google Scholar] [CrossRef] [PubMed]**

22. Piskin, B.; Türkün, M. Stability of Various Sodium Hypochlorite Solutions. *J. Endod.* **1995**, *21*, 253–255. **[Google Scholar] [CrossRef]**

23. R Core Team. *R: A Language and Environment for Statistical Computing*; R Foundation for Statistical Computing: Vienna, Austria, 2021; Available online: **https://www.R-project.org** (accessed on 13 December 2021).

24. Forensic Science Community. *Comparison Study of Disinfectants for Decontamination Project Information Stakeholder Information Photos*; National Forensic Science Technology Center: Largo, FL, USA, 2011; No. 727. **[Google Scholar]**

25. Verdon, T.J.; Mitchell, R.J.; van Oorschot, R.A.H. Swabs as DNA Collection Devices for Sampling Different Biological Materials from Different Substrates. *J. Forensic Sci.* **2014**, *59*, 1080–1089. **[Google Scholar] [CrossRef] [PubMed]**

26. Wilson-Wilde, L.; Yakovchyts, D.; Neville, S.; Maynard, P.; Gunn, P. Investigation into Ethylene Oxide Treatment and Residuals on DNA and Downstream DNA Analysis. *Sci. Justice* **2017**, *57*, 13–20. **[Google Scholar] [CrossRef] [PubMed]**

21 ANALYZING THE SKIN PIGMENTATION AND GENETIC ANCESTRY OF THREE SUBPOPULATIONS FROM PAKISTAN: PUNJABI, PASHTUN, AND BALOCH

1. INTRODUCTION: AN OVERVIEW

Human pigmentation is a highly diverse externally visible characteristic (EVC). Anticipating the emergence of EVCs can yield valuable leads from the outset of a police inquiry, as it offers a "genetic eye witness" account of a suspected wrongdoer. This enables investigators to concentrate their efforts on particular groups of persons and reduce the pool of prospective suspects [1,2,3]. The study of human pigmentation genetics is highly intriguing in the field of forensic genetics because to its ability to predict three significant external visible characteristics (EVCs): hair color, eye color, and skin color [4,5]. The relationship between skin pigmentation and biogeographic heritage is highly correlated [6], and it is influenced by environmental factors such as UV radiation [7]. This makes it challenging to determine the specific genetic variants that cause pigmentation and to accurately forecast it. Forecasting the biogeographic lineage can provide insights on the individual's skin coloring. However, skin color prediction based on the assumption that biogeographic ancestry and skin color are well correlated only offers an indirect prediction, and the assumption is not necessarily true in admixed populations [**8**,**9**]. Studies of admixed individuals from North and South American populations showed that the correlation between ancestry and skin pigmentation varied considerably and explained 20–65% of the skin pigmentation [**3**,**6**,**10**,**11**]. However, people with mixed ancestry are well-suited for investigating the impact of genetic variations linked to skin pigmentation. This is because they exhibit a diverse spectrum of skin colors, and the ancestral proportions of each person can be determined using established methods in population genetics.

The population of Pakistan is a heterogeneous population with several subpopulations and ethnic groups. Pakistan is located at the confluence of the Middle East, Central Asia, and Southeast Asia. This region served as the habitat for diverse historical civilizations, such as the Indus Valley Civilization during the Bronze Age [12]. The region was governed by various dynasties and empires representing diverse religions and civilizations, such as Muslims, Turco-Mongols, Hindus, Indo-Greeks, Sikhs, and Afghans [13]. Therefore, the Pakistani population is highly suitable for investigating biogeographic ancestry and the genetics of skin pigmentation.

We examined the connections between different pigmentary variations, biogeographic heritage, and the measurement of skin pigmentation in three mixed subpopulations from Pakistan, consisting of 299 individuals. We assessed the lineage of the participants using the Precision ID Ancestry Panel from Thermo Fisher Scientific. Additionally, we examined the correlation between skin pigmentation and genetic pigmentary variations using the HuPi AmpliSeq™ Custom panel. We conducted an analysis on the relationship between skin pigmentation and the 36 HIrisPlex-S SNPs [18,19], as well as nine SNPs that have been linked to variations in skin pigmentation in a mixed Brazilian population [3]. Finally, we evaluated the efficacy of the skin color prediction using the HIrisPlex-S prediction model [18,19].

2. MATERIALS AND METHODS

2.1. Samples, DNA Extraction, and DNA Quantification

A total of 107 individuals from Baluchistan, 103 individuals from Khyber Pakhtunkhwa, and 89 individuals from northern Punjab, specifically Punjabis, were included in the study.. Signed informed consent was obtained from all participants, and the samples were anonymized. The study was approved by the Review Board/ Ethical Committee of the University of the Punjab, Pakistan (D/No. 019/DFEMS). Genomic DNA was extracted using the Qiagen BioRobot® EZ1 Workstation and the EZ1 DNA investigator kit (Qiagen, Hilden, Germany). Purified DNA was quantified using the Qubit™ dsDNA HS Assay Kit and a Qubit® 4.0 Fluorometer (Thermo Fisher Scientific, Waltham, MA, USA) according to the manufacturer's recommendations.

2.2. Measurement of Quantitative Skin Pigmentation

The UV-Optimize Scientific 555 (Chromo Light APS, Vedbæk, Denmark) calibrated with a white standard (ISO 2469) was used to measure quantitative skin pigmentation. Quantitative skin pigmentation was determined as the pigment protection factor, PPF, which is a measure for skin pigmentation levels [**20,21**]. The skin pigmentation of each individual was measured in triplicates from skin areas on the lower inner forearm (lower arm), upper inner arm (upper arm), and forehead. The median PPF-value for each measuring site was used for further analysis. All skin measurements were taken from areas without hair, freckles, nevi, or tattoos.

2.3. Typing with the Precision ID Ancestry Panel and the Human Pigmentation (HuPi) AmpliSeq™ Custom Panel

DNA was amplified with the Precision ID Ancestry Panel (Thermo Fisher Scientific, Waltham, MA, USA) and the Human Pigmentation (HuPi) AmpliSeq™ Custom panel [17] in two separate reactions. The Precision ID Ancestry Panel targets 165 ancestry-informative markers (AIMs). The HuPi AmpliSeq™ Custom panel targets 183 SNPs and indels, which were previously found to be associated with human pigmentary variation [17]. Sequencing libraries were prepared with the Ion AmpliSeq™ Library Kit 2.0 (Thermo Fisher Scientific, Waltham, MA, USA) using half volume of all reagents but otherwise following the manufacturer's instructions. Amplification of targets was carried out using approximately 1 ng DNA (1–3 μL) and 24 and 26 cycles for the Precision ID Ancestry Panel and the HuPi AmpliSeq™ Custom panel, respectively. Sequencing libraries were purified using the Biomek 3000 pipetting robot [22]. Purified libraries were quantified using the Qubit™ 4.0 Fluorometer with the Qubit™ dsDNA HS Assay kit (Thermo Fisher Scientific, Waltham, MA, USA) and pooled in equimolar concentrations to a final volume of 28 pM for the Precision ID Ancestry panel and 35–60 pM for the HuPi AmpliSeq™ Custom Panel. Template preparation was performed with the Ion Chef™ instrument (Thermo Fisher Scientific, Waltham, MA, USA) using the Ion S5™ Precision ID Chef Kit (Thermo Fisher Scientific, Waltham, MA, USA) following the manufacturer's recommendations (Thermo Fisher Scientific, Waltham, MA, USA). Sequencing was carried out with Ion 530™ Chips and Ion S5™ Precision ID Sequencing Reagents on the Ion S5™ (Thermo Fisher Scientific, Waltham, MA, USA) with up to 80 samples per chip.

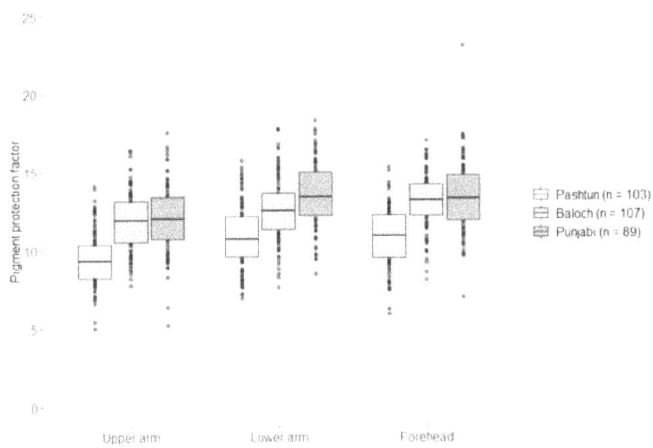

Figure 1. Boxplots of skin pigmentation measurements (pigment protection factor) of the upper arm, lower arm, and forehead in three Pakistani subpopulations. The pigment protection factor was measured in triplicate at each measured skin area.

2.4. Analysis of Sequencing Data

The data were initially analyzed with the Torrent Suite Software v.5.10.1 (Thermo Fisher Scientific, Waltham, MA, USA). For the Precision ID Ancestry panel, data were analyzed with the HID_SNP_Genotyper_v5_2_2 plugin using GRCh37 (hg19) as the reference genome and default parameters (Thermo Fisher Scientific, Waltham, MA, USA). The data obtained with the HuPi AmpliSeq™ Custom Panel were analyzed with the VariantCaller v5.10.1.20 plugin and the GRCh38.p2 (hg38) reference genome. The "Down-sample to coverage" was changed from 400 to 10,000 reads. Otherwise, the analyses were performed with the default parameters (Thermo Fisher Scientific, Waltham, MA, USA). The resulting Excel-files were processed using R v.4.0.2 (R core team, **https://www.R-project.org/**, accessed on 1 November 2020) and an in-house developed script. The data quality control comprised evaluation of the locus balance, heterozygote balance (Hb), and noise levels for each target. Hb was calculated as the number of reads for one nucleotide divided by the number of reads for the other nucleotide in the order A, C, G, and T. Genotypes were accepted when a locus had ≥45 reads and the Hb was between 0.3 and 3. SNPs with 30–44 reads were inspected and accepted if the Hb was between 0.7 and 1.5 or if the noise did not exceed one read for homozygous genotype calls. Genotypes that did not meet the criteria were annotated as NN. For both panels, loci with NNs in more than 15% of the samples were excluded. Moreover, samples with more than 15% missing data (NN) were excluded from further analysis. Hardy–Weinberg Equilibrium (HWE) and pairwise linkage disequilibrium (LD) were calculated with Haploview v. 4.2 [**23**], and pairwise F_{ST}-values were calculated using Arlequin v3.5.2.2 software [**24**].

2.5. Classification of Biogeographic Ancestry

The alleles of the AIM-SNPs in each sample were investigated using STRUCTURE v.2.3.4.21 [**25,26**]. The analyses were carried out using 100,000 steps of burn-in followed by 100,000 repetitions for the MCMC. The 'admixture' and the 'correlated allele frequencies' models were used [**25,26**]. After testing the number of clusters (*K*) from K = 3 to K = 6, K was set to 4, corresponding to the Sub-Saharan African, South-Central Asian, European, and East Asian populations. The results were visualized using CLUMPP v.1.1.222 [**27**] and Distruct v.1.1.23 [**28**]. Principal component analysis (PCA) was carried out using an in-house written Python script. Four SNPs, rs1800414, rs12913832, rs1426654, and rs16891982, were included in both the Precision ID Ancestry Panel and the HuPi AmpliSeq™ Custom panel. One SNP, rs10954737, lacked genotype data in all reference populations. Hence, these five SNPs were not considered in the PCA and STRUCTURE analyses and were not used to classify ancestry.

2.6. Correlations between Biogeographic Ancestry, Pigmentary Variants, and Skin Pigmentation

We excluded monomorphic genetic variants (i.e., variants with only one observed genotype) and variants in LD ($r^2 > 0.8$) with other variants in the HuPi AmpliSeq™ Custom Panel. Hence, the dataset was limited to 102 genetic variants and four metapopulations (Sub-Saharan Africa, South-Central Asia, Europe, and East Asia). Differences in skin pigmentation levels of the lower arm, upper arm, and forehead, as well as differences in skin pigmentation of the three subpopulations, were investigated using Student's *t*-test for paired data or the Welch two-sample *t*-test for unpaired data. The correlations among skin pigmentation levels, genetic variants in the HuPi AmpliSeq™ Custom Panel, and the estimation of the ancestry proportions were carried out using multiple linear regression (MLR). The correlation was evaluated using the adjusted R^2. To select a subset of variants, backwards model selection was performed using MLR with Akaike Information Criterion (AIC) using the *stepAIC* function in the MASS R-package [**30**].

We also evaluated the correlation between skin pigmentation and the genotypes of the 36 HIrisPlex-S SNPs [**18,19**], and between skin pigmentation and nine SNPs (rs1426654, rs1448484, rs16891982, rs4424881, rs10831496, rs6119471, rs12913832, rs10424065, and rs1408799) previously found to explain up to 65% of the skin pigmentation variation in a Brazilian population of primarily European ancestry [**3**].

2.7. Prediction of Skin Colour Using the HIrisPlex-S

Prediction of skin color using the HIrisPlex-S model was carried out with the online web-tool (**https://hirisplex.erasmusmc.nl/**, accessed on 1 Novmber 2020). The HIrisPlex-S model predicts skin color in five categories: Very pale, Pale, Intermediate, Dark, and Dark to Black [**18,19**]. The skin color category with the highest predictive value was used as the skin pigmentation prediction. The HIrisPlex-S skin color prediction and the quantitative skin pigmentation measurements were compared.

2.8. Statistical Methods

The statistical methods used are described in each section. When multiple comparisons were performed, the statistical significance was corrected with the Bonferroni method.

3. RESULTS

3.1. Skin Pigmentation Measurements

Quantitative skin pigmentation measurements were performed for the 299 Pakistani individuals, Baloch (*n* = 107), Pashtun (*n* = 103), and Punjabi (*n* = 89). The lowest pigmentation levels were observed on the upper arm (mean PPF of 11.08), followed by the lower arm (mean PPF of 12.30) and the forehead (mean PPF of 12.43). We observed statistically significant differences between the pigmentation levels on the upper arm and lower arm ($p < 0.05$) and the upper arm and forehead ($p < 0.05$) among all subpopulations. Statistically significant difference between pigmentation levels of the lower arm and forehead was only observed in the Baloch subpopulation ($p < 0.05$).

Individuals from the Pashtun subpopulation showed statistically significant ($p < 0.05$) lower pigmentation levels than individuals in the Baloch and Punjabi subpopulations on all measured areas (**Figure 1**).

3.2. The Precision ID Ancestry Panel

The median number of reads per target was 493 (range: 45–12,105), the median Hb was 1.0 (range: 0.3–3.0) for heterozygous SNPs, and the median level of noise was 0.0% (range: 0–14.3%). One locus, rs2070586, deviated from HWE ($p < {}^{3-}10$) in the Baloch population. The loci rs3811801 and rs671 were monomorphic in all three populations, rs1871534 and rs2814778 were monomorphic in Pashtuns and Punjabis, and rs1800414 was monomorphic in the Baloch. No pairwise LD (pairwise $r^2 > 0.8$) was detected in any of the three populations. The observed values of pairwise F_{ST} genetic distances were statistically significantly different among all subpopulations ($p < {}^{5-}10$).

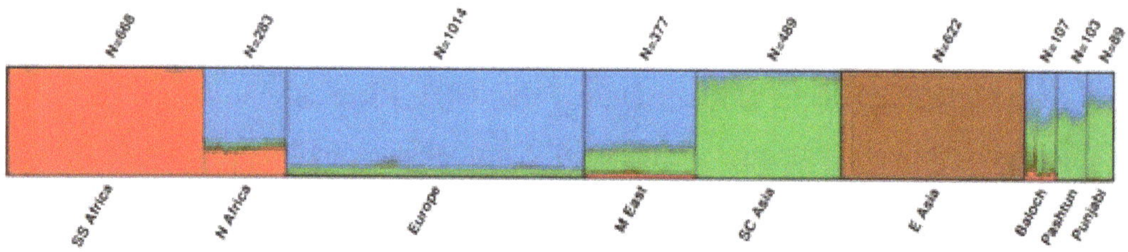

Figure 2. STRUCTURE analysis with K = 4 using 160 ancestry informative markers (AIMs) and six meta-populations. Each cluster (K) is represented by a color. Population abbreviations used: SS Africa: Sub-Saharan Africa; N Africa: North Africa; M East: Middle East; SC Asia: South-Central Asia; E Asia: East Asia.

3.3. Proportion of Ancestry Components

The biogeographic ancestry of each individual was investigated with STRUCTURE and PCA using reference data from 34 populations grouped into six metapopulations: Sub-Saharan African, North African, European, Middle Eastern, South-Central Asian, and East Asian. The results of the STRUCTURE analysis with K = 4 are shown in **Figure 2**. The Baloch, Punjabi, and Pashtun subpopulations were admixed populations with major contributions from South-Central Asian and European populations. The Baloch differed from the other two subpopulations due to approximately 10% Sub-Saharan African genetic contribution, whereas the Sub-

Saharan African contribution was <1% in the Punjabis and Pashtuns. The Baloch had the lowest proportion of South-Central Asian ancestry (44.7%), followed by the Pashtuns (56.6%) and the Punjabis (69.3%).

The first two principal components (PC1 and PC2) in the PCA analysis separated the Sub-Saharan African, European, and East Asian populations from each other. The North Africans clustered closest to the European and Middle Eastern populations, while the South-Central Asian populations clustered between the East Asian and European populations. The positions of the three Pakistani subpopulations, Punjabis, Pashtuns, and the Baloch, overlapped with each other and the South-Central Asian and Middle Eastern populations. The Baloch individuals clustered closer to the African populations than the Pashtuns and Punjabis. In contrast, the Punjabis clustered closer to the South-Central Asian populations than the Pashtuns and Baloch, supporting the STRUCTURE analysis conclusions.

3.4. The HuPi AmpliSeq™ Custom Panel

A total of 163 variants were successfully typed with the HuPi AmpliSeq™ Custom Panel. The median number of reads was 597 (range: 60–1532) per target. The median Hb for heterozygous SNP calls was 1.0 (range: 0.35–2.86). The median level of noise was 0.0% (range: 0–12.7%). The allele frequencies of rs4778241 in *OCA2* deviated statistically significantly from HWE ($p < {}^{3-}10$). Twelve variants were monomorphic, and these data were removed from the subsequent analysis. For the correlation measurements between skin pigmentation and genetic variants, only independent genetic variants were considered (pairwise $r^2 < 0.8$), reducing the number of genetic variants to 102.

3.5. Correlation between Skin Pigmentation and Biogeographic Ancestry

The proportions of Sub-Saharan African, European, South-Central Asian, and East Asian ancestry were compared with the skin pigmentation at each measured site using MLR. We found statistically significant ($p < {}^{3-}10$) adjusted R^2-values for the upper arm: 0.10, lower arm: 0.13, and forehead: 0.12 when comparing the full ancestry profiles with the skin pigmentation. We found statistically significant ($p < 0.05$) negative correlations between the proportion of European ancestry and skin pigmentation at all three measured sites (adjusted R^2: 0.04–0.09) and statistically significant positive correlation ($p < 0.05$) between Sub-Saharan African ancestry and skin pigmentation at the upper arm and forehead (adjusted $R^2 = 0.06$ and 0.07, respectively). In contrast, the proportions of East Asian and South-Central Asian ancestry showed no statistically significant correlation with skin pigmentation. We analyzed the correlation between the proportions of only European and Sub-Saharan African ancestry with skin pigmentation at all three measuring sites using

MLR. The resulting adjusted R^2-values were 0.10 (p < 0.13 ,($^{3-}$10 (p < $^{6-}$10), and 0.12 (p < $^{6-}$10) for the upper arm, lower arm, and forehead pigmentation, respectively.

3.6. Correlations among Skin Pigmentation, Pigmentary Variants, and Biogeographic Ancestry

Each of the independent variants typed with the HuPi AmpliSeq™ Custom Panel was compared with the skin pigmentation at the upper arm, lower arm, and forehead using linear regression. We observed statistically significant (p < 0.05) correlations between skin pigmentation on either the upper arm, lower arm, or forehead and the SNPs rs1042602 in *TYR* and rs10831496 in *GRM5*, rs1426654 in *SLC24A5*, rs16891982 in *SLC45A2*, and rs12913832 in *HERC2* (**Table 1**). Only rs1426654 and rs1042602 were statistically significantly correlated with skin pigmentation at all three sites (p < $^{3-}$10) (**Table 1**).

Table 1. Five SNPs with statistically significant association with skin pigmentation on the upper arm, lower arm, or forehead in 299 Pakistani individuals from three subpopulations (Baloch, Pashtun, and Punjabi).

		Upper Arm		Lower Arm		Forehead	
Gene	SNP	Adjusted R2	Sign.1	Adjusted R2	Sign. 1	Adjusted R2	Sign. 1
TYR	rs1042602	0.13	***	0.11	**	0.071	**
GRM5	rs10831496	0.029	–	0.047	*	0.014	–
HERC2	rs12913832	0.025	–	0.016	–	0.064	*
SLC24A5	rs1426654	0.16	***	0.10	**	0.082	**
SLC45A2	rs16891982	0.075	**	0.028	–	0.058	**

1 Statistical significance (Sign.) after Bonferroni correction: * p < 0.05, ** p < 10−3, *** p < 10−6.

The five SNPs were compared with skin pigmentation using MLR, and the adjusted R^2 ranged from 0.24 to 0.33 (p < $^{6-}$10) (**Table 2**). Subsequently, model selection was performed using stepwise AIC, giving the best model with rs1426654, rs1042602, rs16891982, and rs12913832 for all three sites (upper arm, lower arm, and forehead). When the proportions of European and Sub-Saharan African ancestry were included, the adjusted R^2-values increased slightly (**Table 2, Figure 3**). The cumulative correlations between (1) the four SNPs and ancestry and (2) the skin pigmentation are shown in **Figure 3**. The four SNPs, ancestry, and the upper arm pigmentation showed the highest adjusted R^2-values (**Table 2** and **Figure 3**).

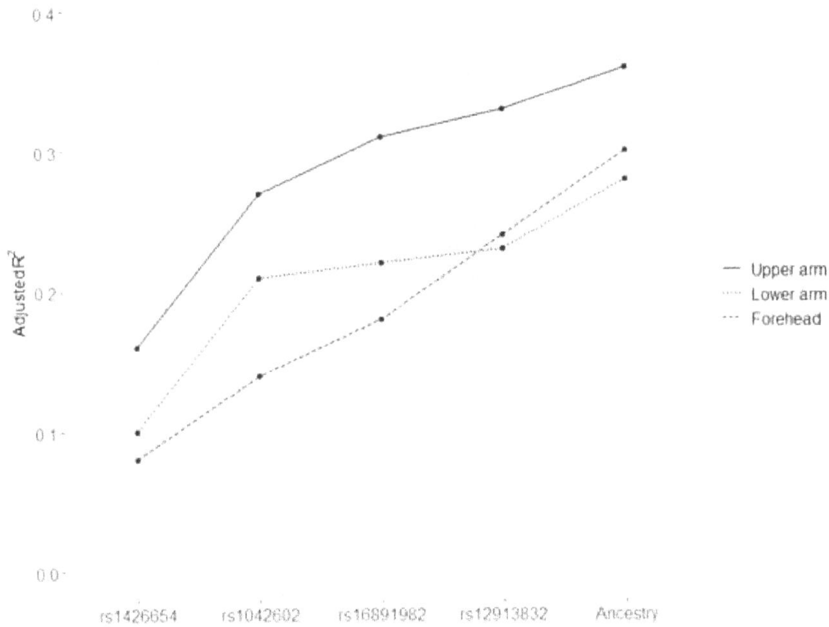

Figure 3. Cumulative correlations between (1) four SNPs and ancestry and (2) upper arm, lower arm, and forehead pigmentation. The SNPs were ranked based on their correlations with skin color (adjusted R^2). Ancestry is the proportion of Sub-Saharan African and European ancestry.

Table 2. Multiple loci correlations with skin pigmentation on the upper arm, lower arm, or forehead in 299 Pakistani individuals from three subpopulations (Baloch, Pashtun, and Punjabi).

	Upper Arm		Lower Arm		Forehead	
	Adjusted R^2	Sign. [1]	Adjusted R^2	Sign. [1]	Adjusted R^2	Sign. [1]
Five SNPs [2]	0.33	***	0.24	***	0.24	***
Best model [3]	0.33	***	0.23	***	0.24	***
Best model + ancestry [4]	0.37	***	0.28	***	0.30	***

[1] Statistical significance (Sign.): *** $p < ^{2}.^{6}$-10 rs1042602, rs10831496, rs12913832, rs1426654, and rs16891982. [3] rs1042602, rs12913832, rs1426654, and rs16891982. [4] Ancestry is the estimated proportion of European and Sub-Saharan African ancestry

3.7. Correlation of Skin Pigmentation with SNPs from Existing Skin Colour Models

The skin pigmentation in the Pakistani individuals was compared with nine SNPs previously found to be associated with skin pigmentation in a Brazilian population of primarily European ancestry [3]. The nine SNPs included rs10831496, rs12913832, rs1426654, and rs16891982, which were associated with skin pigmentation in the Pakistani population (**Table 1**), as well as rs1448484 in *OCA2*, rs4424881 in *APBA2*, rs6119471 in *ASIP*, rs1408799 in *TYRP1*, and rs10424065 in *MFSD12*, which were not associated with skin pigmentation in our study. In the Pakistani population, these nine SNPs explained up to %26 of the skin pigmentation (upper arm). The adjusted R^2-values were 0.26 ($p < 0.15$,(0.05 ($p < 0.05$), and 0.19 ($p < 0.05$) for the upper arm, lower arm, and forehead pigmentation, respectively.

We also correlated the HIrisPlex-S SNPs with the skin pigmentation on the upper arm, lower arm, and forehead using MLR. Four SNPs in the *MC1R*-gene, rs3212355, rs1805006, rs11547464, and rs1110400 were monomorphic and excluded from the analysis. Thus, the correlation was based on 32 of the 36 skin color predictive SNPs in the HIrisPlex-S. The adjusted R^2-values were 0.36 ($p < 0.29$,($^{6-}10$ ($p < {}^{6-}10$), and 0.27 ($p < {}^{6-}10$), for pigmentation on the upper arm, lower arm, and forehead, respectively.

Lastly, we evaluated the performance of the HIrisPlex-S model that is a forensically validated skin color prediction model [**18,19**]. The predicted skin colors of the 299 Pakistani individuals were: Intermediate: 102, Dark: 167, and Black: 30. No individual was predicted to have very pale or pale skin colors. Comparisons between the predicted skin color categories and the quantitative skin pigmentation measurements are shown in **Figure 4**.

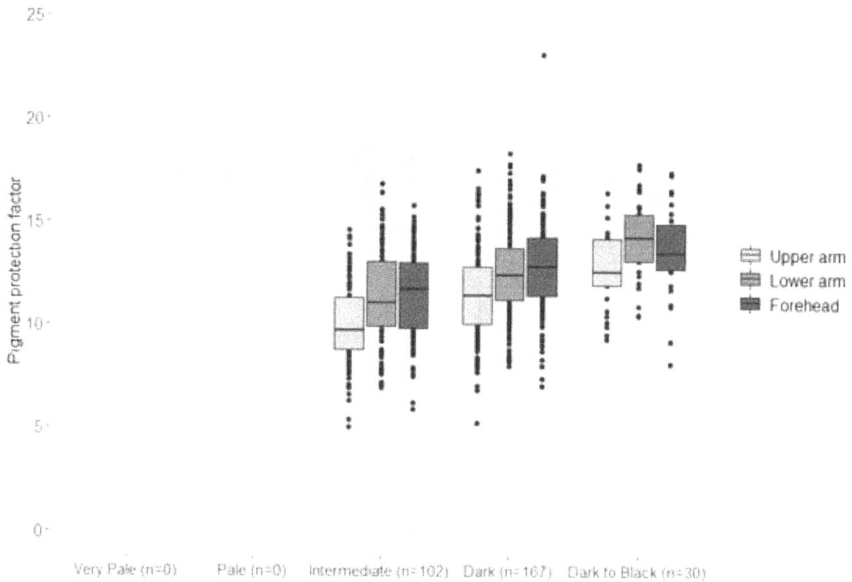

Figure 4. Boxplots showing the skin color category predicted with the HIrisPlex-S prediction model and the pigmentation protection factor on the upper arm, lower arm, and forehead.

4. DISCUSSION

The genetic distances computed between the three subpopulations were found to be statistically significant, as shown in Supplementary Table S3. The Baloch were more geographically separated from Punjabis and Pashtuns compared to the level of separation between Punjabis and Pashtuns. The findings corroborated our prior results obtained using short tandem repeats, which revealed genetic variations between the Baloch population and other subpopulations in Pakistan [31]. The STRUCTURE analysis (Figure 2) revealed that all three subpopulations from Pakistan exhibited admixture, primarily with genetic contributions from South-Central Asia and Europe. Nevertheless, the Baloch population also exhibits a genetic admixture from Sub-Saharan African groups, potentially indicating the presence of African individuals who migrated and settled on the Indian subcontinent [32,33].

The proportion of European and Sub-Saharan African ancestry affected the skin pigmentation, whereas the proportion of South-Central Asian ancestry did not. However, the proportions of European and Sub-Saharan African ancestry only explained approximately 10% of the variation and only influenced the adjusted R^2-values of the MLR model marginally (**Table 2**). This showed that ancestry was a relatively poor predictor of skin pigmentation in the highly admixed Pakistani populations. It also indicated that any association between a locus and skin pigmentation levels would most likely come from causative SNPs or indels, and not be a consequence of population genetic differences.

Of the 163 variants that were previously shown to be associated with pigmentary traits, five SNPs were statistically significantly correlated with the skin pigmentation levels of the investigated individuals from the Pakistani subpopulations (**Table 1**). These five SNPs are located in five genes, *TYR, GRM5, SLC24A5, SLC45A2,* and *HERC2.* Four of the five SNPs, rs1042602 in *TYR*, rs1426654 in *SLC24A5*, rs16891982 in *SLC45A2*, and rs12913832 in *HERC2*, explained 33% of the skin pigmentation variation on the upper arm area and 23–24% of the skin pigmentation variation on the lower arm and forehead that are less protected from UV irradiation (**Table 2**). Three of the SNPs, rs1426654 (ranked 1), rs1042602 (ranked 2), and rs16891982 (ranked 3), were previously found to be associated with skin pigmentation in individuals of South Asian descent [**7,34,35**]. In agreement with our results, the *SLC24A5* SNP rs1426654 had the most pronounced effect on skin pigmentation levels [**7,34,36**]. The *HERC2* SNP rs12913832 (ranked 4) has not previously reported to be associated with skin pigmentation in South Asian populations. However, it was reported to be associated with skin color in other populations [**11,19**]. The fifth SNP, *GRM5* rs10831496 (ranked 5), did not increase the cumulative adjusted R^2-value of the MLR model at any of the measured pigmentation areas. This SNP is highly polymorphic and located in intron 3 of the *GRM5* gene, 340 kbp upstream of the *TYR* gene on chromosome 11. *GRM5* rs10831496 was previously reported to be associated with tanning response after exposure to sunlight [**37**]. In our study, it was associated with skin pigmentation on the lower arm.

The three highest ranking SNPs, rs1426654, rs1042602, and rs16891982, are non-synonymous variants: Thr111Ala in *SLC24A5*, Ser192Tyr in *TYR*, and Phe374Leu in *SLC45A2*, respectively, and they all influence melanin production. In human primary melanocytes that were homozygous for the *SLC24A5* 111Ala variant, the TYR activity, melanin content, and the number of mature melanosomes were decreased [**38**]. The 192Tyr *TYR* variant reduced the TYR activity by **39**] %40] and the number of primary melanocytes, and human skin tissue samples that were homozygous for the 192Tyr variant had reduced TYR activity and expression [**40**]. The *SLC45A2* 374Leu variant was found in increased frequency in non-European populations, and 374Leu homozygous melanocytes had higher TYR activity and melanin content than melanocytes that were homozygous for the 374Phe variant [**38**]. The fourth-ranking SNP, *HERC2* rs12913832, is the most important variant for eye color determination [**41,42,43**].

It is positioned in a key enhancer element of the *OCA2* pigmentary gene [**44**]. Human primary melanocytes that were homozygous for the derived rs12913832 G allele had reduced TYR activity, melanin content, and numbers of mature melanosomes [**38**]. When we analyzed five additional SNPs that were previously found to be associated with pigmentation in an admixed Brazilian population with primarily European and Sub-Saharan African genetic contributions [**3**], the correlation with skin pigmentation of the upper arm was reduced from 33% to 26%. This indicates that the biogeographic background of an individual is important for the selection of markers that eventually will give the best prediction of skin color, which is well in line with the knowledge that some genetic markers are associated with skin pigmentation in certain populations and not in others [**45**]. It also indicates that the estimation of the biogeographic background should be the first step in the estimation of skin pigmentation levels from DNA samples of unknown individuals. The second step should be to investigate causative variants found to be associated with skin pigmentation in the most likely biogeographic background of the collected trace sample and these variants should form the basis for the estimation of skin pigmentation.

In forensics genetic casework, ancestry inference can provide helpful information about an unidentified perpetrator who left a biological trace sample at a crime scene. Several well-established AIM panels that successfully differentiate the major human populations, including Europeans, East Asians, South Central Asians, Native Americans, Sub-Saharan Africans, and Oceanians, have been developed [**14,46,47,48,49**]. Ancestry information may also provide indications on EVCs that are typical for a given population. However, caution must be taken when using ancestry to infer the phenotypes of an individual. An individual of mixed ancestry may have AIM alleles inherited from one parental population, but phenotypic characteristics, e.g., skin color, of the other parental population [**3**,**9**]. Thus, skin color and other EVCs should be estimated using causative genetic variants. We identified four likely causative variants that influence skin pigmentation in Pakistanis and are included in the leading skin color prediction model, HIrisPlex-S [**18**,**19**]. Nevertheless, the variations accounted for only about one-third of the diversity in skin pigmentation, and the other SNPs in the HIrisPlex-S had only a minimal impact on the corrected R^2-values. This highlights the necessity for more genetic research on human skin pigmentation.

5. CONCISE SUMMARY

Skin pigmentation is a highly noticeable and diverse characteristic in people. We conducted a comparison of the genetic variations (alleles) of 163 SNPs and indels using the Human Pigmentation (HuPi) AmpliSeq™ Custom panel. Additionally, we examined the relationship between the biogeographic ancestry and the levels of skin pigmentation on the upper arm, lower arm, and forehead of 299 individuals from three subpopulations in Pakistan: Baloch, Pashtun, and Punjabi. The Precision ID Ancestry Panel was used to estimate the biogeographic ancestry of each individual. The majority of persons have a predominantly mixed heritage, consisting of both South-Central Asian and European ancestry. However, the Baloch individuals exhibited an average amount of Sub-Saharan African ancestry of over 10%, whereas it was less than 1% in the Punjabi and Pashtun populations.

The genetic distances between the Pashtun, Punjabi, and Baloch subpopulations, as determined by the ancestry markers, exhibited statistically significant variations. The skin pigmentation of individuals from the Pashtun subpopulation was considerably lower than that of persons from the Punjabi and Baloch subpopulations, with a p-value of less than 0.05. The study found that the levels of European and Sub-Saharan African ancestry, as well as the presence of five specific SNPs (rs1042602, rs10831496, rs1426654, rs16891982, and rs12913832), were strongly linked to variations in skin pigmentation at the upper arm, lower arm, and forehead among individuals in the Pakistani population. These associations remained statistically significant even after adjusting for multiple testing, with a p-value of less than 10-3. A model incorporating four single nucleotide polymorphisms (SNPs) - rs1426654, rs1042602, rs16891982, and rs12913832 - accounted for 33% of the variation in upper arm skin pigmentation. 37% of the variation in upper arm skin color can be attributed to the four single nucleotide polymorphisms (SNPs) and the proportions of European and Sub-Saharan African ancestry. The findings from our study suggest that the four specific single nucleotide polymorphisms (SNPs), namely rs1426654, rs1042602, rs16891982, and rs12913832, which are situated in the genes SLC24A5, TYR, SLC45A2, and HERC2, respectively, have a crucial role in determining the variation in skin color among the mixed Pakistani subpopulations.

REFERENCES

1. Kayser, M.; Schneider, P.M. DNA-based prediction of human externally visible characteristics in forensics: Motivations, scientific challenges, and ethical considerations. *Forensic Sci. Int. Genet.* **2009**, *3*, 154–161. **[Google Scholar] [CrossRef] [PubMed]**

2. Samuel, G.; Prainsack, B. Forensic DNA phenotyping in Europe: Views "on the ground" from those who have a professional stake in the technology. *New Genet. Soc.* **2019**, *38*, 119–141. **[Google Scholar] [CrossRef] [Green Version]**

3. Andersen, J.D.; Meyer, O.S.; Simão, F.; Jannuzzi, J.; Carvalho, E.; Andersen, M.M.; Pereira, V.; Børsting, C.; Morling, N.; Gusmão, L. Skin pigmentation and genetic variants in an admixed Brazilian population of primarily European ancestry. *Int. J. Legal Med.* **2020**, *134*, 1569–1579. **[Google Scholar] [CrossRef]**

4. Kayser, M.; De Knijff, P. Improving human forensics through advances in genetics, genomics and molecular biology. *Nat. Rev. Genet.* **2011**, *12*, 179–192. **[Google Scholar] [CrossRef] [PubMed]**

5. Maroñas, O.; Phillips, C.; Söchtig, J.; Gomez-Tato, A.; Cruz, R.; Alvarez-Dios, J.; de Cal, M.C.; Ruiz, Y.; Fondevila, M.; Carracedo, Á.; et al. Development of a forensic skin colour predictive test. *Forensic Sci. Int. Genet.* **2014**, *13*, 34–44. **[Google Scholar] [CrossRef]**

6. Parra, E.J.; Kittles, R.A.; Shriver, M.D. Implications of correlations between skin color and genetic ancestry for biomedical research. *Nat. Genet.* **2004**, *36*, S54–S60. **[Google Scholar] [CrossRef] [PubMed]**

7. Stokowski, R.P.; Pant, P.V.K.; Dadd, T.; Fereday, A.; Hinds, D.A.; Jarman, C.; Filsell, W.; Ginger, R.S.; Green, M.R.; Van Der Ouderaa, F.J.; et al. A genomewide association study of skin pigmentation in a South Asian population. *Am. J. Hum. Genet.* **2007**, *81*, 1119–1132. **[Google Scholar] [CrossRef] [Green Version]**

8. Shriver, M.D.; Parra, E.J.; Dios, S.; Bonilla, C.; Norton, H.; Jovel, C.; Pfaff, C.; Jones, C.; Massac, A.; Cameron, N.; et al. Skin pigmentation, biogeographical ancestry and admixture mapping. *Hum. Genet.* **2003**, *112*, 387–399. **[Google Scholar] [CrossRef] [PubMed]**

9. Valenzuela, R.K.; Henderson, M.S.; Walsh, M.H.; Garrison, N.; Kelch, J.T.; Cohen-Barak, O.; Erickson, D.T.; John Meaney, F.; Bruce Walsh, J.; Cheng, K.C. Predicting phenotype from genotype: Normal pigmentation. *J. Forensic Sci.* **2010**, *55*, 315–322. **[Google Scholar] [CrossRef] [Green Version]**

10. Beleza, S.; Johnson, N.A.; Candille, S.I.; Absher, D.M.; Coram, M.A.; Lopes, J.; Campos, J.; Araújo, I.I.; Anderson, T.M.; Vilhjálmsson, B.J.; et al. Genetic Architecture of Skin and Eye Color in an African-European Admixed Population. *PLOS Genet.* **2013**, *9*, e1003372. **[Google Scholar] [CrossRef] [Green Version]**

11. Lona-Durazo, F.; Hernandez-Pacheco, N.; Fan, S.; Zhang, T.; Choi, J.; Kovacs, M.A.; Loftus, S.K.; Le, P.; Edwards, M.; Fortes-Lima, C.A.; et al. Meta-analysis of GWA studies provides new insights on the genetic architecture of skin pigmentation in recently admixed populations. *BMC Genet.* **2019**, *20*, 59. **[Google Scholar] [CrossRef] [Green Version]**

12. Wright, R.P. *The Ancient Indus: Urbanism, Economy, and Society*; Case Studies in Early Societies; Cambridge University Press: Cambridge, UK, 2010; ISBN 9780521576529. **[Google Scholar]**

13. Wynbrandt, J. *A Brief History of Pakistan*; Brief history; Infobase Publishing: New York, NY, USA, 2009; ISBN 9780816061846. **[Google Scholar]**

14. Pereira, V.; Mogensen, H.S.; Børsting, C.; Morling, N. Evaluation of the Precision ID Ancestry Panel for crime case work: A SNP typing assay developed for typing of 165 ancestral informative markers. *Forensic Sci. Int. Genet.* **2017**, *28*, 138–145. **[Google Scholar] [CrossRef] [PubMed]**

15. Mogensen, H.S.; Tvedebrink, T.; Børsting, C.; Pereira, V.; Morling, N. Ancestry prediction efficiency of the software GenoGeographer using a z-score method and the ancestry informative markers in the Precision ID Ancestry Panel. *Forensic Sci. Int. Genet.* **2020**, *44*, 102154. **[Google Scholar] [CrossRef] [PubMed]**

16. Al-Asfi, M.; McNevin, D.; Mehta, B.; Power, D.; Gahan, M.E.; Daniel, R. Assessment of the Precision ID Ancestry panel. *Int. J. Legal Med.* **2018**, *132*, 1581–1594. **[Google Scholar] [CrossRef] [PubMed]**

17. Meyer, O.S.; Andersen, J.D.; Børsting, C. Presentation of the Human Pigmentation (HuPi) AmpliSeq™ custom panel. *Forensic Sci. Int. Genet. Suppl. Ser.* **2019**, *7*, 478–479. **[Google Scholar] [CrossRef]**

18. Walsh, S.; Chaitanya, L.; Breslin, K.; Muralidharan, C.; Bronikowska, A.; Pospiech, E.; Koller, J.; Kovatsi, L.; Wollstein, A.; Branicki, W.; et al. Global skin colour prediction from DNA. *Hum. Genet.* **2017**, *136*, 847–863. **[Google Scholar] [CrossRef] [Green Version]**

19. Chaitanya, L.; Breslin, K.; Zuñiga, S.; Wirken, L.; Pośpiech, E.; Kukla-Bartoszek, M.; Sijen, T.; de Knijff, P.; Liu, F.; Branicki, W.; et al. The HIrisPlex-S system for eye, hair and skin colour prediction from DNA: Introduction and forensic developmental validation. *Forensic Sci. Int. Genet.* **2018**, *35*, 123–135. **[Google Scholar] [CrossRef] [PubMed] [Green Version]**

20. Kongshoj, B.; Thorleifsson, A.; Wulf, H.C. Pheomelanin and eumelanin in human skin determined by high-performance liquid chromatography and its relation to in vivo reflectance measurements. *Photodermatol. Photoimmunol. Photomed.* **2006**, *22*, 141–147. **[Google Scholar] [CrossRef] [PubMed]**

21. Ravnbak, M.H. Objective determination of Fitzpatrick skin type. *Dan. Med. Bull.* **2010**, *57*, B4153. **[Google Scholar]**

22. Farzad, M.S.; Pedersen, B.M.; Mogensen, H.S.; Børsting, C. Development of an automated AmpliSeq™ library building workflow for biological stain samples on the Biomek(®) 3000. *Biotechniques* **2020**, *68*, 342–344. **[Google Scholar] [CrossRef] [Green Version]**

23. Barrett, J.C.; Fry, B.; Maller, J.; Daly, M.J. Haploview: Analysis and visualization of LD and haplotype maps. *Bioinformatics* **2005**, *21*, 263–265. **[Google Scholar] [CrossRef] [Green Version]**

24. Excoffier, L.; Lischer, H.E.L. Arlequin suite ver 3.5: A new series of programs to perform population genetics analyses under Linux and Windows. *Mol. Ecol. Resour.* **2010**, *10*, 564–567. **[Google Scholar] [CrossRef] [PubMed]**

25. Falush, D.; Stephens, M.; Pritchard, J.K. Inference of population structure using multilocus genotype data: Linked loci and correlated allele frequencies. *Genetics* **2003**, *164*, 1567–1587. **[Google Scholar] [CrossRef]**

26. Pritchard, J.K.; Stephens, M.; Donnelly, P. Inference of Population Structure Using Multilocus Genotype Data. *Genetics* **2000**, *155*, 945–959. **[Google Scholar] [CrossRef]**

27. Jakobsson, M.; Rosenberg, N.A. CLUMPP: A cluster matching and permutation program for dealing with label switching and multimodality in analysis of population structure. *Bioinformatics* **2007**, *23*, 1801–1806. **[Google Scholar] [CrossRef] [PubMed] [Green Version]**

28. Rosenberg, N.A. DISTRUCT: A program for the graphical display of population structure. *Mol. Ecol. Notes* **2004**, *4*, 137–138. **[Google Scholar] [CrossRef]**

29. Simayijiang, H.; Børsting, C.; Tvedebrink, T.; Morling, N. Analysis of Uyghur and Kazakh populations using the Precision ID Ancestry Panel. *Forensic Sci. Int. Genet.* **2019**, *43*, 102144. **[Google Scholar] [CrossRef] [PubMed]**

30. Venables, W.N.; Ripley, B.D. *Modern Applied Statistics with S*; Springer: New York, NY, USA, 2002. **[Google Scholar]**

31. Shan, M.A.; Børsting, C.; Morling, N. Forensic application and genetic diversity of 21 autosomal STR loci in five major population groups of Pakistan. *Int. J. Legal Med.* **2021**, *135*, 775–777. **[Google Scholar] [CrossRef]**

32. Korn, A.; Nourzaei, M. "Those were the hungry years": A glimpse of Coastal Afro-Balochi. *J. R. Asiat. Soc.* **2018**, *28*, 661–695. **[Google Scholar] [CrossRef] [Green Version]**

33. Siddiqi, M.H.; Akhtar, T.; Rakha, A.; Abbas, G.; Ali, A.; Haider, N.; Ali, A.; Hayat, S.; Masooma, S.; Ahmad, J.; et al. Genetic characterization of the Makrani people of Pakistan from mitochondrial DNA control-region data. *Leg. Med.* **2015**, *17*, 134–139. **[Google Scholar] [CrossRef] [PubMed]**

34. Mallick, C.B.; Iliescu, F.M.; Möls, M.; Hill, S.; Tamang, R.; Chaubey, G.; Goto, R.; Ho, S.Y.W.; Gallego Romero, I.; Crivellaro, F.; et al. The light skin allele of SLC24A5 in South Asians and Europeans shares identity by descent. *PLoS Genet.* **2013**, *9*, e1003912. **[Google Scholar] [CrossRef] [Green Version]**

35. Jonnalagadda, M.; Faizan, M.A.; Ozarkar, S.; Ashma, R.; Kulkarni, S.; Norton, H.L.; Parra, E. A Genome-Wide Association Study of Skin and Iris Pigmentation among Individuals of South Asian Ancestry. *Genome Biol. Evol.* **2019**, *11*, 1066–1076. **[Google Scholar] [CrossRef] [Green Version]**

36. Lamason, R.L.; Mohideen, M.-A.P.K.; Mest, J.R.; Wong, A.C.; Norton, H.L.; Aros, M.C.; Jurynec, M.J.; Mao, X.; Humphreville, V.R.; Humbert, J.E.; et al. SLC24A5, a Putative Cation Exchanger, Affects Pigmentation in Zebrafish and Humans. *Science* **2005**, *310*, 1782–1786. **[Google Scholar] [CrossRef] [Green Version]**

37. Nan, H.; Kraft, P.; Qureshi, A.A.; Guo, Q.; Chen, C.; Hankinson, S.E.; Hu, F.B.; Thomas, G.; Hoover, R.N.; Chanock, S.; et al. Genome-wide association study of tanning phenotype in a population of European ancestry. *J. Investig. Dermatol.* **2009**, *129*, 2250–2257. **[Google Scholar] [CrossRef] [PubMed] [Green Version]**

38. Cook, A.L.; Chen, W.; Thurber, A.E.; Smit, D.J.; Smith, A.G.; Bladen, T.G.; Brown, D.L.; Duffy, D.L.; Pastorino, L.; Bianchi-Scarra, G. Analysis of cultured human melanocytes based on polymorphisms within the SLC45A2/MATP, SLC24A5/NCKX5, and OCA2/P loci. *J. Investig. Dermatol.* **2009**, *129*, 392–405. **[Google Scholar] [CrossRef] [Green Version]**

39. Chaki, M.; Sengupta, M.; Mondal, M.; Bhattacharya, A.; Mallick, S. Molecular and functional studies of tyrosinase variants among Indian oculocutaneous albinism type 1 patients. *J. Investig. Dermatol.* **2011**, *131*, 260–262. **[Google Scholar] [CrossRef]**

40. Jagirdar, K.; Smit, D.J.; Ainger, S.A.; Lee, K.J.; Brown, D.L.; Chapman, B.; Zhen Zhao, Z.; Montgomery, G.W.; Martin, N.G.; Stow, J.L. Molecular analysis of common polymorphisms within the human Tyrosinase locus and genetic association with pigmentation traits. *Pigment Cell Melanoma Res.* **2014**, *27*, 552–564. **[Google Scholar] [CrossRef] [PubMed] [Green Version]**

41. Eiberg, H.; Troelsen, J.; Nielsen, M.; Mikkelsen, A.; Mengel-From, J.; Kjaer, K.W.; Hansen, L. Blue eye color in humans may be caused by a perfectly associated founder mutation in a regulatory element located within the HERC2 gene inhibiting OCA2 expression. *Hum. Genet.* **2008**, *123*, 177–187. **[Google Scholar]** **[CrossRef]** **[PubMed]**

42. Sturm, R.A.; Duffy, D.L.; Zhao, Z.Z.; Leite, F.P.N.; Stark, M.S.; Hayward, N.K.; Martin, N.G.; Montgomery, G.W. A single SNP in an evolutionary conserved region within intron 86 of the HERC2 gene determines human blue-brown eye color. *Am. J. Hum. Genet.* **2008**, *82*, 424–431. **[Google Scholar]** **[CrossRef]** **[Green Version]**

43. Liu, F.; van Duijn, K.; Vingerling, J.R.; Hofman, A.; Uitterlinden, A.G.; Janssens, A.C.J.W.; Kayser, M. Eye color and the prediction of complex phenotypes from genotypes. *Curr. Biol.* **2009**, *19*, R192–R193. **[Google Scholar]** **[CrossRef]** **[PubMed]** **[Green Version]**

44. Visser, M.; Kayser, M.; Palstra, R.-J. HERC2 rs12913832 modulates human pigmentation by attenuating chromatin-loop formation between a long-range enhancer and the OCA2 promoter. *Genome Res.* **2012**, *22*, 446–455. **[Google Scholar]** **[CrossRef]** **[PubMed]** **[Green Version]**

45. Pavan, W.J.; Sturm, R.A. The genetics of human skin and hair pigmentation. *Annu. Rev. Genomics Hum. Genet.* **2019**, *20*, 41–72. **[Google Scholar]** **[CrossRef]**

46. Kosoy, R.; Nassir, R.; Tian, C.; White, P.A.; Butler, L.M.; Silva, G.; Kittles, R.; Alarcon-Riquelme, M.E.; Gregersen, P.K.; Belmont, J.W.; et al. Ancestry informative marker sets for determining continental origin and admixture proportions in common populations in America. *Hum. Mutat.* **2009**, *30*, 69–78. **[Google Scholar]** **[CrossRef]** **[PubMed]** **[Green Version]**

47. Kidd, K.K.; Speed, W.C.; Pakstis, A.J.; Furtado, M.R.; Fang, R.; Madbouly, A.; Maiers, M.; Middha, M.; Friedlaender, F.R.; Kidd, J.R. Progress toward an efficient panel of SNPs for ancestry inference. *Forensic Sci. Int. Genet.* **2014**, *10*, 23–32. **[Google Scholar]** **[CrossRef]** **[Green Version]**

48. Phillips, C.; Parson, W.; Lundsberg, B.; Santos, C.; Freire-Aradas, A.; Torres, M.; Eduardoff, M.; Børsting, C.; Johansen, P.; Fondevila, M.; et al. Building a forensic ancestry panel from the ground up: The EUROFORGEN Global AIM-SNP set. *Forensic Sci. Int. Genet.* **2014**, *11*, 13–25. **[Google Scholar]** **[CrossRef]** **[Green Version]**

49. Phillips, C.; McNevin, D.; Kidd, K.K.; Lagacé, R.; Wootton, S.; de la Puente, M.; Freire-Aradas, A.; Mosquera-Miguel, A.; Eduardoff, M.; Gross, T.; et al. MAPlex—A massively parallel sequencing ancestry analysis multiplex for Asia-Pacific populations. *Forensic Sci. Int. Genet.* **2019**, *42*, 213–226. **[Google Scholar]** **[CrossRef]** **[PubMed]** **[Green Version]**

22 AN ASSESSMENT OF OPENARRAY™ AS A GENOTYPING TECHNIQUE FOR FORENSIC DNA PHENOTYPING AND HUMAN IDENTIFICATION

1. INTRODUCTION: AN OVERVIEW

Short tandem repeats (STRs) are genetic variations commonly employed in forensic investigations to identify perpetrators and in cases involving connection testing [1,2]. Nevertheless, the multiplexing capability of Short Tandem Repeats (STRs) is comparatively limited due to the requirement for capillary electrophoresis and the inability to accept a significant number of markers. It is widely recognized in the genomic era that STRs are not the sole contributor to genetic diversity in humans [3]. Single-nucleotide polymorphisms (SNPs) are the prevailing form of genetic variation in individuals and can offer further insights, such as the assumed ancestral origin of a DNA sample and phenotypic traits like eye and hair color [4]. SNPs can be utilized in forensic DNA analysis due to their numerous possible markers, capacity to be automated, and the possibility of reducing the required fragment length to only 60-80 base pairs [5,6,7].

Forensic analysis typically entails comparing DNA profiles obtained from biological samples gathered at a crime scene with things and people believed to be linked to the crime. The efficacy of forensic DNA analysis has been bolstered by the establishment of DNA databases, which enable the identification of unidentified perpetrators and serial offenders by connecting many criminal incidents [1,8,9]. Nevertheless, in criminal cases where there is no match with a STR profile, law enforcement can benefit from several types of genetic studies.

Within this framework, the examination of single nucleotide polymorphisms (SNPs) holds great potential for investigating variations in physical traits (forensic DNA phenotyping, FDP) and determining the geographical origin of individuals [1,10]. The primary benefit of this approach is to reduce the pool of potential contributors of trace evidence at the crime scene to a smaller group of individuals who are very likely to possess the externally observable traits and biogeographic ancestry that were deduced from the DNA found at the crime scene [11]. Forensic DNA phenotyping is officially authorized and governed by legislation in two European Union member states, namely the Netherlands and Slovakia. It is also carried out in accordance with existing rules in six more countries, namely Poland, Czech Republic, Sweden, Hungary, Austria, and Spain, as well as in the United Kingdom [11].

In the era of genomics, several methodologies have been devised to examine numerous genetic variations for forensic applications, including phenotypic determination, ancestral analysis, and identification. The utilization of TaqMan chemistry in real-time PCR is a widely recognized technology that is regularly employed in the operational procedures of forensic laboratories for the purpose of quantifying DNA obtained from biological materials [12,13]. The OpenArray™ system, developed by Thermo Fisher Scientific in Waltham, MA, USA, utilizes TaqMan chemistry to enable the simultaneous and high-throughput analysis of samples and single nucleotide polymorphisms (SNPs) [14,15]. In this study, we conducted a preliminary investigation to assess the suitability of OpenArray™ technology (Thermo Fisher Scientific, Waltham, MA, USA) for the analysis of SNPs in human identification and phenotype inference. In order to achieve this objective, we chose a comprehensive set of 41 SNPs from the HIrisPlex-S tool, which is known for predicting eye, hair, and skin color. Additionally, we selected 19 SNPs from the Precision ID Identity SNP Panel (Thermo Fisher Scientific, Waltham, MA, USA) and ForenSeq DNA Signature Prep Kit (Verogen Inc, San Diego, CA, USA) for individual identification purposes. A total of 314 samples were analyzed to assess the genotyping accuracy of OpenArray™ by comparing it with resequencing data. The platform's sensitivity was evaluated by employing four consecutive dilutions of DNA, containing 10, 5, 1, and 0.5 ng respectively [16,17,18].

2. MATERIALS AND METHODS

2.1. Selection of SNPs

In this work we selected 41 SNPs for eye, hair, and skin color prediction from the HIrisPlex-S system and 19 SNPs from a universal individual identification SNPs panel validated for forensic purposes [**16**,**17**,**18**,**19**]. The SNPs for human identification were selected based on their chromosomal location (each SNP is mapped to a different chromosome to avoid linkage and linkage disequilibrium effects). These genetic variants were included in a customized panel of 60 SNPs able to provide both human identification and phenotyping information in the same analytical session. The TaqMan probe related to rs1805008 could not be developed. Therefore, rs11538871 was chosen because of its complete linkage disequilibrium (LD) with rs1805008 [**20**].

The selected SNPs are summarized in **Table 1**.

Table 1. Single-nucleotide polymorphism (SNP) details. FDP: forensic DNA phenotyping; ID: SNPs for forensic identification [16,17,18,19].

Reference SNP Number	Assay Name	Call Rate (%)	Concordance Rate (%)	FDP/ID
rs3114908	C__27458355	100%	100%	skin
rs1800414	C___8866240_10	100%	100%	skin
rs10756819	C_____69643_10	97.8%	97.5%	skin
rs2238289	C__26683035_10	97.8%	97.5%	skin
rs17128291	C__34508041_10	100%	100%	skin
rs6497292	C__31479145_10	97.8%	100%	skin
rs1129038	C____489033_10	97.8%	100%	skin
rs1667394	C__26683195_10	97.8%	100%	skin
rs1126809	C_175675702_10	100%	100%	skin
rs1470608	C___8866144_10	97.8%	100%	skin
rs1426654	C___2908190_10	100%	100%	skin
rs6119471	C__30035872_10	100%	100%	skin
rs1545397	C__11751858_10	100%	100%	skin
rs6059655	C_189135912_10	100%	100%	skin
rs12441727	C___1289851_10	97.8%	100%	skin
rs8051733	C__43670381_20	100%	100%	skin
rs3212355	C__32390816_10	100%	100%	hair
rs312262906/ rs796296176	C_339644346_10	100%	100%	hair
rs11547464	C__27541634_10	100%	100%	hair
rs885479	C___7519060_20	100%	100%	hair
rs1805008 (rs11538871)	C_175977481_10	100%	100%	hair
rs1805005	C___7519054_10	100%	100%	hair
rs1805006	C___7519055_20	100%	100%	hair
rs1805007	C___2033404_30	97.8%	100%	hair
rs1805009	C___1010501_30	100%	100%	hair
rs201326893	C_190249836_20	100%	100%	hair
rs2228479	C___2033405_20	97.8%	100%	hair
rs1110400	ANKCDY2	97.8%	100%	hair
rs28777	C___2390574_10	100%	100%	hair
rs12821256	C__26988870_10	100%	100%	hair

Reference SNP Number	Assay Name	Call Rate (%)	Concordance Rate (%)	FDP/ID
rs4959270	C__29334795_10	97.8%	100%	hair
rs1042602	C___8362862_10	95.7%	100%	hair
rs2402130	C__27204528_20	100%	100%	hair
rs2378249	C__16003674_10	100%	100%	hair
rs683	C___3119206_10	97.8%	100%	hair
rs16891982	C___2842665_10	100%	100%	eye
rs12203592	C__31918199_10	100%	100%	eye
rs1800407	C___8866200_10	97.8%	97.5%	eye
rs12913832	C__30724404_10	97.8%	100%	eye
rs12896399	C___3244615_10	100%	100%	eye
rs1393350	C___9491300_10	100%	100%	eye
rs10488710	C___2450075_30	97.8%	100%	ID
rs2920816	C__16156638_10	97.8%	100%	ID
rs6955448	C__29220288_10	100%	100%	ID
rs1058083	C___1619935_1_	95.7%	97.4%	ID
rs221956	C____824925_10	100%	100%	ID
rs13182883	C___2556113_10	100%	100%	ID
rs1821380	C__11673733_10	100%	100%	ID
rs6811238	C__11245682_10	100%	100%	ID
rs430046	C___9603287_10	100%	100%	ID
rs576261	C____788229_10	97.8%	100%	ID
rs10092491	C___2049946_10	95.7%	100%	ID
rs560681	C___1006721_1_	100%	100%	ID
rs1736442	C___3285337_1_	91.3%	97.3%	ID
rs4364205	C__26449463_10	95.7%	100%	ID
rs445251	C___2997607_10	100%	100%	ID
rs7041158	C__29060279_10	97.8%	100%	ID
rs13218440	C___9371416_10	100%	100%	ID
rs9786184	C__29554891_10	100%	100%	ID
rs2032652	C___2259382_10	100%	100%	ID

2.2. Ethical Committee Approval

These investigations were carried out following the rules of the Declaration of Helsinki of 1975, revised in 2013. According to Point 23 of this declaration, approval from the ethical committee of Policlinico Tor Vergata (protocol number: Prot. CE/ PROG.177/20) was obtained.

2.3. Reference DNA Samples

A total of 314 reference DNA samples were used. In particular, 114 genomic DNA samples were obtained from the 1000 Genomes Project (Coriell Institute for Medical Research, Camden, NJ, USA) and 200 blood samples were obtained from the Laboratory of Genomic Medicine UILDM (Italian Muscular Dystrophy Association) Santa Lucia Foundation [**21**]. The 114 samples from 1000 Genomes were selected from a small town near Florence in the Tuscany region of Italy. The genotype data of genomic DNA samples from 1000 Genomes were extracted from the Ensembl genome browser [**22**]. An additional four volunteer donors were used for the sensitivity testing.

2.4. Biological Samples from Real Cases

We analyzed a total of 21 samples of biological evidence that had single STR profiles generated during the routine workflow and that were assigned to their reference perpetrator. This evidence was from different tissues (10 from saliva, 2 from semen, 7 from blood, 2 from urine) and was attributed to 8 different subjects from different tissues [**23,24**]. All samples were marked with a specific internal code, as shown in **Table 2** [**2**]. The forensic samples were loaded in the same plate twice with the purpose of verifying the reproducibility of genotyping assignation.

Table 2. Details of forensic biological samples. Subject A (SubjA), Subject B (SubjB), Subject C (SubjC), Subject D (SubjD), Subject E (SubjE), Subject F (SubjF), Subject G (SubjG), and Subject H (SubjH).

Sample	Source	Subject	[ng/μL]	Call Rate (%)	Predicted Phenotype	Real Phenotype
20-352	Saliva	SubjA	9.483	97.5	blue eyes; brown hair/dark hair; pale skin	confirmed
20-367	Blood	SubjA	1.263	91.7		
20-368	Blood	SubjB	2.322	93.3	blue eyes; blond hair/light hair; intermediate skin	confirmed
20-369	Blood	SubjC	6.609	97.5	brown eyes; brown hair/dark hair; dark skin	confirmed
20-361	Saliva	SubjD	18.493	95.8	brown eyes; brown hair/ dark hair; intermediate skin	confirmed
20-370	Blood	SubjD	2.178	94.2		

<table>
<tr><th>Sample</th><th>Source</th><th>Subject</th><th>[ng/µL]</th><th>Call Rate (%)</th><th>Predicted Phenotype</th><th>Real Phenotype</th></tr>
<tr><td>20-364</td><td>Saliva</td><td>SubjE</td><td>36.327</td><td>96.7</td><td rowspan="2">brown eyes; red hair; intermediate skin</td><td rowspan="2">confirmed</td></tr>
<tr><td>20-371</td><td>Blood</td><td>SubjE</td><td>1.657</td><td>95.0</td></tr>
<tr><td>20-144</td><td>Blood</td><td>SubjF</td><td>0.078</td><td>16.7</td><td rowspan="4">blue eyes; blond hair/light hair; intermediate skin</td><td rowspan="4">confirmed</td></tr>
<tr><td>20-145</td><td>Semen</td><td>SubjF</td><td>7.022</td><td>92.5</td></tr>
<tr><td>20-146</td><td>Urine</td><td>SubjF</td><td>0.009</td><td>4.2</td></tr>
<tr><td>20-147</td><td>Saliva</td><td>SubjF</td><td>5.534</td><td>95.0</td></tr>
<tr><td>20-148</td><td>Blood</td><td>SubjG</td><td>4.994</td><td>96.7</td><td rowspan="4">brown eyes; brown hair/ light hair; intermediate skin</td><td rowspan="4">confirmed</td></tr>
<tr><td>20-149</td><td>Semen</td><td>SubjG</td><td>0.642</td><td>52.5</td></tr>
<tr><td>20-150</td><td>Urine</td><td>SubjG</td><td>0.012</td><td>0.8</td></tr>
<tr><td>20-151</td><td>Saliva</td><td>SubjG</td><td>20.448</td><td>99.2</td></tr>
<tr><td>20-065</td><td>Saliva</td><td>SubjH</td><td>18.562</td><td>97.5</td><td rowspan="5">brown eyes; brown hair/ dark hair; intermediate skin</td><td rowspan="5">confirmed</td></tr>
<tr><td>20-069</td><td>Saliva</td><td>SubjH</td><td>19.640</td><td>97.5</td></tr>
<tr><td>20-071</td><td>Saliva</td><td>SubjH</td><td>13.438</td><td>98.3</td></tr>
<tr><td>20-072</td><td>Saliva</td><td>SubjH</td><td>12.512</td><td>97.5</td></tr>
<tr><td>20-074</td><td>Saliva</td><td>SubjH</td><td>10.603</td><td>97.5</td></tr>
</table>

2.5. DNA Purification and Quantification

In accordance with the quality standards of forensic laboratories that require the use of automatized instruments, the genomic DNA of 225 samples were extracted using the Maxwell®16 MDx Instrument (Promega, Madison, WI, USA) and the DNA IQ Casework Pro Kit for Maxwell®16 (Promega, Madison, WI, USA).

Each DNA sample was quantified using the Quantifiler®Trio DNA Quantification Kit (Thermo Fisher Scientific, Waltham, MA, USA).

2.6. OpenArray™ Technology

OpenArray™ Technology (Thermo Fisher Scientific, Waltham, MA, USA) is a high-throughput real-time PCR genotyping method that allows for rapid screening of several TaqMan assays in many samples [**13**]. This real-time method involves the use of an array composed of 3072 through-holes running on the QuantStudio 12K Flex Real Time PCR System (Thermo Fisher Scientific, Waltham, MA, USA) with an OpenArray™ block. The OpenArray™ system is composed of a specific plate (OpenArray™ plate) enabling 192 samples to be typed in a single run [**14**].

For each sample, 5 ng (volume of 3 µL) of extracted DNA from the reference samples and 3 µL of 2× TaqMan OpenArray™ Genotyping Master Mix were manually loaded into 384 well-plates according to the manufacturer's instructions (Thermo

Fisher Scientific, Waltham, MA, USA). A negative control was obtained by adding 3 μL of pure distilled water to the Master Mix. The QuantStudio 12K Flex OpenArray™ AccuFill System transferred the previously generated mix to the TaqMan OpenArray™ plate. The amplification was performed using the QuantStudio 12K Flex Real Time PCR System (Thermo Fisher Scientific, Waltham, MA, USA) instrument, and the results were analyzed using TaqMan Genotyper Software (Thermo Fisher Scientific, Waltham, MA, USA) [**14**]. For each SNP, the call rate and concordance rate were calculated as described in **Section 2.7** [25].

2.7. Sensitivity

For sensitivity studies, genotyping results from four volunteer donors were assessed from serial dilutions of DNA consisting of 10 ng, 5 ng, 1 ng, and 0.5 ng. These four volunteer donors provided a total of one sample (buccal swab) each.

2.8. Resequencing Analysis

The genotyping accuracy was assessed through comparison with resequencing data. Resequencing data were available from the Ensembl genome browser for the 1000 Genomes Project samples and from an internal database for the Laboratory of Genomic Medicine UILDM Santa Lucia Foundation samples. The PCR sequence reactions were performed with BigDye Terminator v3.1 (Thermo Fisher Scientific, Waltham, MA, USA) and the purification with BigDyeXTerminator (Thermo Fisher Scientific, Waltham, MA, USA) according to the manufacturer's instructions. The samples were run on ABI3130xl (Thermo Fisher Scientific, Waltham, MA, USA) and the obtained sequences were read by Sequencing Analysis software v5.2 (Thermo Fisher Scientific, Waltham, MA, USA) [**13**].

2.9. Statistical Analysis

Upon completion of the analytical assay, the TaqMan Genotyper Software provided the call rate, which represents the percentage of successful genotyping. The call rate is the proportion of single nucleotide polymorphisms (SNPs) that received a genotype call from the program, out of the total number of SNPs that were typed. As per the software's guidelines, a call rate is considered successful if it exceeds 90%. The concordance rate refers to the ratio of SNPs identified by OpenArray™ that match the resequencing data. The determination of this value was made by comparing the allele calls produced by OpenArray™ technology with the genotypes obtained from resequencing data [25]. Concordance refers to the proportion of single nucleotide polymorphisms (SNPs) that yield identical outcomes when comparing data obtained from both OpenArray™ and resequencing methods.

For phenotypic characterization of biological forensic samples (n = 17), we constructed a statistical prediction model using HIrisPlex-S webtool systems, available at **https://hirisplex.erasmusmc.nl/** (accessed on 20 January 2021). Each phenotype was predicted by applying established algorithms for the color of the eyes, hair, and skin, which returns a *p*-value (prediction probability) [**19**]. In terms of eye color, the highest *p*-value is assigned to the most probable phenotype [**19,26**]. For the interpretation of hair color, we considered the hair color prediction guide developed by Walsh S. et al. [**23,27**]. This guide examines categorical hair color probabilities in combination with light/dark hair color shade probabilities as obtained from genotype data [**23,27**]. For the interpretation of skin color, we used the guide developed by L. Chaitanya et al. [**16**]. This guide considers the highest *p*-value in combination with the second highest *p*-value [**16,28**]. For example, if in the pale category we obtained a *p*-value greater than 0.7, this phenotypic characteristic depends on the effect of the second highest category if this *p*-value is greater than 0.15. In particular, it will appear darker if the second most likely category is intermediate, whereas it will appear lighter if the second most likely category is pale [**16**].

Prior visual phenotyping was performed by two blind assessments (two independent individuals from our laboratory not involved in this study) to verify the validity of the software prediction.

In order to verify the discrimination power of 17 human identification SNPs (not including rs9786184 and rs2032652 on the Y chromosome), we determined the random match probability (RMP) [**29**]. The random match probability is the estimated frequency of a genetic profile in the reference population and is calculated using allele frequencies from that population group. Therefore, the RMP is a useful value for evaluating the discriminating power of a DNA profiling system, and it indicates the probability of obtaining a match between two distinct and unrelated individuals [**29**].

3. RESULTS

3.1. Genotyping of the Control Sample and Concordance

The OpenArray™ System (Thermo Fisher Scientific, Waltham, MA, USA) successfully analyzed 100% of the reference DNA samples. The average call rate and the average concordance rate were 96.9 ± 1.72 and 99.8 ± 0.56, respectively [**25**]. The call rate was also calculated for each SNP. The average SNP call rate was 98.9% ± 1.67, with values from 91.3–100% (**Table 1**).

3.2. Sensitivity Study

The results of the sensitivity study are shown in **Table 3**. A call rate higher than or equal to 90% was obtained for all samples. The results obtained demonstrated that the system is able to correctly type DNA evidence in quantities starting from 0.5 ng.

Table 3. The call rate and concordance rate obtained from the sensitivity study.

Sample	Call Rate (%)	Concordance Rate (%)
SAMPLE_1_10 ng	100	100
SAMPLE_1_5 ng	100	100
SAMPLE_1_1 ng	100	100
SAMPLE_1_0.5 ng	100	100
SAMPLE_2_10 ng	95	100
SAMPLE_2_5 ng	98.3	100
SAMPLE_2_1 ng	96.7	100
SAMPLE_2_0.5 ng	96.7	100
SAMPLE_3_10 ng	96.7	100
SAMPLE_3_5 ng	95	100
SAMPLE_3_1 ng	90	98.21
SAMPLE_3_0.5 ng	91.7	100
SAMPLE_4_10 ng	96.7	100
SAMPLE_4_5 ng	95	100
SAMPLE_4_1 ng	91.7	98.28
SAMPLE_4_0.5 ng	95	100

3.3. Genotyping of Biological Evidence from Real Cases

A total of 21 single biological evidence samples from different tissues (10 from saliva, 2 from semen, 7 from blood, 2 from urine), previously assigned to their reference perpetrator, were analyzed using the OpenArray™ System. This evidence was attributed to eight different subjects [**23,24**] (**Table 2**). A total of 17 samples presented call rates higher than 90%, with only 4 samples showing call rates below 90% (**Table 2**). As expected, these 4 samples were characterized by an extremely low starting concentration (**Table 2**).

We typed two replicates of each sample to verify the genotype call reliability. We obtained concordant results between the two replicates for all samples.

3.3.1. Predicting Human Appearance from Forensic Samples

Once the correct allelic call was ascertained for all subjects, it was possible to carry out predictive analysis on the HIrisPlex-S System software [**19**].

Table 2 reports the phenotype predictions given by the HIrisPlex-S System software [**19**].

3.3.2. RMP Calculation

Calculation of the random match probability (RMP) was performed. The calculation considered 17 SNPs located on the autosomal chromosomes. The frequencies relating to the European (EUR), African (AFR), and East Asian (EAS) populations of the 1000 Genomes Project, available on the Ensembl genome browser, were used [22].

Table 4 shows the RMP values obtained for each subject.

Table 4. The random match probability (RMP) values relating to the eight subjects. European (EUR), African (AFR), and East Asian (EAS).

Subject	RMP_{EUR}	RMP_{AFR}	RMP_{EAS}
SubjA	2.67×10^{-8}	1.48×10^{-8}	1.56×10^{-8}
SubjB	5.60×10^{-9}	1.57×10^{-9}	3.51×10^{-9}
SubjC	1.15×10^{-7}	2.49×10^{-8}	$4.51 \times 10_{-8}$
SubjD	$4.24 \times 10_{-7}$	1.91×10^{-7}	1.37×10^{-7}
SubjE	$2.89 \times 10_{-8}$	1.51×10^{-8}	$2.81 \times 10_{-8}$
SubjF	1.80×10^{-9}	6.07×10^{-10}	3.27×10^{-10}
SubjG	$2.23 \times 10_{-7}$	8.95×10^{-8}	1.19×10^{-7}
SubjH	4.26×10^{-8}	2.30×10^{-8}	6.44×10^{-8}

These numbers indicate that the studied SNPs do not possess enough discriminatory ability to definitively identify an individual in the global population. However, they can be utilized to validate genetic compatibility with suspects, which can then be confirmed using standard STR analysis [30,31].

4. DISCUSSION

As far as we know, this is the initial utilization of the OpenArray™ System in the forensic domain. We developed a set of 60 single nucleotide polymorphisms (SNPs) for the purpose of estimating phenotypes and identifying individuals [2]. The SNPs used for predicting eye, hair, and skin color were directly chosen from the HIrisPlex-S system. The SNPs used for human identification were selected from subsets that had already been validated, namely the Precision ID Identity SNP Panel (Thermo Fisher Scientific, Waltham, MA, USA) and the ForenSeq DNA Signature Prep Kit (Verogen Inc, San Diego, CA, USA) [16,17,18,19]. The SNPs were initially assessed on 314 reference samples, with an average call rate of 96.9% and a concordance percentage approaching 99.8%.

Sensitivity testing performed on serial dilutions demonstrated that samples with 0.5 ng of total input DNA can be correctly typed.Of 21 biological evidence samples from different tissues (10 from saliva, 2 from semen, 7 from blood, 2 from urine), a total of 17 were correctly typed using OpenArray™, revealing a call rate higher than 90%.

Regarding identification purposes, since most SNPs are biallelic, they are less informative for identity testing than STR analyses [**32**]. Thus, many more SNPs will be needed to achieve the same level of discrimination afforded by the commonly used STR loci [**33**]. However, our system is not intended to replace the present STR analysis for human identification. Despite the reduced discrimination power of this system compared to STR based kits, the OpenArray™ System can be used to exclude suspects, to prioritize samples for downstream analyses providing well-established information about the prediction of eye color, hair color, and skin pigmentation.

It should be used primarily as a tool for phenotype prediction, while identification should also be supported by traditional methods. A total of 17 SNPs for human identification were used in this study, yielding a mean RMP of approximately 10^{-8}.It is important to outline that our work was intended to provide a new tool, based on previously validated SNPs, for phenotype prediction and prioritizing samples for downstream analyses for human identification [**16,17,18,19,34,35**]. The HIrisPlex-S panel and the related software were previously validated elsewhere [**16,19**]. Thus, our work was aimed at verifying the ability of OpenArray™ to correctly type the selected SNPs.

The main strengths of the OpenArray™ System are the speed of the analysis (less than four hours against the current 24–48 h), the standardized interpretation of the results, the low cost for the analysis of a single sample (about 10 euros, in the region of 20% of the current average) and the opportunity to provide both a first evaluation of biological compatibility and phenotypic information. Moreover, TaqMan chemistry is a well-known technology routinely used in the workflow of forensic laboratories to quantify the DNA extracted from biological samples.

An existing limitation is the requirement to employ an additional system based on conventional STRs for the assessment of biological compatibility during sample analysis. In addition, the OpenArray™ System is not utilized in the workflow of forensic genetics laboratories, hence restricting its accessibility for regular utilization.

5. AT A GLANCE

To the best of our knowledge, this study represents the first application of the OpenArray™ System as a genotyping tool in forensic genetics. In particular, the results obtained suggest that the OpenArray™ System is robust and reliable and is suitable for use in forensic genotyping. Herein, we developed a panel for phenotypic prediction and human identification. Although the 19 SNPs used provide useful information for personal identification, a greater number of SNPs is required for the reliable identification of a subject.

The quality of the results obtained from biological evidence also promotes the application of this system in forensic laboratories to assist investigators in resolving cases. Finally: these data suggest the applicability of this system as a genotyping method in the forensic field. In particular, it could be suitable as a first line method to verify the presence of genetic compatibility between evidence and suspects and to achieve rapid investigative information also providing well-founded predictions of eye color, hair color, and skin pigmentation. More studies will be needed for further validation of this technology and to consider opportunities to implement this custom array with more SNPs to lower the RMP and to include markers for studies of ancestry and lineage.

6. CONCISE SUMMARY

An OpenArray™ technology custom plate was utilized to assess 60 single-nucleotide polymorphisms (SNPs) that have been confirmed to accurately predict eye color, hair color, skin pigmentation, and personal identification. The SNPs were chosen from pre-validated subsets, specifically the Hirisplex-s, Precision ID Identity SNP Panel, and ForenSeq DNA Signature Prep Kit. The concordance rate and call rate for each single nucleotide polymorphism (SNP) were determined by evaluating a total of 314 DNA samples that were sequenced. The assay's sensitivity was evaluated by creating a series of dilutions including 10.0, 5.0, 1.0, and 0.5 ng.

The OpenArray™ platform achieved a call rate of 96.9% on average and a concordance rate of nearly 99.8%. Serial dilutions were used to conduct sensitivity testing, which revealed that a sample containing 0.5 ng of total input DNA may be accurately identified. The 19 selected SNPs for human identification exhibited an average random match probability (RMP) of 10–8. An examination of 21 instances of biological evidence from 8 individuals, which produced individual short tandem repeat profiles within the regular workflow, proved the practicality of this approach in actual scenarios. A total of seventeen samples were accurately classified, indicating a call rate exceeding 90%. The phenotypic prediction demonstrated same accuracy to that reported in the matching validation data. Although the discrimination strength of this system is lower than that of STR based kits, the OpenArray™ System can still be utilized to exclude suspects and prioritize samples for further analysis. It offers reliable information regarding the prediction of eye color, hair color, and skin pigmentation. Additional research will be required to further confirm the effectiveness of this technology and to explore the possibility of incorporating a larger number of SNPs into the custom array in order to achieve a reduced RMP. Furthermore, it is important to consider including markers for ancestry and lineage studies.

REFERENCES

1. Kayser, M. Forensic DNA Phenotyping: Predicting human appearance from crime scene material for investigative purposes. *Forensic Sci. Int. Genet.* **2015**, *18*, 33–48. **[Google Scholar] [CrossRef] [PubMed]**

2. Ragazzo, M.; Carboni, S.; Caputo, V.; Buttini, C.; Manzo, L.; Errichiello, V.; Puleri, G.; Giardina, E. Interpreting Mixture Profiles: Comparison between Precision ID GlobalFiler™ NGS STR Panel v2 and Traditional Methods. *Genes* **2020**, *11*, 591. **[Google Scholar] [CrossRef] [PubMed]**

3. Cascella, R.; Strafella, C.; Caputo, V.; Errichiello, V.; Zampatti, S.; Milano, F.; Potenza, S.; Mauriello, S.; Novelli, G.; Ricci, F.; et al. Towards the application of precision medicine in Age-Related Macular Degeneration. *Prog. Retin. Eye Res.* **2018**, *63*, 132–146. **[Google Scholar] [CrossRef] [PubMed]**

4. Giardina, E.; Spinella, A.; Novelli, G. Past, present and future of forensic DNA typing. *Nanomedicine* **2011**, *6*, 257–270. **[Google Scholar] [CrossRef] [PubMed]**

5. Giardina, E.; Pietrangeli, I.; Martone, C.; Zampatti, S.; Marsala, P.; Gabriele, L.; Ricci, O.; Solla, G.; Asili, P.; Arcudi, G.; et al. Whole genome amplification and real-time PCR in forensic casework. *BMC Genom.* **2009**, *10*, 159. **[Google Scholar] [CrossRef] [Green Version]**

6. Oldoni, F.; Podini, D. Forensic molecular biomarkers for mixture analysis. *Forensic Sci. Int. Genet.* **2019**, *41*, 107–119. **[Google Scholar] [CrossRef]**

7. Schwark, T.; Meyer, P.; Harder, M.; Modrow, J.-H.; von Wurmb-Schwark, N. The SNPforID Assay as a Supplementary Method in Kinship and Trace Analysis. *Transfus. Med. Hemother.* **2012**, *39*, 187–193. **[Google Scholar] [CrossRef] [Green Version]**

8. Machado, H.; Granja, R. DNA Databases and Big Data. In *Forensic Genetics in the Governance of Crime*; Palgrave Pivot: Singapore, 2020; pp. 57–70. **[Google Scholar] [CrossRef] [Green Version]**

9. Amankwaa, A.O. Trends in forensic DNA database: Transnational exchange of DNA data. *Forensic Sci. Res.* **2019**, *5*, 8–14. **[Google Scholar] [CrossRef] [Green Version]**

10. Butler, J.M.; Willis, S. Interpol review of forensic biology and forensic DNA typing 2016–2019. *Forensic Sci. Int. Synerg.* **2020**, *2*, 352–367. **[Google Scholar] [CrossRef]**

11. Schneider, P.M.; Prainsack, B.; Kayser, M. The Use of Forensic DNA Phenotyping in Predicting Appearance and Biogeographic Ancestry. *Dtsch. Arztebl. Int.* **2019**, *51-52*, 873–880. **[Google Scholar] [CrossRef]**

12. Butler, J.M. DNA Quantitation. In *Advanced Topics in Forensic DNA Typing: Methodology*; Butler, J.M., Ed.; Academic Press: Cambridge, MA, USA, 2012; pp. 49–67. **[Google Scholar] [CrossRef]**

13. Cascella, R.; Strafella, C.; Ragazzo, M.; Manzo, L.; Costanza, G.; Bowes, J.; Hüffmeier, U.; Potenza, S.; Sangiuolo, F.; Reis, A.; et al. *KIF3A* and *IL4*- are disease-specific biomarkers for psoriatic arthritis susceptibility. *Oncotarget* **2017**, *8*, 95401–95411. **[Google Scholar] [CrossRef] [PubMed] [Green Version]**

14. Broccanello, C.; Gerace, L.; Stevanato, P. QuantStudio™ 12K Flex OpenArray® System as a Tool for High-Throughput Genotyping and Gene Expression Analysis. In *Quantitative Real-Time PCR Methods in Molecular Biology*; Biassoni, R., Raso, A., Eds.; Humana: New York, NY, USA, 2020; Volume 2065, pp. 199–208. **[Google Scholar] [CrossRef]**

15. Pomeroy, R.; Duncan, G.; Sunar-Reeder, B.; Ortenberg, E.; Ketchum, M.; Wasiluk, H.; Reeder, D. A low-cost, high-throughput, automated single nucleotide polymorphism assay for forensic human DNA applications. *Anal. Biochem.* **2009**, *395*, 61–67. **[Google Scholar] [CrossRef] [PubMed]**

16. Chaitanya, L.; Breslin, K.; Zuñiga, S.; Wirken, L.; Pośpiech, E.; Kukla-Bartoszek, M.; Sijen, T.; Knijff, P.; Liu, F.; Branicki, W.; et al. The HIrisPlex-S system for eye, hair and skin colour prediction from DNA: Introduction and forensic developmental validation. *Forensic Sci. Int. Genet.* **2018**, *35*, 123–135. **[Google Scholar] [CrossRef] [PubMed] [Green Version]**

17. Pakstis, A.J.; Speed, W.C.; Fang, R.; Hyland, F.C.; Furtado, M.R.; Kidd, J.R.; Kidd, K.K. SNPs for a universal individual identification panel. *Hum. Genet.* **2010**, *127*, 315–324. **[Google Scholar] [CrossRef]**

18. Ochiai, E.; Minaguchi, K.; Nambiar, P.; Kakimoto, Y.; Satoh, F.; Nakatome, M.; Miyashita, K.; Osawa, M. Evaluation of Y chromosomal SNP haplogrouping in the HID-Ion AmpliSeq™ Identity Panel. *Leg. Med.* **2016**, *22*, 58–61. **[Google Scholar] [CrossRef]**

19. HIrisPlex-S Eye. Hair and Skin Colour DNA Phenotyping Webtool. Available online: **https://hirisplex.erasmusmc.nl/** (accessed on 28 September 2020).

20. rs1805008 SNP. Available online: **http://www.ensembl.org/Homo_sapiens/Variation/HighLD?db=core;r=16:89919236-89920236;v=rs1805008;vdb=variation;vf=22711199#373537_tablePanel** (accessed on 30 November 2020).

21. 1000 Genomes Project Consortium; Abecasis, G.R.; Altshuler, D.; Auton, A.; Brooks, L.D.; Durbin, R.M.; Gibbs, R.A.; Hurles, M.E.; McVean, G.A. A map of human genome variation from population-scale sequencing. *Nature* **2010**, *467*, 1061–1073. **[Google Scholar] [CrossRef] [Green Version]**

22. Yates, A.D.; Achuthan, P.; Akanni, W.; Allen, J.; Allen, J.; Alvarez-Jarreta, J.; Amode, M.R.; Armean, I.M.; Azov, A.G.; Bennett, R.; et al. Ensembl 2020. *Nucleic Acids Res.* **2020**, *48*, D682–D688. **[Google Scholar] [CrossRef]**

23. Walsh, S.; Chaitanya, L.; Clarisse, L.; Wirken, L.; Draus-Barini, J.; Kovatsi, L.; Maeda, H.; Ishikawa, T.; Sijen, T.; de Knijff, P.; et al. Developmental validation of the HIrisPlex system: DNA-based eye and hair colour prediction for forensic and anthropological usage. *Forensic Sci. Int. Genet.* **2014**, *9*, 150–161. **[Google Scholar] [CrossRef]**

24. Cascella, R.; Stocchiy, L.; Strafella, C.; Mezzaroma, I.; Mannazzu, M.; Vullo, V.; Montella, F.; Parruti, G.; Borgiani, P.; Sangiuolo, F.C.; et al. Comparative analysis between saliva and buccal swabs as source of DNA: Lesson from *HLA-B*57:01* testing. *Pharmacogenomics* **2015**, *16*, 1039–1046. **[Google Scholar] [CrossRef] [Green Version]**

25. Ragazzo, M.; Melchiorri, S.; Manzo, L.; Errichiello, V.; Puleri, G.; Nicastro, F.; Giardina, E. Comparative Analysis of ANDE 6C Rapid DNA Analysis System and Traditional Methods. *Genes* **2020**, *11*, 582. **[Google Scholar] [CrossRef]**

26. Walsh, S.; Liu, F.; Ballantyne, K.N.; van Oven, M.; Lao, O.; Kayser, M. IrisPlex: A sensitive DNA tool for accurate prediction of blue and brown eye colour in the absence of ancestry information. *Forensic Sci. Int. Genet.* **2011**, *5*, 170–180. **[Google Scholar] [CrossRef] [PubMed]**

27. Walsh, S.; Liu, F.; Wollstein, A.; Kovatsi, L.; Ralf, A.; Kosiniak-Kamysz, A.; Branicki, W.; Kayser, M. The HIrisPlex system for simultaneous prediction of hair and eye colour from DNA. *Forensic Sci. Int. Genet.* **2013**, *7*, 98–115. **[Google Scholar] [CrossRef] [PubMed] [Green Version]**

28. Walsh, S.; Chaitanya, L.; Breslin, K.; Muralidharan, C.; Bronikowska, A.; Pospiech, E.; Koller, J.; Kovatsi, L.; Wollstein, A.; Branicki, W.; et al. Global skin colour prediction from DNA. *Hum. Genet.* **2017**, *136*, 847–863. **[Google Scholar] [CrossRef] [PubMed] [Green Version]**

29. Butler, J.M. Statistical Interpretation Overview. In *Advanced Topics in Forensic DNA Typing: Interpretation*; Butler, J.M., Ed.; Academic Press: Cambridge, MA, USA, 2014; pp. 213–237. **[Google Scholar] [CrossRef]**

30. Sanchez, J.J.; Phillips, C.; Børsting, C.; Balogh, K.; Bogus, M.; Fondevila, M.; Harrison, C.D.; Musgrave-Brown, E.; Salas, A.; Syndercombe-Court, D.; et al. A multiplex assay with 52 single nucleotide polymorphisms for human identification. *Electrophoresis* **2006**, *27*, 1713–1724. **[Google Scholar] [CrossRef]**

31. Kidd, K.K.; Soundararajan, U.; Rajeevan, H.; Pakstis, A.J.; Moore, K.N.; Ropero-Miller, J.D. The redesigned Forensic Research/Reference on Genetics-knowledge

base, FROG-kb. *Forensic Sci. Int. Genet.* **2018**, *33*, 33–37. **[Google Scholar]** **[CrossRef]** **[PubMed]** **[Green Version]**

32. Yang, J.; Lin, D.; Deng, C.; Li, Z.; Pu, Y.; Yu, Y.; Li, K.; Li, D.; Chen, P.; Chen, F. The advances in DNA mixture interpretation. *Forensic Sci. Int.* **2019**, *301*, 101–106. **[Google Scholar]** **[CrossRef]**
33. King, J.L.; Churchill, J.D.; Novroski, N.M.M.; Zeng, X.; Warshauer, D.H.; Seah, L.H.; Budowle, B. Increasing the discrimination power of ancestry- and identity-informative SNP loci within the ForenSeq™ DNA Signature Prep Kit. *Forensic Sci. Int. Genet.* **2018**, *36*, 60–76. **[Google Scholar]** **[CrossRef]**
34. Palencia-Madrid, L.; Xavier, C.; de la Puente, M.; Hohoff, C.; Phillips, C.; Kayser, M.; Parson, W. Evaluation of the VISAGE Basic Tool for Appearance and Ancestry Prediction Using PowerSeq Chemistry on the MiSeqFGx System. *Genes* **2020**, *11*, 708. **[Google Scholar]** **[CrossRef]**
35. Breslin, K.; Wills, B.; Ralf, A.; Ventayol Garcia, M.; Kukla-Bartoszek, M.; Pospiech, E.; Freire-Aradas, A.; Xavier, C.; Ingold, S.; de La Puente, M.; et al. HIrisPlex-S system for eye, hair, and skin color prediction from DNA: Massively parallel sequencing solutions for two common forensically used platforms. *Forensic Sci. Int. Genet.* **2019**, *43*, 102152. **[Google Scholar]** **[CrossRef]**

23 OPTIMIZING DNA SAMPLING METHODS FOR BURGLARY INVESTIGATIONS

1. INTRODUCTION: AN OVERVIEW

Burglaries, which involve property damage, trespassing, and theft, are a significant form of property crime. The victims are left with a heightened sense of insecurity, further exacerbating the impact. Burglaries are the most prevalent type of crime investigated with forensic genetics in Switzerland and likely in many other nations. Firstly, burglaries are a common form of property crime, accounting for 24,010 recorded instances in Switzerland in 2020, which represents around 9% of all registered property offenses [1]. Secondly, burglaries are especially conducive to collecting DNA evidence compared to other high-volume crimes like fraud. In the department of Forensic Molecular Biology at the Institute of Forensic Medicine, where this study took place, "touch" DNA traces from burglaries and attempted burglaries account for over 50% of the total number of traces studied. Considering the duration of time spent in a household, it is reasonable to assume that most of the DNA evidence found on the premises can be attributed to the people who live there. Nevertheless, despite the substantial quantity of cases, we seldom acquire reference samples from residents in order to prevent the establishment of their profiles based on DNA traces collected at the crime scene and subsequently transferred to the Swiss DNA database for comparison using CODIS. Considering an average of 2–3 individuals per household, or maybe more for workspaces, and factoring in burglaries in commercial settings, conducting routine sampling of reference swabs from all occupants would result in significant financial and administrative expenses.

Nevertheless, disregarding the fact that a significant proportion of resident profiles ultimately become registered in CODIS as trace profiles presents a few challenges: (a) The police run the risk of spoiling investigations through the presentation of inadmissible evidence. The following real case might serve as an example to illustrate this particular problem: Two persons, a man and a woman living together, were assaulted by three men. From the jacket of the woman, a DNA mixture containing a female and a male component was recovered and sent to CODIS. The female component belonged to the female victim, as inferred by crosschecking all samples in the same case. The male component, however, resulted in profile matches with DNA traces from two burglaries. The police therefore started investigations against the male victim of the assault. The Bern Cantonal High Court, however, prohibited all the evidence gathered in this case from being used, since the investigation was launched based on inadmissible evidence given by two random and case unrelated DNA matches that could have been avoided by taking reference samples [2]. The verdict prevented the assailants from being brought to justice and the affaire became a cold case. (b) A lack of prior exclusion might prevent crime victims from contacting the police. The risk of being associated with a past crime or a juvenile sin, e.g., illegal graffiti spraying, might discourage people from contacting the police, once they have become victims themselves. In Switzerland, trace DNA sampling is legitimate in cases of property damage having caused more than 10,000 CHF of pecuniary damage, a sum that is easily reached through graffiti spraying. We can imagine a 16-year-old graffiti sprayer, who becomes a victim in a more serious crime and who refrains from contacting the police, because he fears that through incidental DNA matches, his past or present spraying activities might be revealed to the police. (c) The database becomes inflated. DNA trace profiles caused by residents will normally not generate hits in forensic databases, at least not with profiles of convicted individuals. Therefore, the traces will remain in the database for years and create candidate matches with newly added suspect profiles or trace profiles from other cases. DNA mixture profiles frequently lead to candidate matches that have to be verified by the submitting lab. The lab than decides whether the "candidate match" is considered as a true match ("HIT"), or not. For mixture profiles, most of the verifications result in "NO HIT" verdicts. The control of those candidate matches is time consuming and constitutes a possible source of error, justifying a cleanup of the database from profiles from crime-authorized people.

Based on rigorous research on DNA transmission and persistence [3,4,5,6], it is anticipated that a significant proportion of the DNA trace profiles obtained during burglary investigations will originate from the residents. Nevertheless, there is a lack of data regarding the proportion of profiles in actual case scenarios that originate from residents. In order to have a better understanding of these statistics, we specifically examined burglaries that occurred in private residences, as the level of access is typically more limited compared to stores or other commercial settings. We obtained reference samples from the individuals and matched them to the DNA profiles obtained from the crime scene evidence. Subsequently, we sent them to the Swiss DNA database for comparison.

2. MATERIALS AND METHODS

2.1. Sampling

The Bern Cantonal Police conducted sampling for crime scene traces in burglaries in accordance with their customary procedure. The police officers independently and spontaneously determined at the location where DNA evidence from the perpetrator may potentially be found, and collected samples accordingly. Reference samples from residents of private households were collected between November 2020 and April 2021. Residents present during the crime scene investigation were asked for reference samples. Consenting residents were handed out a written agreement form with explanations concerning the procedure, the modalities for the usage of their DNA profiles and information on why prior exclusion of their profiles would be in their favor. The police officer further registered the number of persons living in the household and whether there had been any visitors at the house within the last 7 days prior to the burglary. For all inhabitants not present at the apartment or house during the crime scene sampling, the police officer left a set of documents, buccal swabs, and a prepaid reply envelope on-site, so the absent inhabitants could send their reference swabs directly to the DNA laboratory by mail. All samples were given a pseudonymized family role (such as "father" or "man") under which they were subsequently processed in the DNA laboratory. Buccal swabs were collected with Sarstedt Forensic Swab L. (Sarstedt AG, Nümbrecht, Germany), crime scene traces were collected using the forensiX Cardboard Evidence Collection Kit (Art. Nr. 9021040; Prionics, Schlieren, Switzerland). One or two partially moistened swabs were used for the sampling of one trace.

2.2. Analysis

DNA from buccal swabs was extracted by Chelex 100 (Bio-Rad Laboratories, Hercules, CA, USA). Samples were placed in 1500 µL of 20% Chelex 100, put for two times 30 s at 5900 rpm on a Precellys®24 homogenizer (Bertin instruments, Montigny-le-Bretonneux, France), incubated at 100 °C for 10 min and subsequently centrifuged at 13,000 rpm on an Eppendorf Minispin centrifuge (Eppendorf, Hamburg, Germany). DNA extracted from buccal swabs was quantified using a Qubit fluorometer (Thermo Fisher, Waltham, MA, USA). DNA profiles were established in 12.5 µL reaction volume by multiplex-PCR using the AmpFlSTR® NGM Select Express™ (Thermo Fisher, Waltham, MA, USA) and PowerPlex® ESI17 Fast kits (Promega, Madison, WI, USA). DNA from trace samples was extracted with the AutoMateExpress™ device and the PrepFilerExpress™ Kit (both Thermo Fisher, Waltham, MA, USA), with an elution volume of 50 µL, representing our standard lab procedure for swabs from touched surfaces [7]. DNA was quantified by Real-Time-PCR (qPCR) using the Quantifiler® HP Kit from Thermo Fisher on a 7500 RT PCR System (Thermo Fisher, Waltham, MA, USA). DNA profiles were established by multiplex-PCR using the AmpFlSTR® NGM Select™ and NGMDetect™ Kits (Thermo Fisher, Waltham, MA, USA) in a total reaction volume of 25 µL (at least two independent amplifications). A maximum of 0.5 ng DNA was amplified per reaction, using the maximum sample volume of 10 µL for samples with DNA-concentrations below 50 pg/µL. In line with our standard operating procedures for casework, all samples with a DNA concentration below 20 pg/µL were amplified with 32 instead of 30 PCR cycles. Capillary electrophoresis was run either on a 3130 xl or on a 3500 xl genetic analyzer (Thermo Fisher, Waltham, MA, USA). Signal interpretation was performed with Genemapper ID-X, v1.4 (Thermo Fisher, Waltham, MA, USA). All peaks above 50 rfu (3130 xl) and 100 rfu (3500 xl) were considered as true alleles.

2.3. Criteria for a Submission to Swiss CODIS Database

The Swiss DNA database uses the CODIS (combined DNA index system) software v7.0. Switzerland has defined by law the 16 loci included in the AmpFlSTR® NGMSelect/NGMDetect and PowerPlex® ESI17/ESX17 kits as database core loci [**8**]. More or less all offences leading to ordinary criminal proceedings are recordable, except misdemeanors penalized with a fine. Familial searches are permitted for all offences penalized with a prison sentence of more than three years and can be ordered by the public prosecutor. If reference samples are collected, the generated profiles are used only for the exclusion of people with lawful right of access to the scene. Reference profiles are not registered in CODIS. Profile exclusion is performed locally in the laboratory, before sending the trace profile to CODIS. The entry criteria for a regular database search in Switzerland are a minimum of 6 loci for single or major component profiles and 8 loci for two person mixtures. The sex locus Amelogenin can be entered, but is not searched for in CODIS. Apparent major components are selected, if no two-person mixture can be clearly distinguished. A submission was considered when the lab internal quality conditions were met, i.e., a minimum ratio of 3:1 for major to minor component and heterozygote peak balance of at least 60%. Profiles that did not fulfill our lab or the search criteria for CODIS, although they could possibly still be interpreted, e.g., by probabilistic genotyping [**9,10**], were not investigated any further. It is important to note that our study targets the investigative phase of burglary investigations and not the evaluative phase [**11**].

3. RESULTS

3.1. Samples Obtained

We had a bad timing for our study, since we registered in the relevant timeframe between November 2020 and April 2021 an unusual low number of burglaries in private households. This was presumably due to the COVID-19 pandemic. As part of the measures taken in the fight against the pandemic, e.g., restrictions in cross-border traffic or the obligation of working in the home office, people spent more time at home. In 2020, we temporarily observed a workload drop of about 80%. From November to April, our lab registered 226 burglaries in private housings; attempts and sneak-in thefts not included. In 55 of those cases, so approximately one-quarter, we obtained the reference samples from the inhabitants. We analyzed 144 trace samples in total, what corresponds to 2.6 traces per burglary. We obtained a total of 118 references samples (50 women, 47 men, and 21 children), so in average about 2.14 reference samples per case, corresponding well with the average household size in Switzerland of 2.2 persons [**12**].

3.2. Sample Categories

Samples were classified into three categories: (a) furniture, e.g., handles from cupboards, pieces of furniture that had been moved by the offenders; (b) portable items, e.g., keys or jewelry; and (c) samples taken around the entry of the offender, e.g., door or window. The majority, about 65%, of the samples were taken in the vicinity of the entry (**Figure 1**a). The samples around the entry were further divided into subcategories (**Figure 1**b). Half of them (n = 46) were collected from window glass or glass fragments.

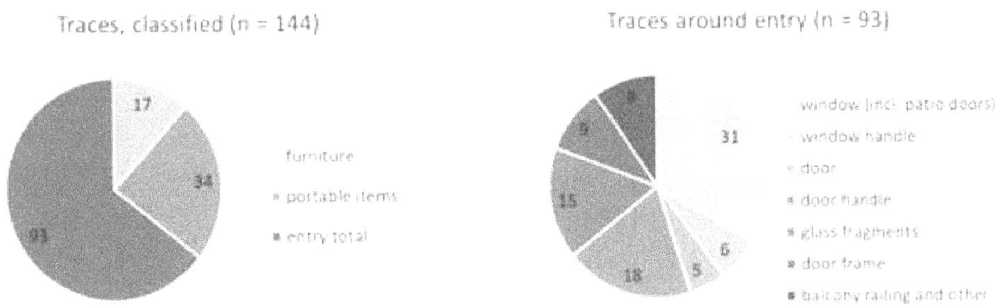

(a) (b)

Figure 1. (a) All samples classified in three categories; (b) samples from entry divided in several sub-classes.

3.3. Typing Success

We were able to establish DNA profiles suitable for submission to CODIS from 48 traces, what corresponds to an overall success rate of 33% (**Figure 2**).

Figure 3 shows the distribution of the total amount of DNA extracted from each sample of all 144 samples (**Figure 3**a) and of the 48 samples that resulted in CODIS-suitable profiles (**Figure 3**b). We were able to establish DNA profiles for database entry for all traces with a total DNA amount of more than 1 ng. Around 85% of the traces with less than 0.5 ng total DNA did not result in CODIS-suitable profiles.

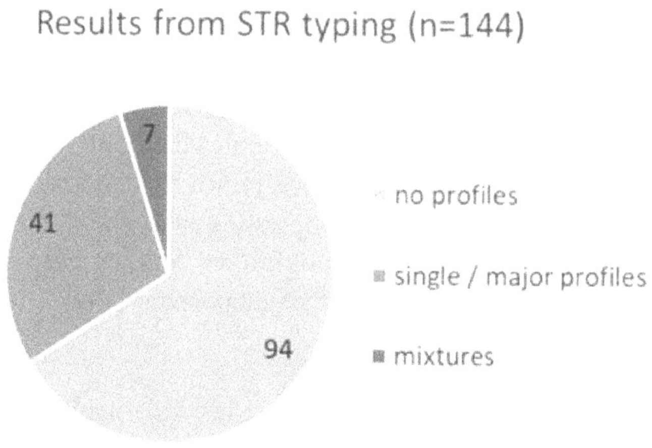

Figure 2. The STR typing results for all 144 crime scene samples summarized. The category "no profiles" includes all profiles not fulfilling the CODIS entry criteria.

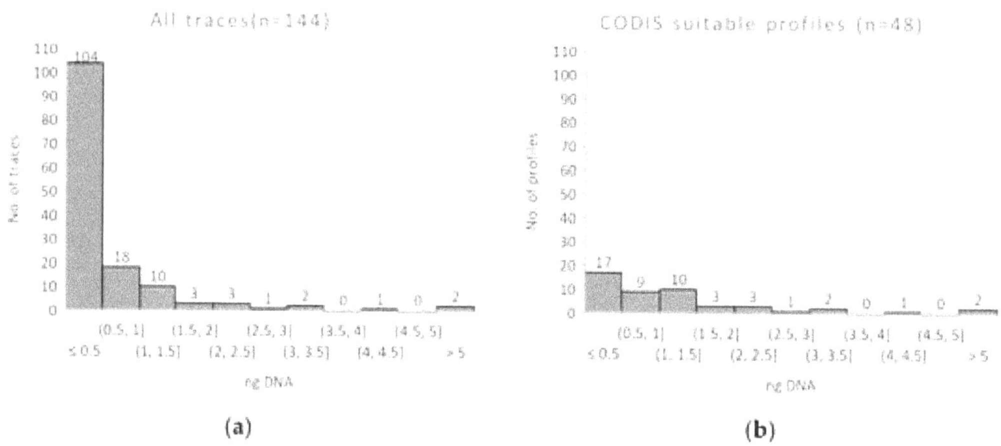

Figure 3. Histograms of the distributions of DNA amounts in ng obtained from: (a) all samples and (b) samples for which we were able to establish CODIS-suitable DNA profiles.

The DNA typing success rate of samples taken from furniture and portable items reaches 70% and 50%, respectively, whereas only about 30% of traces taken around the entry resulted in CODIS-suitable profiles (**Figure 4**).

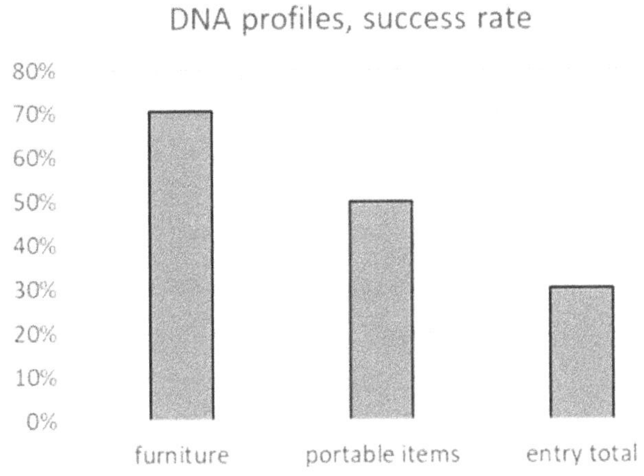

Figure 4. Fraction of samples of the respective class that resulted in CODIS-suitable DNA profiles.

3.4. Comparison with Reference Samples

The 48 CODIS-suitable profiles include interpretable components from 57 DNA contributors. We obtained 41 single or major component profiles and 7 two-person mixtures. Two of the major component profiles showed a reproducible (at least three PCR amplifications conducted) minor component with more than five but less than eight alleles. Therefore, those three minor components are not suitable for CODIS as a mixture, but would be sufficient according to the Swiss CODIS criteria for interpretation, the reason why we included them in our interpretation. Of those 57 trace donors, 17 were detected on portable items, 12 on furniture and 28 on samples around the entry, half of which were from windows glass (**Figure** 5a). Of those 57 trace contributors, 43 (75%) were identified as crime scene authorized persons, i.e., inhabitants.

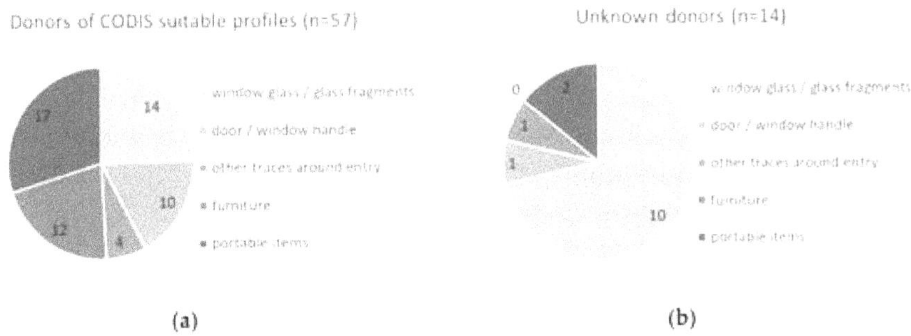

Figure 5. (a) Donors of the 48 CODIS-suitable profiles, according to trace class. (b) Unknown donors according to trace class. There were 0 unknown donors on furniture.

If we have a closer look on the different sample categories, we can see that traces from window glass clearly outperform the rest of the samples (**Figure 5**b). Whereas 71% of the interpretable traces detected on window glass and glass fragments are from unknown contributors, and therefore from possible suspects, less than 10% of the contributors from all other traces were from unknown individuals. The ratio of unknown vs. known contributors is depicted in **Figure 6.**

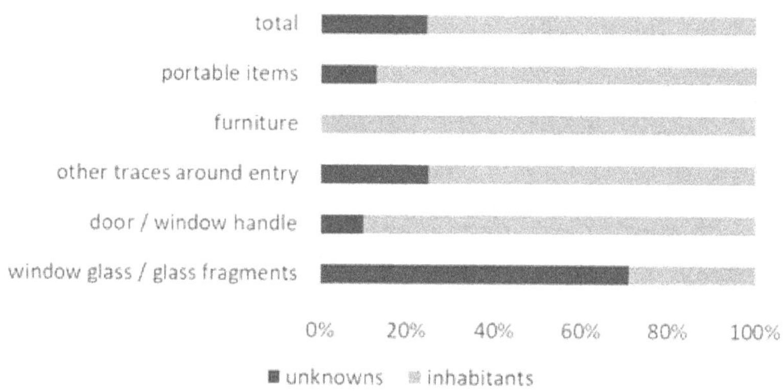

Figure 6. Ratios between unknown contributors and inhabitants detected on the different types of traces.

4. DISCUSSION

The success rate of DNA typing for contact traces in burglaries, which is 33%, aligns with the general typing efficiency of approximately 25% that we have found in our laboratory in previous years. Furthermore, it aligns with the success rate reported by another Swiss laboratory, which likewise operates within the same national and legal environment [13]. Their average success rate for "touch" DNA traces of 26% is slightly lower than the one for the samples from our study, but we have to take into account the higher sensitivity of the multiplex kits nowadays, since the study from 2008 used AmpFlSTR® SGM Plus™ and we used the more sensitive NGMSElect™ and NGMDetect™ kits.

Around 70% of the profiles established from window glass and glass fragments were from unknown individuals. A low level of background DNA on windows has been reported in a previous study, which suggests windows as premium choice for DNA sampling in burglary cases. However, no numbers were presented by the authors of this study from 2008 [**14**]. The results from our study confirm that windows have a comparatively low background of resident DNA and are therefore most suited for DNA trace analysis. We can see that the police officers of the Bern Cantonal Police are already focusing on the most promising sampling spots, since 32% of the traces we received were taken from windows or glass fragments.

Even though the data shown in **Figure 6** seem to be very strong in favor of the sampling of window glass, we have to keep one limitation in mind, concerning the pre-selection effectuated by the police officer handling the case. This pre-selection is twofold: First, the officer will select the traces appearing most promising to him or her at the crime scene. Second, he or she will make a subsequent selection as to which samples to send for analysis to the DNA laboratory. If the analysis of the first samples sent to the lab do not result in interpretable DNA profiles, the case-managing officer might decide to send additional samples for analysis. Therefore, we encounter some risk of bias if samples from window glass are prioritized and no other samples are analyzed in the case. If we assume that the unknown profiles we can recover are maybe from "less prudent" burglars, than we might also possibly expect more of their profiles on other items in the apartment. However, we would not see those suspect profiles, because we do not analyze additional samples after having successfully established a profile from window glass.

In the years from 2009 to 2017, between 30 to 40% of all the DNA trace profiles submitted every year to the Swiss DNA database originated from burglary cases [**15**]. From our study, we would expect that at least 75% of those profiles originate from residents. The number of non-suspect profiles could even be higher, because we cannot be sure that we obtained reference samples from all people authorized to the crime scene, e.g., visitors or first responding local police officers, and DNA might also be transferred and subsequently picked up [**3**,**4**,**5**,**6**].

In typical casework, it is not customary for us to collect reference samples from individuals. Out of the 144 samples in this investigation, we would typically have submitted 44 distinct profiles to CODIS (excluding redundant profiles from the same instance). Following the comparison with the reference samples, we ultimately submitted a total of 11 profiles, consisting of 10 single/major component profiles and 1 blend. As of 10 December 2021, six of these profiles resulted in matches with profiles from individuals listed in the Swiss DNA database. Five of these matches were derived from samples collected from window glass, highlighting the significant value of this sort of trace in advancing burglary investigations. Currently, the Swiss DNA database holds approximately 100,000 trace DNA profiles. Hence, based on the findings of our investigation, we anticipate that around 25,000 trace DNA profiles recorded in the database are genuinely derived from residents. The introduction states that if a resident is not eliminated from a trace DNA profile, there is a danger of being linked to further offenses through a trace-to-trace match in CODIS. Nevertheless, the absence of previous exclusion also hampers the privacy of the residents in a broader sense. Currently, the normal STR profiles provide minimal personal information. Nevertheless, increasingly invasive methods, such as forensic DNA phenotyping (FDP), are becoming more prevalent [16]. While we view the likelihood of this development as low, there is a possibility that privacy concerns could arise for authorized individuals at crime scenes if FDP is used for volume crimes in the future.

5. AT A GLANCE

The analysis revealed that over 70% of the profiles obtained from window glass near the location of the break-in were attributed to unidentified individuals. In contrast, less than 10% of the remaining traces were associated with unknown profiles, which may potentially be linked to suspects. Thus, it is crucial to give priority to collecting samples from (damaged) windows when investigating a crime scene in order to provide leads for the investigation. In order to reduce the presence of DNA profiles from personnel permitted to access crime scenes in CODIS, it is advisable to consider collecting reference samples for traces that were not obtained from window glass.

6. CONCISE SUMMARY

Burglaries rank as the most prevalent form of mass crime in Switzerland when it comes to crime scene investigations utilizing forensic DNA analysis. Approximately 33% of the DNA trace profiles recorded in the Swiss DNA database are linked to incidents of burglary. However, while collecting potential DNA traces at a crime scene, it is uncertain if the sampled DNA came from the burglar or from a resident of the house. Due to the prevalence of burglaries, crime scene investigators typically refrain from collecting reference samples from all residents for reasons of cost and administration. Hence, the presumed likelihood that a DNA profile attributed to an individual permitted to be present at the crime scene is ultimately submitted to a DNA database for the purpose of comparison must be considered. As far as we know, no study has been conducted to assess the proportion of these non-offender profiles that enter DNA databases. In order to elucidate this inquiry, we gathered reference specimens from individuals who had experienced recent burglaries in their personal residences. By juxtaposing the profiles derived from these reference samples with the profiles created from trace DNA, we can demonstrate that the majority of the DNA samples obtained in burglary investigations pertain to the inhabitants. Although the study had a small number of cases, likely because of a decrease in crime during the pandemic, we demonstrate that collecting trace DNA near the point of entry, specifically from window and door glasses, is highly effective for obtaining DNA samples from the perpetrator rather than the resident.

REFERENCES

1. PolizeilicheKriminalstatistik (PKS). *Jahresbericht 2020 der Polizeilich Registrierten Straftaten*; Swiss Federal Statistical Office: Neuchâtel, Switzerland, 2021.

2. Bern Cantonal High Court. *BK 2015 62—DNA-Analyse (Leitentscheid)*; Judgment of the Bern Cantonal High Court; Bern Cantonal High Court: Bern, Switzerland, 2015.

3. Meakin, G.; Jamieson, A. DNA transfer: Review and implications for casework. *Forensic Sci. Int. Genet.* **2013**, *7*, 434–443. **[Google Scholar] [CrossRef]**

4. van Oorschot, R.A.H.; Szkuta, B.; Meakin, G.E.; Kokshoorn, B.; Goray, M. DNA transfer in forensic science: A review. *Forensic Sci. Int. Genet.* **2019**, *38*, 140–166. **[Google Scholar] [CrossRef] [PubMed]**

5. Burrill, J.; Daniel, B.; Frascione, N. A review of trace "Touch DNA" deposits: Variability factors and an exploration of cellular composition. *Forensic Sci. Int. Genet.* **2019**, *39*, 8–18. **[Google Scholar] [CrossRef] [PubMed] [Green Version]**

6. Gosch, A.; Courts, C. On DNA transfer: The lack and difficulty of systematic research and how to do it better. *Forensic Sci. Int. Genet.* **2019**, *40*, 24–36. **[Google Scholar] [CrossRef] [PubMed]**

7. Zieger, M.; Schneider, C.; Utz, S. DNA recovery from gelatin fingerprint lifters by direct proteolytic digestion. *Forensic Sci. Int.* **2019**, *295*, 145–149. **[Google Scholar] [CrossRef]**

8. DNA-Analyselaborverordnung EJPD. Available online: **https://www.fedlex.admin.ch/eli/cc/2014/605/de** (accessed on 20 December 2021).

9. Bright, J.-A.; Richards, R.; Kruijver, M.; Kelly, H.; McGovern, C.; Magee, A.; McWhorter, A.; Ciecko, A.; Peck, B.; Baumgartner, C.; et al. Internal validation of STRmix™—A multi laboratory response to PCAST. *Forensic Sci. Int. Genet.* **2018**, *34*, 11–24. **[Google Scholar] [CrossRef] [PubMed] [Green Version]**

10. Bleka, Ø.; Storvik, G.; Gill, P. EuroForMix: An open source software based on a continuous model to evaluate STR DNA profiles from a mixture of contributors with artefacts. *Forensic Sci. Int. Genet.* **2016**, *21*, 35–44. **[Google Scholar] [CrossRef] [PubMed] [Green Version]**

11. ENFSI Guideline for Evaluative Reporting in Forensic Science. 2015. Available online: **https://enfsi.eu/wp-content/uploads/2016/09/m1_guideline.pdf** (accessed on 22 December 2021).

12. Swiss Federal Statistical Office. Households. 2020. Available online: **https://www.bfs.admin.ch/bfs/de/home/statistiken/bevoelkerung/stand-entwicklung/haushalte.html** (accessed on 22 December 2021).

13. Castella, V.; Mangin, P. DNA profiling success and relevance of 1739 contact stains from caseworks. *Forensic Sci. Int.-Genet. Suppl. Ser.* **2008**, *1*, 405–407. **[Google Scholar] [CrossRef]**

14. Raymond, J.J.; Walsh, S.J.; van Oorschot, R.A.H.; Gunn, P.R.; Evans, L.; Roux, C. Assessing trace DNA evidence from a residential burglary: Abundance, transfer and persistence. *Forensic Sci. Int. Genet. Suppl. Ser.* **2008**, *1*, 442–443. [**Google Scholar**] [**CrossRef**]
15. Killias, M.; Gut, M.; Biberstein, L.; Walser, S. *DNA-AnalysenimStrafverfahren: Entwicklung, Umfang und Wirkungen*; Killias Research & Consulting: Zürich, Switzerland, 2018. [**Google Scholar**]
16. Scudder, N.; McNevin, D.; Kelty, S.F.; Walsh, S.J.; Robertson, J. Massively parallel sequencing and the emergence of forensic genomics: Defining the policy and legal issues for law enforcement. *Sci. Justice J. Forensic Sci. Soc.* **2018,** *58*, 153–158. [**Google Scholar**] [**CrossRef**]

24 THE FORCE PANEL: A COMPREHENSIVE SNP MARKER SET DESIGNED FOR VERIFYING LEADS IN INVESTIGATIVE GENETIC GENEALOGY AND FOR BROADER FORENSIC APPLICATIONS

1. INTRODUCTION: AN OVERVIEW

Forensic DNA databases were constructed based on the utilization of non-coding DNA markers for the purpose of human identification, specifically focusing on short tandem repeat (STR) loci found in autosomal and Y-chromosomal targets [1]. Forensic DNA profiling has developed separately and concurrently with other areas of genetics, using a specific collection of DNA markers that have little relevance outside the medico-legal system. However, the achievement of technological independence was quickly compromised by the use of investigative genetic genealogy (IGG) and high-density single nucleotide polymorphism (SNP) genotyping. These techniques have been employed to solve prominent criminal cases (such as the Golden State Killer) and to identify unidentified human remains [2]. Existing direct-to-consumer (DTC) single nucleotide polymorphism (SNP) databases offer a rapidly expanding and extensive collection of genetic information. These databases can potentially reveal the genetic connections of individuals whose family members have engaged in consumer DNA testing and have chosen to be part of genealogical DNA databases like GEDmatch [3,4]. The IGG method entails identifying an expanded DNA profile from a sample obtained from a crime scene or other forensic setting utilizing high-density SNP genotyping microarrays [5] or whole-genome sequencing (WGS) [6]. Next, the SNP genotype dataset is queried in a publicly accessible DNA database to detect matches with substantial DNA overlap, suggesting a familial connection. Genealogists analyze the matches to recreate family trees, identifying branches that may indicate potential suspects, which are then shared with law authorities. After the "investigative lead" is determined, the subsequent step in DNA profiling involves utilizing the forensic DNA core markers (known as STRs) to validate the identification. The DNA of the individual under investigation is collected for analysis, and the STR profile of the suspect is compared to the STR profile generated from the sample collected at the crime scene. The ultimate DNA-based identification relies on the court-acceptable STR results, which serve as the conclusive forensic evidence in the case.

Confirmatory STR testing is possible when the DNA extracted from a forensic sample is of the expected quality for forensic analysis. However, specific types of samples may contain DNA that has undergone degradation, rendering them inappropriate for STR analysis. This is particularly true for hair shafts without roots and skeletal remains that have aged. Confirmatory STR testing is constrained in cases where direct/reference samples are unavailable and when inferring distant relationships, such as in the identification of historical remains. When faced with such situations, mitochondrial DNA (mtDNA) analysis is frequently employed to get a DNA profile.

This is because mtDNA has a greater number of copies compared to nuclear DNA (nDNA), resulting in a better likelihood of successfully obtaining an mtDNA target. Due to its non-recombining nature and the fact that it is passed down exclusively from mothers to their offspring, mtDNA is a lineage marker that is shared among maternal relatives. However, its usefulness in forensic applications is restricted because its ability to differentiate individuals is rather low. Hence, mitochondrial DNA (mtDNA) is most suitable for cases involving deteriorated DNA that may be supported by additional forms of evidence, whether DNA-related or otherwise. Novel techniques for sequencing the entire mitochondrial genome in a high-throughput manner, known as massively parallel sequencing (MPS), have been successfully utilized on DNA samples that have undergone degradation.

These methods, such as mini-amplicon PCR, primer extension capture (PEC), and hybridization capture, have been thoroughly tested and proven to be suitable for forensic applications. These technologies are highly sensitive and can accurately provide mtDNA data from even the most difficult forensic samples. These MPS procedures have also been modified for SNP profiling of DNA specimens that have undergone degradation [11,12,13,14,15]. Custom hybridization capture [14] can easily enrich large SNP multiplexes that are suited for extended kinship analysis. Nevertheless, these expansive panels may not be appropriate for regular forensic casework because of the markers that could be connected and/or correlated with pathogenic variations.

2. MATERIALS AND METHODS

2.1. FORCE Panel Design

A SNP panel for kinship analysis was developed to optimize sequencing efficiency by minimizing the number of SNPs while maximizing kinship resolution. The selection of kinship SNPs was based on the criteria outlined in Table 1. (1) The study used three commonly used Illumina SNP genotyping chips from San Diego, CA, USA. (2) The minor allele frequency (MAF) of the selected SNPs ranged from 0.2 to 0.8 in the major 1000 Genomes continental populations, including African (AFR), Admixed American (AMR), East Asian (EAS), European (EUR), and South Asian (SAS). (3) The selected SNPs were at least 0.5-cm apart in terms of genetic distance. (4) The linkage disequilibrium (LD) metric r2 between the SNPs was smaller than 0.1. (5) The frequency difference between the 1000 Genomes five continental populations was less than 0.35. X-SNPs were chosen for situations where X-chromosomal data is needed, using the same criteria as the autosomal kinship SNPs. Pre-existing SNP marker sets were added to the FORCE panel to offer cross-compatibility with commercial SNP assays (**Table 2**). Identity-informative SNPs (iiSNPs) were compiled from the ForenSeq DNA Signature Prep Kit (Primer Mix A; San Diego, CA, USA), the Precision ID Identity Panel (Thermo Fisher Scientific, Waltham, MA, USA), and a published QIAseq panel (QIAGEN) [**19**]. The ancestry- and phenotype-informative SNPs (aiSNPs and piSNPs, respectively) were compiled from the ForenSeq kit (Primer Mix B), the Precision ID Ancestry Panel (Thermo Fisher Scientific), and markers identified by the VISAGE consortium [**20**]. To increase the Y haplogroup resolution beyond the 34 Y-SNPs in the Precision ID Identify Panel, 884 Y-SNPs targeted by the amplicons of the Ralf et al. [**16**]. AmpliSeq panel (Thermo Fisher Scientific) were included in the test. The identified SNPs from the FORCE panel were compared to the genes from the American College of Medical Genetics (ACMG) secondary findings (SF) 2.0 to determine their clinical significance [21]. SNPs mentioned the ACMG SF 2.0 were removed from the panel. The panel excluded tri-allelic SNPs and did not incorporate autosomal microhaplotypes in order to streamline the data processing.

Table 1. Kinship SNP selection criteria and resulting viable SNPs.

Kinship Panel Criterion	Number of Remaining SNPs
1. Included on all major SNP array chips (**Table 2**)	116,544
2. MAF 0.2–0.8 in 1000 Genomes populations	34,851
3. Minimum 0.5 cm distance	5426
4. LD metric $r^2 < 0.1$	5376
5. Maximum 35% frequency difference between 1000 Genomes populations	3937

Table 2. Commercial and/or published SNP panels utilized for the FORCE panel selection. iiSNP = identity-informative SNP; aiSNP = ancestry-informative SNP; piSNP = phenotype-informative SNP; X-SNP = X-chromosomal SNP; Y-SNP = Y-chromosomal SNP.

Marker Type	SNP Panels (# SNPs)
Kinship SNP/X-SNP	Infinium Global Screening Array (654,027), Infinium Omni Express (710,000), Infinium CytoSNP-850K (850,000)
iiSNP	ForenSeq DNA Signature Prep Kit: Primer Mix A (94)
	Precision ID Identity Panel (90)
	QIAseq Investigator 140 SNP panel (140)
aiSNP	ForenSeq DNA Signature Prep Kit: Primer Mix B (56)
	Precision ID Ancestry Panel (165)
	VISAGE panel (115)
piSNP	ForenSeq DNA Signature Prep Kit: Primer Mix B (24)
	VISAGE panel (41)
Y-SNP	Precision ID Identity Panel (34)
	AmpliSeq (884)

A custom myBaits-20,000 kit (Arbor Biosciences, Ann Arbor, MI, USA) was designed for hybridization capture of the selected SNPs. The design included four baits per autosomal SNP (right flank, left flank, and each allele) and two baits per X/Y SNP (each allele). A BLAST [**22**] analysis was then performed to filter out non-specific baits. A total of 228 baits failed to meet the BLAST analysis requirements. Most of these failed baits (139/228) came from the AmpliSeq Y-SNP marker set [**16**], including 61 Y-SNPs for which both baits failed. The remaining failed baits targeted autosomal markers (74 SNPs with one failed bait, six SNPs with two failed baits, and one SNP with three failed baits). The failed baits were removed from the kit, thus eliminating 61 of the 884 Y-SNPs. Overall, there were 97 SNP markers with at least one failed bait and thus may show reduced coverage in the SNP capture results. The α version of the myBaits FORCE capture assay tested in this study included 5446 SNPs, targeted by 19,526 unique baits. There were 24 SNPs included in the α panel but ultimately excluded from analysis due to the outlined criteria. The excluded SNPs were six kinship SNPs with clinical relevance, 15 triallelicaiSNPs (one of which also had clinical relevance), one Y-SNP from the Precision ID Identity Panel with no Y haplogroup association, and two uninformative autosomal SNPs. A summary of the marker types comprising the 5422 FORCE panel SNP targets analyzed in this study is presented in **Figure 1**

2.2. Simulated Panel Performance

The work employed a simulation method to analyze the anticipated distributions of likelihood ratios (LR) for different kinship case scenarios, using the kinship SNPs contained in the FORCE panel. Five distinct kinship case scenarios were chosen, encompassing pairwise relatedness ranging from second to sixth degrees (namely, half siblings, first cousins, first cousins once removed, second cousins, and second cousins once removed). The simulations were conducted using the entire kinship SNP set, which initially consisted of 3937 SNPs, but later eliminated six of them. Additionally, simulations were conducted using 75% and 25% of the kinship single nucleotide polymorphisms (SNPs) to replicate scenarios when only incomplete SNP profiles are accessible.

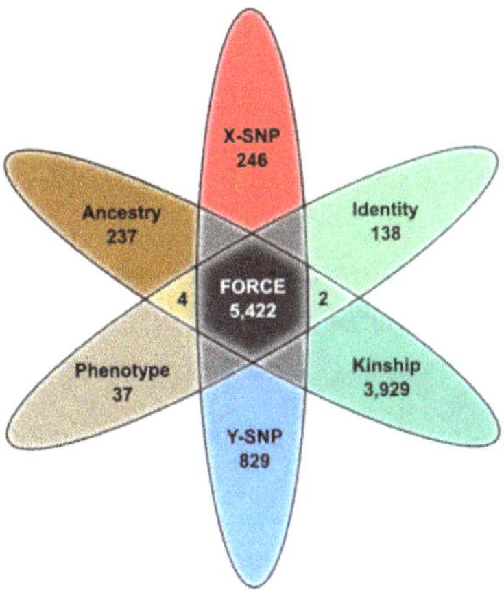

Figure 1. FORCE panel SNP marker summary (*n* = 5422).

The simulations were performed using the software Merlin [**23**]. Merlin was used both to create the pedigree-based DNA data and to calculate likelihoods for the tested hypotheses. European allele frequencies from [**18**] were used and genetic linkage was accounted for using genetic position information from Rutgers map [**24**]. For each simulated kinship case scenario, the likelihood ratio was calculated as LR = Pr (DNA|H1)/Pr (DNA|H2), where H1 was the hypothesis representing the related case scenario, and H2 was the hypothesis representing the alternative, in which the tested individuals are unrelated. For each case scenario and true hypothesis, 10,000 simulation events were performed. A front-end script written in R [**25**] was used to format and manage input files, and R was also used to handle the outputs from Merlin and for plotting the distribution curves (R package *ggplot2*).

2.3. SNP Capture Assay Laboratory Testing

2.3.1. Sample Selection

The samples used in this investigation were obtained from non-probative Defense Personnel Accounting Agency (DPAA) cases that had been previously analyzed by the Armed Forces Medical Examiner System's Armed Forces DNA Identification Laboratory (AFMES-AFDIL). The instances pertain to bone samples obtained from five male military personnel who were previously identified as participants in World War II (WWII), together with their corresponding family reference specimens. This investigation included one or more DNA extracts from a single skeletal part per service member, as well as two or three family reference samples for each service member (Table 3). The selection of the five cases was based on a variety of sample quality and the extent of association. Furthermore, additional non-probative samples were assessed to determine the agreement and/or performance of the kinship SNPs, in addition to the case samples used for testing (Table 4). The samples consisted of seven reference samples, a bone sample from the 19th century belonging to an individual previously identified as sample 25, JB55 [12], another bone sample from the 19th century related with JB55 (sample 26), and four control DNAs.

Table 3. Non-probative case descriptions. Each of the five cases includes one or more DNA extracts from a degraded bone sample (~70 years) of a previously identified World War II (WWII) service member, along with two or three family reference specimens (buccal swabs). DOR = degree of relatedness.

Case	WWII Context	Sample	Sample Type	Relationship to Service Member (DOR)
A	USS *Oklahoma*	1	Buccal swab	Grandniece (3)
		2	Buccal swab	Nephew (2)
		3	Left femur	Self
B	USS *Oklahoma*	4	Buccal swab	Nephew (2)
		5	Buccal swab	Great grandnephew (4)
		6	Buccal swab	Great grandniece (4)
		7.1	Left femur	Self
		7.2		
C	Austria	8	Buccal swab	First cousin twice removed, male (5)
		9	Buccal swab	Nephew (2)
		10	Buccal swab	Daughter (1)
		11.1	Long bone	Self
		11.2		

Case	WWII Context	Sample	Sample Type	Relationship to Service Member (DOR)
D	Italy	12	Buccal swab	Sister (1)
		13	Buccal swab	Grandniece (3)
		14.1	Right parietal	Self
		14.2		
		14.3		
E	Tarawa	15	Buccal swab	Son (1)
		16	Buccal swab	Daughter (1)
		17.1	Right tibia	Self
		17.2		

Table 4. Additional non-probative samples tested: seven reference samples with self-reported phenotypic and ancestry information (24–18), two historical bone samples ~200 years postmortem (25 and 26), and four control DNA samples (2800M, K562, NA12877, and NA12878).

Sample	Sample Type	Sex
18	Buccal swab	Female
19	Buccal swab	Female
20	Buccal swab	Male
21	Buccal swab	Female
22	Buccal swab	Female
23	Buccal swab	Male
24	Buccal swab	Male
25 (JB55)	Femoral bone (~200 years)	Male
26	Unknown bone (~200 years)	Unknown
2800M	Whole blood	Male
K562	Cell line	Female
NA12877	Cell line	Male
NA12878	Cell line	Female

All living sample donors provided informed consent for their samples to be used in research and quality improvement activities. The use of these reference and bone samples was approved by the Defense Health Agency Office of Research Protections (Protocol # DHQ-20-2073).

2.3.2. DNA Extraction and Quantitation

We acquired the DNA extracts and reagent blanks (RBs) that were previously collected from the five skeletal samples and JB55 [12]. In the earlier processing of these samples, DNA was isolated from bone powder weighing between 0.2 and 1.0 grams. The samples were initially incubated at 56 °C overnight in a buffer solution consisting of 0.5 M EDTA, 1% lauroylsarcosine, and 200 µL of proteinase K with a concentration of 20 mg/mL. The DNA extraction process was carried out using either an organic technique or a double QIAquick PCR Purification Kit method (QIAGEN, Hilden, Germany), as detailed in reference [26]. All organic extracts underwent purification using the QIAGEN MinElute PCR Purification Kit, and were subsequently eluted in 50 to 100 µL of Tris-EDTA solution (10 mM Tris, pH 7.5, 0.1 mM EDTA). A bone sample of unidentified origin, dating back to the same historical period as JB55, was subjected to a modified Dabney DNA extraction procedure, along with an accompanying RB. The Dabney procedure employed the technique of utilizing silica columns with a high capacity [27]. However, the process of digestion was carried out overnight at a temperature of 56 °C.

DNA was isolated from the buccal swabs using the QIAGEN EZ1 DNA Investigator Kit automated on an EZ1 Advanced XL instrument following the manufacturer's Trace or Tip Dance Protocol. The final elution volume for all samples ranged from 50 to 100 µL of Tris-EDTA. DNA was quantified with the dsDNA High Sensitivity (HS) Assay Kit on the Qubit 2.0 or Qubit 3.0 Fluorometer (Thermo Fisher Scientific, Waltham, MA, USA) to determine input into library preparation.

2.3.3. Library Preparation

The samples were partitioned into three groups for laboratory manipulation and sequencing. Each group consisted of both negative and positive controls for the library, which were used to assess the success of the library preparation and capture process. The initial sample set (bone) consisted of ten DNA extracts from the bone samples collected during WWII, as well as one historical bone sample (26) and two related RBs. The positive control for the bone set consisted of 1 nanogram of fragmented K562 control DNA, as previously explained in reference [10]. JB55 underwent individual processing, distinct from the other bone samples, inside its own group. This group consisted of a related RB, a library negative control, and a K562 positive control, which were also utilized in the bone sample set. Finally, the reference sample collection included of 12 family references, seven supplementary reference-type samples, and three control DNAs (2800M, NA12877, and NA12878). The reference sample set includes RBs that were associated with a library negative control.

The bone libraries were prepared in a dedicated low copy DNA laboratory using the KAPA Hyper Prep Kit (Roche Sequencing, Wilmington, MA, USA). Library preparation followed the manufacturer's recommendations using 15-μM 8-bp dual-indexed adapters. A target DNA extract volume of 50 μL was used for library preparation, with a maximum DNA input of 1 μg. Some of the previously existing DNA extracts had limited volume available; therefore, Tris-EDTA was added to create a total volume of 50 μL. Reference sample libraries were prepared with the KAPA HyperPlus Kit (Roche Sequencing), as fragmented DNA is required for successful library preparation. Library preparation was performed in accordance with the manufacturer's protocol with a 20-min fragmentation and ligation of dual-indexed 8-bp adapters (Integrated DNA Technologies, Coralville, IA, USA). The library PCR reactions for the bone samples utilized KAPA HiFi HotStart Uracil + ReadyMix (Roche Sequencing) to accommodate for uracils present in the native DNA molecule from cytosine deamination. The reference/control sample libraries were amplified with the KAPA HiFi HotStartReadyMix (Roche Sequencing). Library amplification was completed following the manufacturer's recommendations, using a total of 12 cycles of PCR for each of the sample sets (bone and reference/control). Samples were purified using a 1.0× AMPure XP (Beckman Coulter, Indianapolis, ID, USA) bead-based cleanup and were then eluted in 20 μL of Tris-EDTA. The quality of the libraries was checked on the 2100 Bioanalyzer instrument (Agilent Technologies, Santa Clara, CA, USA) using the Agilent DNA 7500 Kit. For the bone sample and control DNA libraries, an aliquot of the purified library product was reserved for WGS.

2.3.4. Hybridization Capture

Hybridization capture with the α version of the FORCE capture panel was completed using the myBaits 1-20K Custom DNA Hybridization Capture kit (Arbor Biosciences, Ann Arbor, MI, USA). The procedure followed the manufacturer's recommended v5 protocol using a single capture and 5 μL of custom SNP baits. Samples were allowed to incubate overnight for approximately 24 h at 62 °C with a heated lid (72 °C) using a Veriti thermal cycler (Thermo Fisher Scientific, Waltham, MA, USA). PCR amplification was achieved using duplicate PCR reactions to make use of the entire capture product. The post-capture PCR reactions were completed with 19 PCR cycles using the KAPA HiFi HotStartReadyMix following the manufacturer's recommendations. The two amplified capture products were combined for each sample and then purified with the MinElute PCR Purification Kit, eluting in 25 μL of Tris-EDTA.

2.3.5. Normalization, Pooling, and High Throughput Sequencing

Purified capture product was quantified using the 2100 Bioanalyzer instrument with the Agilent DNA 7500 Kit to determine DNA molarity. As in library preparation and hybridization capture, the three sample sets (bone, JB55, and reference) were pooled separately to minimize the impact of crosstalk between samples of disparate quality during sequencing [**10**]. The bone set positive control (K562) was omitted from the bone sample sequencing pool and instead added to the reference sample pool for sequencing. The JB55 sample pool included JB55 and its associated RB, as well as an additional library from a different JB55 DNA extract. (Since both JB55 libraries produced similar results; only the data from the DNA extract that also underwent WGS are presented below.) The JB55 library controls were not sequenced; but they were assessed with the Bioanalyzer for contamination and quality control purposes.

Pooling was achieved by individually normalizing captured libraries to the same concentration as the least concentrated sample (nM) in the set, then adding equal volume of all samples to the pool along with maximum RB and negative control volumes. The captured library pools were quantified using the Agilent DNA 7500 Kit on the 2100 Bioanalyzer. The bone and reference captured library pools were diluted to 1 pM with a 5% spike-in of PhiX Control V3 (Illumina) for sequencing on a NextSeq 550 (Illumina) using NextSeq 500/550 High Output Kits, v2.5 (150 cycles). Paired-end sequencing (75 × 75 cycles) was performed for the bone sample pool (14 captured libraries), while single-end data (150 cycles) were generated for the reference and control DNA samples (26 captured libraries). The JB55 pool was diluted to 12 pM including 5% PhiX control V3, and paired-end sequencing (75 × 75 cycles) was completed on a VerogenMiSeqFGx in RUO mode using a MiSeq Reagent Kit, v3 (150 cycles).

To gauge capture efficiency for the non-probative bone sample and control DNA libraries, baseline WGS data were produced on an Illumina NextSeq 550. WGS was not performed for any of the reference samples. Sequencing was completed with NextSeq 500/550 High Output Kits, v2.5 (150 cycles) using the same approach as the captured libraries (75 × 75 cycles for the bone sample pool and 1 × 150 cycles for the control DNA pool). For direct comparison with the JB55 capture data, WGS of one JB55 library was performed on a MiSeqFGx using RUO mode and paired-end (75 × 75 cycles) sequencing on a v3 cartridge (150 cycles).

2.3.6. Sequence Data Analysis

FASTQ files were imported into the CLC Genomics Workbench v12.0.1 (QIAGEN) and all sample data were analyzed with a custom workflow. The workflow, which was applied to all samples, included five steps. (1) Reads were trimmed based on quality (including ambiguous bases) as well as adapter read-through (for paired reads only). (2) Trimmed reads were then mapped to the GRCh38 human reference genome using stringent mapping parameters designed to prevent off-target mapping of exogenous DNA (e.g., bacterial DNA). Modifications to the default mapping parameters were a length fraction of 0.85 and stringency fraction of 0.95, and non-specific reads (i.e., reads that produce an equal alignment score for multiple mapping regions) were ignored. (3) Duplicate mapped reads were removed with 20% maximum representation of minority sequence. (4) Local realignment of unaligned ends was performed to improve indel alignment in the mapping. (5) The alleles observed at each of the 5422 FORCE SNPs was determined using the Identify Known Mutations from Mappings tool, ignoring broken pairs in the bone set sequence data. Various analysis metrics such as the total number of reads, percent mapped, and percent duplicate reads were reported throughout the workflow using different CLC Genomics Workbench quality control (e.g., QC for Targeted Sequencing) tools.

A custom Microsoft Excel (Redmond, WA, USA) template was developed for genotype generation based on the CLC Identify Known Mutations from Mappings output. The targeted SNP had to be covered by a minimum of 10 reads for analysis. Furthermore, in order for a FORCE SNP to be called (i.e., used for downstream analyses), genotypes required allele frequency of 90% or greater to be called as a homozygote and heterozygous genotypes needed at least 30% MAF. SNPs meeting the minimum coverage requirement (10X) but with imbalanced alleles (10–30% MAF) that could not be confidently called homozygote or heterozygote were noted as "imbalanced" and excluded from further analyses. The called SNPs were separated by type (e.g., kinship SNPs, aiSNPs, piSNPs) for analysis and interpretation as described below.

2.3.7. SNP Concordance Assessment

The genotypes derived from the FORCE capture data of the four control DNAs (2800M, K562, NA12877, and NA12878) and one of the historical bone samples (JB55) were compared to previously published data. For 2800M, data from the ForenSeq DNA Library Preparation Kit were used for comparison [**28**], taking into account reverse strand orientation at 52 markers. For the remaining control DNA samples, published variant call format (VCF) files from WGS data were used for the concordance assessment ([**29**] for K562, [**30,31**] for NA12877 and NA12878). The VCF files for K562 and NA12877 included only those positions in which a variant was detected; thus, the genotype was assumed to be homozygous for the reference allele at SNPs missing from the VCF for the purpose of this comparison. A preliminary assessment identified a few issues in the published control DNA datasets that impacted concordance. There was an error in the K562 VCF for rs10892689, and the genotype for this SNP was determined by visual inspection of the associated BAM [**29**]. The NA12878 VCF included all SNP positions regardless of the detection of a non-reference allele. Five FORCE SNPs were missing from the NA12878 VCF, and the associated BAM [**30,31**] was used to confirm the genotype at those positions. Additionally, strand orientation was corrected in the published genotypes at two SNP markers (rs6670984 and rs4922532).

For JB55, published genotypes produced with the Precision ID Ancestry Panel (n = 165) and targeted Y-SNP typing (n = 4) as described in [**12**] were used for comparison to the FORCE profile. The genotype with the highest coverage was utilized when discordant data were observed between the two Precision ID replicates analyzed in the previously published study. Additionally, there were 16 SNPs without sufficient coverage (<10X) in either Precision ID replicate; these markers were excluded from the concordance assessment.

Supplementary to called SNPs, "imbalanced" FORCE genotypes were also compared to the published data. Imbalanced genotypes were observed at SNPs with sufficient coverage (≥10X), but where the allele frequency thresholds for homozygote/heterozygote calling were not met. When the minor allele(s) was observed with base quality <20 and <10% forward reverse balance, indicating sequencing error, the artifact (minor) allele was removed, and the observed genotype noted for the imbalanced SNP was modified. This was only performed for the concordance assessment. No imbalanced SNPs were utilized in downstream analyses (e.g., ancestry predictions, kinship comparisons).

When discordance was observed between the FORCE and published genotype, the FORCE BAM and the BAM associated with the published data were reviewed to confirm the genotypes at the SNP marker. If the genotypes were still inconsistent after further review, the cause of the discrepancy was investigated.

2.3.8. Biogeographical Ancestry Predictions

The aiSNPs were used to conduct biogeographical ancestry estimates using two distinct methods: principal component analysis (PCA) and naïve Bayes analysis. A script written in the R programming language [25] was created to perform Principal Component Analysis (PCA) using the prcomp package [32]. The SNP profiles obtained from the 1000 Genomes project [18] were used to train the PCA model. The script incorporates the naïve Bayes [33] analysis approach by computing the likelihood Pr (DNA|Population X) for each major continental population group (EUR, EAS, AMR, SAS, and AFR). The likelihood estimates utilized allele frequencies obtained from the 1000 Genomes project [18]. Due to the absence of allele frequencies for one aiSNP (rs10954737) in the 1000 Genomes dataset, this particular marker was not used in the analysis. Ancestry estimates were instead based on 240 other aiSNPs. The posterior probabilities were computed under the assumption of a uniform prior. The aiSNPs that have been previously identified provide a distinct differentiation between the major demographic groupings EUR, AFR, EAS, and SAS [34].

2.3.9. Phenotype Predictions

The 41 piSNPs included in the FORCE panel are consistent with the markers of the HirisPlex-S System [**35**]. Using the Excel template described in **Section 2.3.6**, the genotypes from sequence data were converted into the correct allele orientation and assigned "0", "1", "2", or "N/A" based upon a customized R script [**36**]. The converted genotypes were then submitted to the HirisPlex-S System webtool for phenotype prediction [**37**]. The Excel template organizes the converted genotypes of the piSNPs either for easy manual entry or for batch analysis using the file upload functionality. The FORCE data were uploaded as a single file and the HirisPlex-S output was downloaded. The results were compared to self-reported eye color and hair color for the seven additional non-probative reference samples (18–24; **Table 4**).

2.3.10. Y Haplogroup Predictions

Y haplogroups were predicted based upon the 829 Y-SNPs included in the FORCE panel. Called Y-SNPs were compared to the Yleaf WGS HG38 Position file [**38,39**] within the Excel template and variants (non-reference alleles) were denoted with the corresponding Y haplogroup. Identified haplogroups were reviewed manually and the most refined Y haplogroup was reported. The results were further inspected to ensure that the variants observed were expected based upon the predicted Y haplogroup (e.g., variants associated with ancestral haplogroups of the reported haplogroup). Though not utilized in this study, Yleaf offers automated Y haplogroup predictions [**38**]. Yleaf can analyze both FASTQ and BAM files; however, current functionality does not allow for duplicate removal, and therefore it is recommended to utilize FORCE BAM files after mapped duplicate removal for haplogroup predictions performed with the Yleaf software.

2.3.11. Kinship Statistics (Kinship SNP and X-SNP)

Pairwise kinship calculations (kinship SNPs). Likelihoods were calculated based on observed kinship SNP data given seven different relationship hypotheses (parent/offspring, full siblings, half siblings, first cousins, first cousins once removed, second cousins, and second cousins once removed) and one hypothesis representing no genetic relationship between the two compared individuals (i.e., unrelated). From these likelihoods, likelihood ratios (LRs) were calculated for each related hypothesis using the unrelated hypothesis as the alternative hypothesis (i.e., LR = Pr (DNA|H1)/Pr (DNA|H2), where H1 was the hypothesis representing one of the related relationships given above and H2 was the unrelated hypothesis). These kinship calculations were performed in a pairwise fashion between all reference samples as well as between each "Unknown" (i.e., WWII bone sample) and each reference sample. Posterior probabilities were calculated assuming a flat prior. The likelihoods were calculated using Merlin with the preferences described above

(**Section 2.2**). R was again used to format and manage input files (e.g., filter SNPs when having partial SNP profiles) as well as to handle the outputs from Merlin and for the plotting of distribution curves.

Pedigree kinship calculations (kinship SNPs). In addition to the pairwise kinship calculation, kinship statistics were also calculated in a pedigree fashion. Here, SNP profiles for all reference individuals, in each reference family, were included in the likelihood calculations. In this approach each unknown sample was tested as the missing person in each reference pedigree. For each such pedigree, the LR was calculated as LR = Pr (DNA|"H1: The tested unknown sample comes from the missing person")/Pr (DNA|"H2: The tested unknown sample comes from an individual unrelated to the reference pedigree"). The likelihoods were calculated using Merlin with the preferences described above (**Section 2.2**). As previously noted, a front-end script was written in R to format input files and to handle the outputs from Merlin. Kinship predictions with strong statistical support were defined as those that met the following two criteria: LR ≥ 10,000 (log 10 LR ≥ 4) and posterior probability ≥ 99.99% given equal priors for the tested hypotheses.

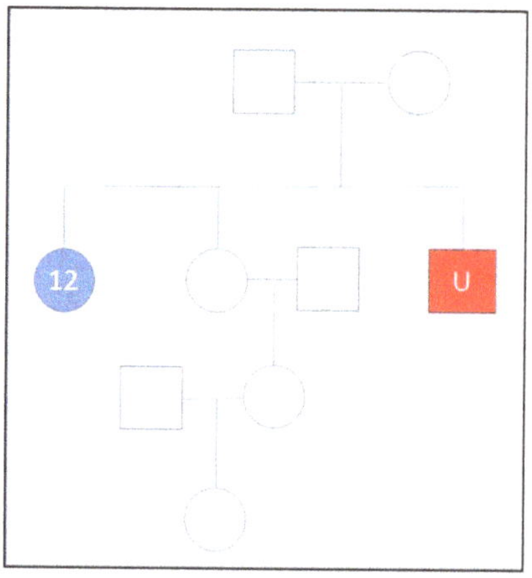

Figure 2. Pedigree of unknown (sample 14.1) and the reference sample (12) from family D utilized in X-SNP kinship assessment.

Kinship calculations using X-SNPs. In order to demonstrate the possibility to calculate kinship statistics solely based on X-SNPs, two samples were selected from family D: sample 12 ("reference") and sample 14.1 ("unknown"). These two individuals are full siblings (**Figure 2**) and their X-SNP genotypes were used to calculate the LR = Pr (DNA|"Sample 12 and sample 14.1 come from two full siblings)/ Pr (DNA|"Sample 12 and sample 14.1 come from two unrelated individuals"). The X-specific version of Merlin ("MINX") calculated the likelihoods of the two hypotheses, using European allele frequencies and genetic positions obtained from Rutgers map [**18,24**]. As previously noted, R was used to format input files and to handle the outputs from Merlin.

3. RESULTS AND DISCUSSION

3.1. Simulated Panel Performance

The kinship simulations unequivocally indicate that the kinship single nucleotide polymorphisms (SNPs) have the potential to resolve the majority of kinship case situations up to the level of first cousins. This includes cases when only incomplete SNP profiles are available, as illustrated in Figure 3. In addition, this collection of kinship single nucleotide polymorphisms (SNPs) can provide valuable information in situations involving more distant familial connections, such as second cousins. However, the likelihood ratio (LR) distribution curves for these cases, both for real relatives and true non-relatives, overlap..

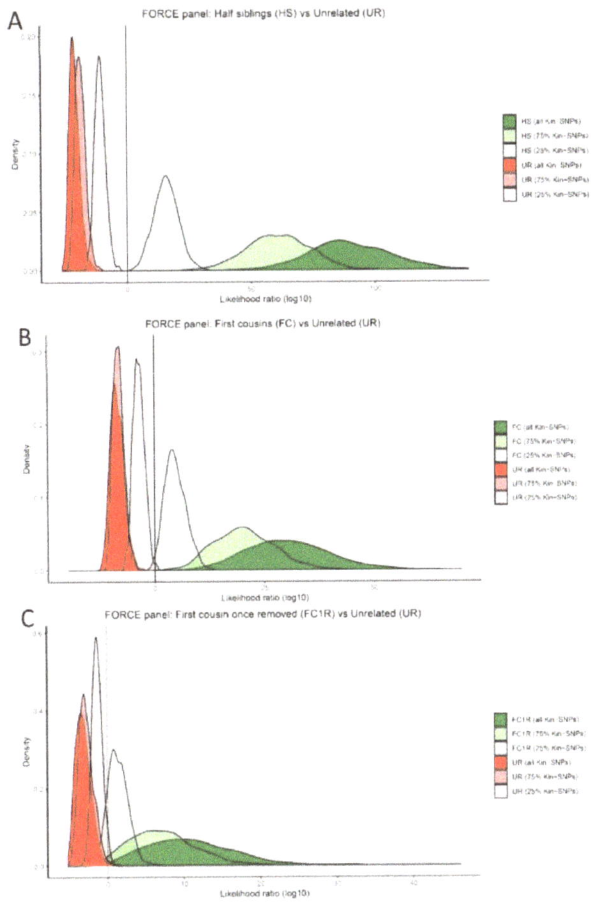

Figure 3. Simulation results for pairwise kinship predictions. Simulations were performed between an unknown sample at varying levels of kinship SNP recovery (%75 ,%100, and %25) compared to a complete reference sample profile. (A) Second degree (half siblings) vs. unrelated (UR); (B) third degree (first cousins) vs. UR; (C) fourth degree (first cousins once removed) vs. UR. Simulation results for additional relationships (first degree and fifth degree) are presented in Figure S2.

Table 5. Summary of the number and proportion of called SNPs after capture of the FORCE panel targets. The results are broken down by sample type. High quality includes family references, additional non-probative reference samples, and control DNAs.

Sample Type	Count	Maximum Possible SNPs	Called SNPs * Count			Percentage		
			Minimum	Maximum	Average	Minimum	Maximum	Average
Male, high quality	11	5422	5133	5395	5355	94.7%	99.5%	98.8%
Female, high quality	12	4593	4396	4582	4552	95.7%	99.8%	99.1%
Bone (all male)	12	5422	11	5361	2407	0.2%	98.9%	44.4%
Control blank	7	5422	0	26	4	0.0%	0.5%	0.1%

* ≥10X coverage, homozygous genotype ≥ 90% allele frequency; heterozygous genotype ≥ 30% minor allele frequency.

The bone sample results varied considerably, ranging from 11–5361 called SNPs and an average of 44.4% of the possible 5422 SNPs across the 12 DNA extracts. To compare the number of called SNPs per bone, the results from each DNA extract were averaged for bone samples with multiple DNA extracts (**Figure 4**). Four of the seven bone samples produced 3719 or more SNPs (two USS *Oklahoma* (3 and 7) and the two historical bones (JB55 and sample 26)). These four samples were extracted with either PCIA followed by MinElute purification or the Dabney DNA extraction method. In contrast, the bone samples that were extracted with QIAquick produced fewer called SNPs. The three QIAquick replicates from the Italy bone sample (14) yielded 5121–282 called SNPs, for an average of 2308. The Austria (11) and Tarawa (17) bone samples, which were also extracted with QIAquick, produced only 50 and 14 called SNPs on average, respectively. The Tarawa bone sample was likely chemically treated with formalin, which was the standard postmortem procedure during WWII. Such chemically treated samples have been shown to produce poor DNA sequencing results [**10**]. The bone from Austria may have also received the same standard WWII postmortem chemical treatment. It is important to note that the variable results obtained from the bone samples may be due to a combination of differing factors, including separate case contexts, the DNA extraction method utilized, as well as the remaining DNA extract volume available for library preparation. The sample selection strategy for this study had prioritized known, non-probative samples with available DNA extracts from both the bone sample and associated family reference samples. Therefore, it was not possible to control for DNA extract preparation and other variables that may have a considerable impact on SNP capture results. An investigation into DNA extraction methods for downstream SNP capture success is an area of future research.

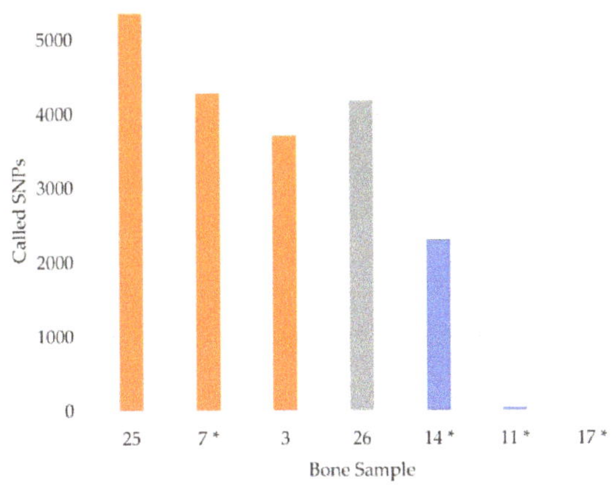

Figure 4. The number of called SNPs per bone sample (maximum 5422). An asterisk (*) represents the average number of called SNPs for a bone sample with multiple DNA extracts. The color indicates the DNA extraction method: orange = phenol chloroform isoamyl alcohol (PCIA) extracts purified with MinElute; gray = Dabney; blue = QIAquick.

In the absence of capture, the bone DNA extracts produced 2–1539 SNPs at 1X coverage and zero SNPs met the 10X calling threshold. The proportion of reads aligning to the FORCE panel regions (i.e., mapped read specificity) was significantly increased by capture, from an average of 0.03% in the shotgun data to 11.4% in the capture data. The capture procedure thus resulted in a 380-fold increase in mapped read specificity over shotgun sequencing. Of note, shotgun sequencing of the four control DNAs produced only partial 10X FORCE SNP profiles. Thus, the capture enrichment was necessary for all samples, including references and control DNAs, in order to meet the 10X coverage requirement for SNP calling. The improved SNP profiling of the captured libraries demonstrates the efficacy of the enrichment step in obtaining on-target FORCE panel SNPs.

3.3. Concordance with Published SNP Data

Overall, 14,519 (99.97%) SNPs were concordant between the 14,524 called FORCE genotypes and published data of the four control DNA samples (**Figure 5**). Complete concordance (100%) was observed for 2800M across the 170 SNPs overlapping called FORCE and expected ForenSeq genotypes. High concordance (99.98%) was also observed for both NA12877 and NA12878 when considering the called FORCE genotypes for each sample (5383 and 4576 SNPs, respectively). Though slightly lower, FORCE data of K562 produced concordant genotypes with published data at 99.93% of the 4395 compared SNPs.

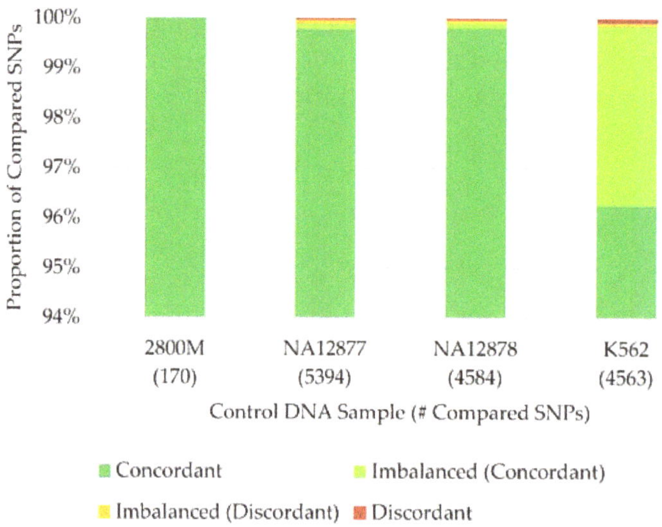

Figure 5. SNP concordance of FORCE capture data compared to published genotypes from the four control DNA samples.

The five discordant called FORCE genotypes were all kinship SNPs, and each discordance was observed in only one control DNA sample. The greatest proportion of discordant calls (60%) was observed in the K562 data. Two of three discordant called SNPs produced high-coverage homozygote genotypes in the FORCE data, while heterozygote genotypes were observed in the published K562 BAM but with allele imbalance (MAF ~20%). The third K562 discordance may be the result of incorrect genotyping based on the published data, as all reads at the SNP location in the published BAM were forward-oriented and unpaired. The greater discordance observed in the K562 data compared to the other control samples is likely due to well-known immortalization effects [**29**] as well as lot-to-lot variability of this cell line [**40**]. For NA12877, there was one difference between the called FORCE SNP and the published data at rs229352. Insufficient coverage (<10X) at this same SNP was observed for many reference/control samples (n = 7). Interestingly, NA12878 and K562 were expected to be homozygous for the alternate allele (GG) at rs229352 based on the published data (this marker is not included in ForenSeq and therefore the expected genotype for 2800M is unknown). Analysis of the rs229352 bait design identified that the alternate allele (G) or left flank did not map under the stringent mapping parameters utilized in the developed workflow. Furthermore, review of the published BAM showed a -5bp insertion located 30 bp upstream ('5) of the SNP on the reads with the G allele at rs229352. For NA12878, the discordant called SNP (rs662185) was identified as a poor performing SNP in **Section 3.2**, with nearly half (47.8%) of the reference/control samples failing to produce sufficient coverage (<10X) and an additional 17.4% with imbalanced genotypes. In NA12878, significantly lower coverage (10X) was observed at rs662185 compared to the average coverage of the other FORCE SNPs (>250X for called SNPs). Similar to rs229352, the published NA12878 BAM showed a 4-bp deletion associated with the G allele at rs662185 located 5 bp downstream (3′) of the SNP. Additionally, evaluation of the three baits for this marker (the right flank bait failed during design) revealed that the alternate allele bait included in the kit was incorrect (T instead of G). Inspection of the baits for all targets identified an additional 522 markers with the incorrect alternate allele. However, these errors in the bait design appeared to have little to no impact on the SNP performance, as only seven (1.3%) of the 523 SNPs with the wrong alternate allele bait performed poorly (<75% called) in reference/control samples. Therefore, the discordances observed at rs229352 and rs662185 are likely due to the indel clusters associated with the alternate allele, which may impact the probe hybridization efficacy and/or prevent mapping under the stringent parameters applied.

Though not utilized for other analyses in this study, the concordance of the 187 observed genotypes at imbalanced FORCE SNPs was assessed for the four control DNA samples. It is noted that 95.19% (178) of the observed imbalanced genotypes were consistent with the published data. Unsurprisingly, a majority (168 of 187; ~90%) of the imbalanced genotypes were detected in K562. As stated above, stochastic variation is expected in the K562 cell line, resulting in allele imbalance and loss of heterozygosity [**29**]. Furthermore, these cell line artifacts may be exacerbated by the relatively low input (1 ng) into library preparation [**10**]. Of the nine discordances found in the imbalanced genotypes, rs169250 accounted for three instances. The imbalanced FORCE genotypes at rs169250 were discordant for three of the control DNA samples (NA12877, NA12878, and K562; this marker is not targeted by ForenSeq and therefore no published genotype was available for 2800M). Further investigation of the BAMs revealed that rs169250 is located in the middle of a 12-bp polycytosine stretch, and thus the complexity of the region for sequencing and/or alignment likely caused discordant genotypes. As a result, future designs of the FORCE panel may exclude rs169250 to ensure successful and accurate genotyping. The remaining imbalanced FORCE genotypes for NA12877 (n = 3) and NA12878 (n = 1) that were discordant with the published data appear to be the result of non-specific mapping in the FORCE data. The last two discordances involving imbalanced FORCE genotypes were found in K562. One appeared to be the result of non-specific mapping in the published BAM (rs241408), while the other (rs7117433) could not be resolved and may be due to cell line artifacts previously discussed.

In the end, considering both called and imbalanced FORCE genotypes, a total of 99.90% (14,697 out of 14,711) SNPs were in agreement between the FORCE and published control DNA datasets. There were, however, a total of 69 genotypes could not be compared out of a maximum of 14,780 possible SNPs because no call was possible for the FORCE or published data (<10X). Most (68) of the failed SNPs were observed in the FORCE data, while one SNP for K562 had low coverage (5X) in the published BAM that prevented the associated published genotype from being confirmed. The number of failed SNPs varied by sample: two of 172 SNPs for 2800M, nine of 4593 SNPs for NA12878, 28 of 5422 SNPs for NA12877, and 30 of 4593 SNPs for K562. Of note, FORCE data from all four control DNA samples failed to produce called genotypes for 16 SNPs. Thirteen of these SNPs failed to produce sufficient coverage for any samples, as previously noted in **Section 3.2**. The three additional poor performing SNPs included two X-SNPs and one Y-SNP. Most (~70%) reference samples failed to produce called genotypes at the two X-SNPs, with no called data in the bone samples. The Y-SNP produced greater reference/control samples success (>65% of male references/controls with a called genotype), but 0% SNP recovery for the bone samples.

Published and FORCE data for JB55, the only bone sample evaluated for concordance, enabled the comparison of 153 SNPs. Sixteen of the 169 overlapping SNPs were excluded from comparison due to insufficient coverage (<10X) in the published Precision ID data. JB55 had 151 of 153 (98.69%) concordant genotypes, and two discordant aiSNPs were observed. The FORCE genotype (CC) for the first discordant marker (rs1879488) was inconsistent with the highest coverage Precision ID replicate from the published study (AC with 395X in replicate 2); however, the FORCE results were concordant with the replicate with slightly lower coverage (CC with 236X in replicate 1). The second difference was observed at a SNP with low coverage in both Precision ID replicates (8X and 16X), potentially the result of non-specific amplification and/or mapping. In the case of these two discordant SNPs, stricter parameters such as a higher coverage threshold or heterozygote threshold than what was used for the previous JB55 study (≥6X and ≥10% MAF) [**12**] may have improved the concordance for JB55.

3.4. Ancestry, Phenotype, and Y Haplogroup Predictions

The completeness of the reference and control sample profiles allowed for ancestry, phenotype, and Y-chromosomal haplogroup predictions, which were consistent with available metadata. The predictive results from the bone sample DNA extracts were much more variable due to the wide range of SNPs obtained. As an example of the bone sample data, the ancestry of sample 7.2 was predicted to be 100% European from 210 called aiSNPs (**Figure 6**). Phenotype predictions were also possible for this bone sample (7.2) using 33 called piSNPs, which estimated blue eyes (93.2% probability), blond-brown hair (66.0% and 27.4% probabilities, respectively), light hair shade (92.8% probability), and pale-intermediate skin (54.1% and 41.6% probabilities, respectively). The Y haplogroup was predicted to be R1a1a1b1a2b~ from 370 Y-SNPs, which was the same Y haplogroup that was predicted for the paternal nephew (sample 4). These predictions were possible even with the partial SNP profile (4165, or 76.8%, of possible 5422 called SNPs) generated from this degraded USS *Oklahoma* sample. Thus, hybridization capture was effective in producing sufficient SNPs from the multiple marker types included within the FORCE panel to allow for predictive information to be gained.

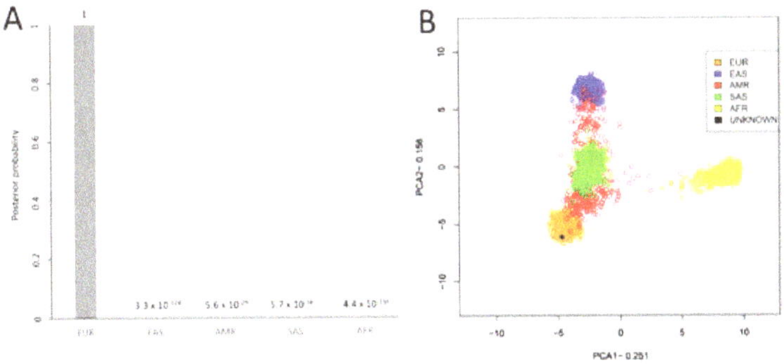

Figure 6. An illustration of ancestry predictions, here exemplified with predictions for sample 7.2. Individuals from the 1000 Genomes project (grouped into the five different continental groups: EUR (European), EAS (East Asian), AMR (American), SAS (South Asian), and AFR (African)) were used as reference data. (A) Posterior probabilities using a naïve Bayes approach; (B) Prediction using a PCA.

3.5. Kinship Predictions

The Figure 7 displays the results of kinship prediction for the reference samples. The initial degree (parent/offspring and complete sibling) estimations yielded log 10 likelihood ratios ranging from 287.7 to 439.3, accompanied by posterior probabilities exceeding 99.99%. This indicates a rather robust statistical backing for these estimations. The log 10 LRs for the third degree reference to reference predictions were 34.4 and 37.8 with over 99.99% posterior probability, indicating very strong statistical support. The two fifth degree relationship predictions produced log 10 LRs of 3.8 and 6.5, with 99.98% and 99.99% posterior probabilities, respectively. Thus, the kinship analysis was capable of correctly predicting fifth degree relatives from reference quality FORCE panel profiles.

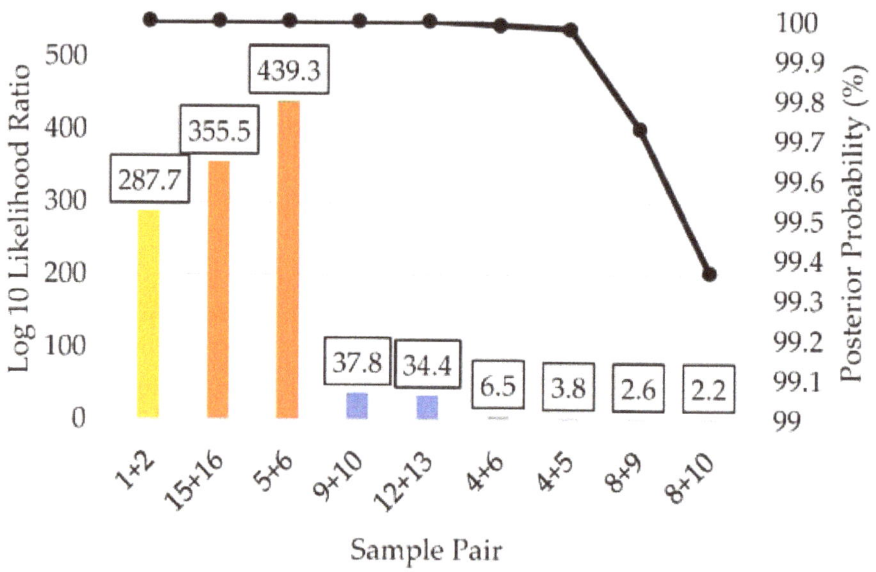

Figure 7. Pairwise kinship prediction log 10 likelihood ratio (LR) (left y-axis; bars) and posterior probabilities (right y-axis; black line) of reference-to-reference comparisons. All predictions compared the expected (known) relationships vs. unrelated. Parent/offspring = yellow (sample pair 2 + 1); full siblings = orange (sample pairs 16 + 15 and 6 + 5); third degree = blue (sample pairs 10 + 9 and + 12 13); fifth degree = gray (sample pairs 6 + 4 and 5 + 4); and sixth degree = purple (sample pairs 9 + 8 and 10 + 8).

The two sixth degree relationship predictions were supported with over 99% posterior probability, but with log 10 LRs of 2.2 and 2.6, indicating moderate statistical support. Considering the four examples of fifth-sixth degree relationships, log 10 LRs corresponding to moderate statistical support [41] were obtained in a majority (75%) of the fifth-sixth degree comparisons. Interestingly, sample 4 (fifth degree comparison) had the lowest SNP recovery of the references (95%) and sample 8 (sixth degree comparison) also had reduced SNP recovery (98%). All but one of the remaining references had >99% complete FORCE profiles (sample 13 at 98%). Therefore, the ability to obtain stronger statistical support for these more distant degrees of relatedness (as in the case of samples 4 and 8) could be possible when complete FORCE panel profiles are obtained. Regardless, the simulation studies indicated considerable overlap in expected versus unrelated relationships when testing 5th degree relatives and beyond. When pairwise kinship predictions were performed in a blind search fashion allowing up to either third degree or sixth degree relatedness, zero incorrect (false positive) relationships with strong statistical support were observed.

The Figure 7 displays the results of kinship prediction for the reference samples. The initial degree (parent/offspring and complete sibling) estimations yielded log 10 likelihood ratios ranging from 287.7 to 439.3, accompanied by posterior probabilities exceeding 99.99%. This indicates a rather robust statistical backing for these estimations.

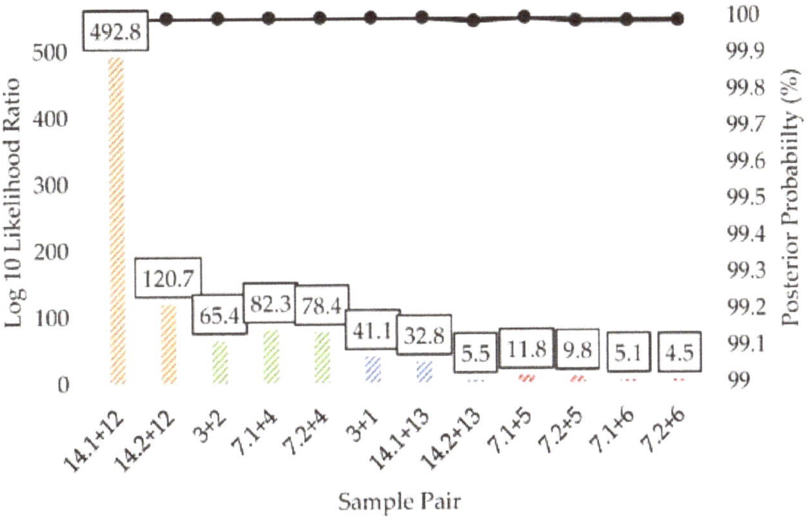

Figure 8. Pairwise kinship prediction log 10 likelihood ratio (LR) (left y-axis; bars) and posterior probabilities (right y-axis; black line) of unknown to reference comparisons. Results are presented for each bone sample DNA extract that produced %25~) 1000) or more kinship SNPs. All predictions compared the expected (known) relationships vs. unrelated. Full siblings = orange (sample pairs 12 + 14.1 and 12 + 14.2); second degree = green (sample pairs 4 + 7.1 ,2 + 3, and + 7.2 4); third degree = blue (sample pairs 13 + 14.1 ,1 + 3, and 13 + 14.2); fourth degree = red (sample pairs 6 + 7.1 ,5 + 7.2 ,5 + 7.1, and 6 + 7.2).

Unknown to reference pairwise kinship predictions from bone sample DNA extracts that produced 1000 (~25%) or more kinship SNPs are presented in **Figure 8**. Similar to what was observed in the reference to reference pairwise comparisons, the unknown to reference full sibling predictions had extremely strong statistical support. The log 10 LRs of these first degree relationship predictions were 120.7 and 492.8, with posterior probabilities over 99.99%. The three second degree relationship predictions produced log 10 LRs between 65.4 and 82.3 and posterior probabilities over 99.99%, also indicating extremely strong statistical support. Two of the three third degree relationship predictions (for sample pairs 3 + 1 and 14.1 + 13) were consistent with the reference to reference predictions. These two had log 10 LRs of 32.8 and 41.1 with posterior probabilities over 99.99%, indicating extremely strong statistical support. The other third degree prediction (14.2 + 13) also had a posterior probability over 99.99% but with a log 10 LR of 5.5, which was much lower than that of the other two third degree predictions. This LR indicating strong statistical support was made from a bone sample profile with only 1324 (33.7%) kinship SNPs. It is likely that a higher LR would have been obtained if sample 14.2 had greater SNP recovery. The four fourth degree relationship predictions had log 10 LRs from 4.5 to 11.8, and all had posterior probabilities over 99.99%. These fourth degree unknown to reference predictions would be classified as having strong to very strong support.

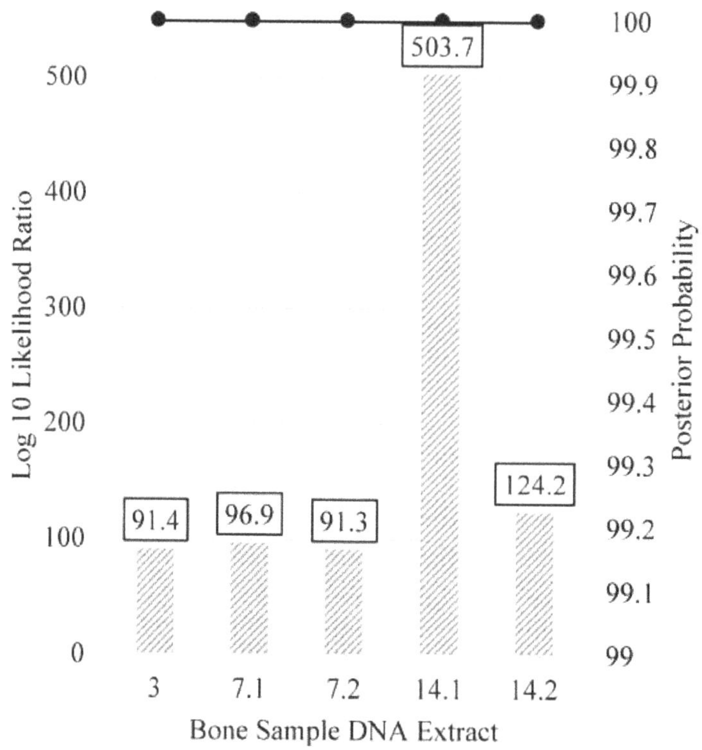

Figure 9. Pedigree scenario kinship prediction results of all bone samples with more than %25~) 1000) kinship SNPs. All predictions compared the unknown when situated in the expected (known) pedigree vs. unrelated. The log 10 likelihood ratio (LR) (left y-axis; bars) and posterior probabilities (right y-axis; black line) are shown.

The five bone sample DNA profiles with >25% kinship SNPs were utilized for kinship predictions in a pedigree scenario, as presented in **Figure 9**. All of the predicted relationships indicated relatedness between the unknown and the expected family members. These pedigree scenario relationship predictions were supported by log 10 LRs greater than 91.3 (and up to 503.7) with posterior probabilities exceeding 99.99%. Thus, the pedigree scenarios yielded extremely strong statistical support when bone sample profiles yielded ~25% or more kinship SNPs.

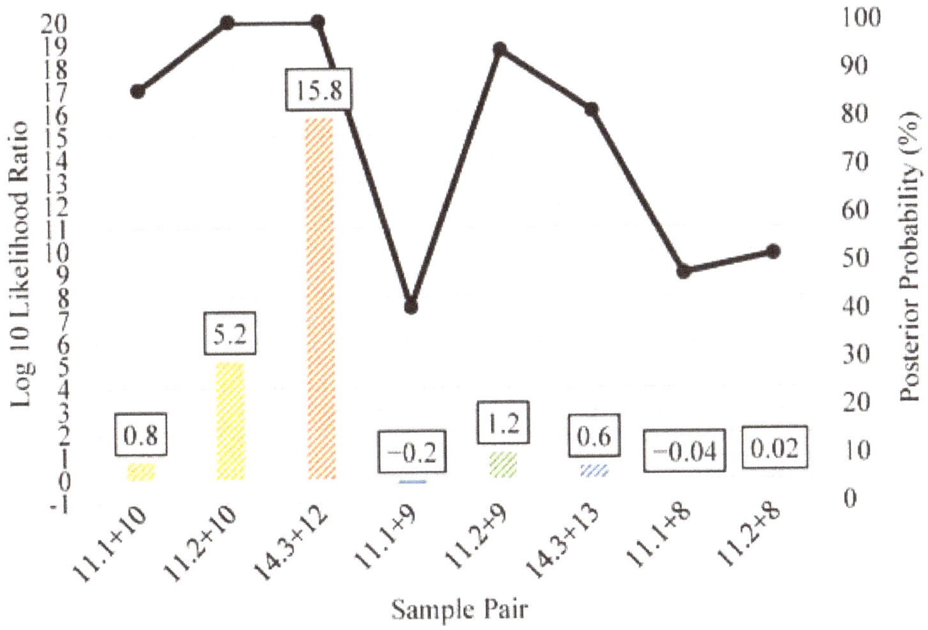

Figure 10. Pairwise kinship prediction log 10 likelihood ratio (LR) (left y-axis; bars) and posterior probabilities (right y-axis; black line) of unknown to reference comparisons. Results are presented for each bone sample DNA extract that produced 247–10 kinship SNPs (all results <%25 called SNPs). Samples 17.1 and 17.2 yielded <15 kinship SNP calls that were inconsistent with the expected parent/offspring relationships; therefore 17.1 and 17.2 and were excluded altogether. All predictions compared the expected (known) relationships vs. unrelated. Parent/offspring = yellow (sample pairs 10 + 11.1 and 10 + 11.2), full siblings = orange (sample pair 12 + 14.3); second degree = green (sample pairs 11.1 9 + and 9 + 11.2); third degree = blue (sample pair 13 + 14.3); and fifth degree = gray (sample pairs 8 + 11.1 and 8 + 11.2).

When bone sample profiles produced minimal SNP data (10–247 kinship SNPs), the kinship predictions were expectedly weaker overall (**Figure 10**). Only two bone sample profiles yielded predictions that would indicate strong statistical support, and both involved first degree relationships. Unknown (bone) sample 11.2 had 73 kinship SNPs, and its parent/offspring relationship with sample 10 was strongly supported by a log 10 LR of 5.2 and a posterior probability > 99.99%. Sample 14.3 had 247 kinship SNPs, and its full sibling relationship with sample 12 was very strongly supported by a log 10 LR of 15.8 and posterior probability > 99.99%. The remaining six relationship pairs shown in **Figure 10** had low log 10 LRs (1.2 or less) and posterior probabilities (<%95). The pedigree scenario kinship predictions did not significantly improve the log 10 LRs or posterior probabilities of the bone sample profiles with 247–10 called SNPs (**Figure 11**). Additionally, samples 17.1 and 17.2 yielded <15 kinship SNPs each, with some SNPs that were inconsistent with the profiles of the associated child reference (one son and one daughter of the unknown) in these parent/offspring relationships. As a result of the poor SNP recovery, bone samples 17.1 and 17.2 were excluded altogether from the pairwise and pedigree scenario kinship predictions. Regardless of the number of kinship SNPs in the bone sample profiles, there were no incorrect (false positive) relationship predictions between unknowns and unrelated reference samples with strong statistical support in blind search scenarios.

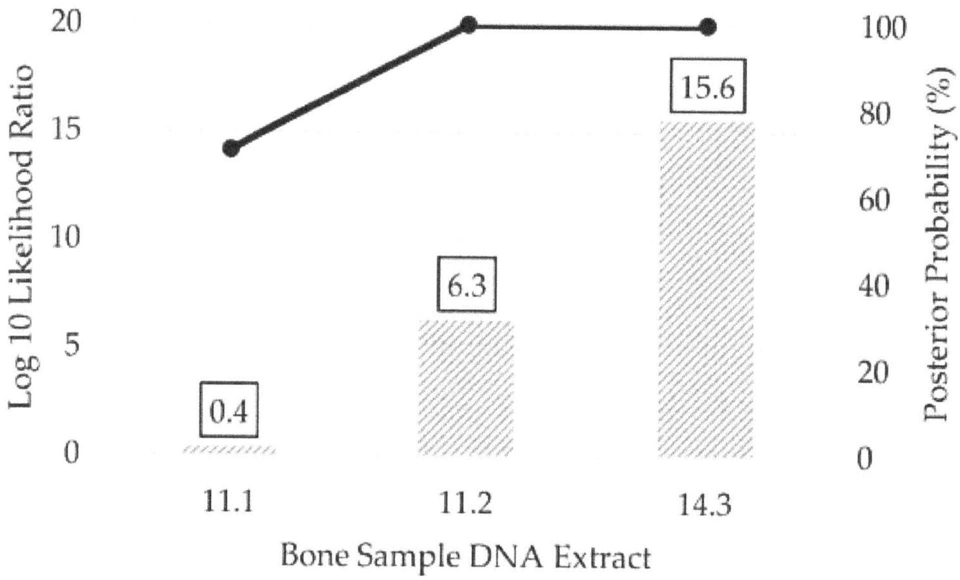

Figure 11. Pedigree scenario kinship prediction results of bone sample DNA extracts that produced 247–10 kinship SNPs. All predictions compared the expected (known) pedigree vs. unrelated. The log 10 likelihood ratio (LR) (left y-axis; bars) and posterior probabilities (right y-axis; black line) are shown. Samples 17.1 and 17.2 yielded <15 kinship SNP calls that were inconsistent with the expected parent/offspring relationships; therefore, 17.1 and 17.2 were excluded altogether.

In addition to the autosomal single nucleotide polymorphisms (SNPs), a total of 246 X-SNPs were specifically selected in the panel for the purpose of kinship testing in certain case scenarios. The X-SNPs included in the panel were utilized to establish a genetic link between individual 12 and the remains of individual 14.1 in family D. The log 10 likelihood ratio (LR) for the hypothesis of full sibling relationship compared to unrelatedness was determined to be 6.7, with a posterior probability exceeding 99.99%. Therefore, the X-SNP data provided strong evidence in support of the case example. X-chromosomal data is particularly valuable in kinship testing due to its unique mechanism of inheritance. When comparing two individuals who are claimed to be paternal half-sisters or a claimed paternal grandmother and granddaughter, the ability to exclude a biological relationship using X-chromosomal markers is not nonexistent, as with autosomal markers [17].

4. AT A GLANCE

This study showcases the effectiveness of the 5422-SNP FORCE panel for forensic purposes. The purpose of this panel was to specifically omit clinically significant markers in order to address privacy concerns associated with DNA databanking. The SNPs incorporated in the FORCE panel are present on microarray genotyping chips utilized for genetic genealogy, facilitating cross-compatibility with externally generated data that will be valuable in validation studies. In addition, comprehensive reference databases are accessible for all markers included in the kit, in contrast to other marker categories like microhaplotypes. The FORCE panel comprises kinship single nucleotide polymorphisms (SNPs) encompassing forensic identity-informative SNPs (iiSNPs), ancestry-informative SNPs (aiSNPs), phenotype-informative SNPs (piSNPs), Y-chromosome SNPs (Y-SNPs), and X-chromosome SNPs (X-SNPs). The hybridization capture approach successfully retrieved SNP profiles from both high-quality reference and historical bone samples. An extremely high level of agreement (>99.9%) was found when comparing FORCE profiles with data that had been previously published. To enhance efficiency, future versions of the FORCE panel should exclude the limited number of underperforming SNPs found in this study. It was feasible to make predictions about ancestry and physical characteristics, and these predictions matched the information provided by the reference sample donors. Moreover, it was feasible to accurately predict the Y-haplogroup at a high resolution using 829 Y-SNPs. The study demonstrated that the innovative collection of 3931 SNPs, which serve as markers for genetic relatedness, is highly successful in accurately predicting relationships up to the 4th degree. The results were supported by robust statistical analysis. Moreover, documented instances involving previously recognized World War II servicemen and corresponding family reference samples yielded kinship forecasts that align with anticipated ties. Future endeavors involving the FORCE panel will involve a collaborative investigation between multiple laboratories, incorporating different methodologies to enhance the identification of single nucleotide polymorphisms (SNPs). Prior to use the FORCE panel in regular casework, this will be relevant.

REFERENCES

1. Butler, J.M.; Hill, C.R. Biology and Genetics of New Autosomal STR Loci Useful for Forensic DNA Analysis. *Forensic Sci. Rev.* **2012**, *24*, 15–26. **[Google Scholar]**
2. Kling, D.; Phillips, C.; Kennett, D.; Tillmar, A. Investigative genetic genealogy: Current methods, knowledge and practice. *Forensic Sci. Int. Genet.* **2021**, *52*, 102474. **[Google Scholar] [CrossRef] [PubMed]**
3. Erlich, Y.; Shor, T.; Pe'er, I.; Carmi, S. Identity inference of genomic data using long-range familial searches. *Science* **2018**, *362*, 690–694. **[Google Scholar] [CrossRef] [Green Version]**
4. GEDmatch. Available online: **https://pro.gedmatch.com/user/login** (accessed on 11 November 2021).
5. Greytak, E.M.; Moore, C.; Armentrout, S.L. Genetic genealogy for cold case and active investigations. *Forensic Sci. Int.* **2019**, *299*, 103–113. **[Google Scholar] [CrossRef]**
6. Tillmar, A.; Fagerholm, S.A.; Staaf, J.; Sjolund, P.; Ansell, R. Getting the conclusive lead with investigative genetic genealogy—A successful case study of a 16 year old double murder in Sweden. *Forensic Sci. Int. Genet.* **2021**, *53*, 102525. **[Google Scholar] [CrossRef] [PubMed]**
7. Cuenca, D.; Battaglia, J.; Halsing, M.; Sheehan, S. Mitochondrial Sequencing of Missing Persons DNA Casework by Implementing Thermo Fisher's Precision ID mtDNA Whole Genome Assay. *Genes* **2020**, *11*, 1303. **[Google Scholar] [CrossRef]**
8. Holt, C.L.; Stephens, K.M.; Walichiewicz, P.; Fleming, K.D.; Forouzmand, E.; Wu, S.F. Human Mitochondrial Control Region and mtGenome: Design and Forensic Validation of NGS Multiplexes, Sequencing and Analytical Software. *Genes* **2021**, *12*, 599. **[Google Scholar] [CrossRef] [PubMed]**
9. Eduardoff, M.; Xavier, C.; Strobl, C.; Casas-Vargas, A.; Parson, W. Optimized mtDNA Control Region Primer Extension Capture Analysis for Forensically Relevant Samples and Highly Compromised mtDNA of Different Age and Origin. *Genes* **2017**, *8*, 237. **[Google Scholar] [CrossRef] [Green Version]**
10. Marshall, C.; Sturk-Andreaggi, K.; Daniels-Higginbotham, J.; Oliver, R.S.; Barritt-Ross, S.; McMahon, T.P. Performance evaluation of a mitogenome capture and Illumina sequencing protocol using non-probative, case-type skeletal samples: Implications for the use of a positive control in a next-generation sequencing procedure. *Forensic Sci. Int. Genet.* **2017**, *31*, 198–206. **[Google Scholar] [CrossRef] [Green Version]**

11. Bose, N.; Carlberg, K.; Sensabaugh, G.; Erlich, H.; Calloway, C. Target capture enrichment of nuclear SNP markers for massively parallel sequencing of degraded and mixed samples. *Forensic Sci. Int. Genet.* **2018**, *34*, 186–196. **[Google Scholar] [CrossRef]**

12. Daniels-Higginbotham, J.; Gorden, E.M.; Farmer, S.K.; Spatola, B.; Damann, F.; Bellantoni, N.; Gagnon, K.S.; de la Puente, M.; Xavier, C.; Walsh, S.; et al. DNA Testing Reveals the Putative Identity of JB55, a 19th Century Vampire Buried in Griswold, Connecticut. *Genes* **2019**, *10*, 636. **[Google Scholar] [CrossRef] [Green Version]**

13. Marshall, C.; Sturk-Andreaggi, K.; Gorden, E.M.; Daniels-Higginbotham, J.; Sanchez, S.G.; Basic, Z.; Kruzic, I.; Andelinovic, S.; Bosnar, A.; Coklo, M.; et al. A Forensic Genomics Approach for the Identification of Sister Marija CrucifiksaKozulic. *Genes* **2020**, *11*, 938. **[Google Scholar] [CrossRef]**

14. Gorden, E.M.; Greytak, E.M.; Sturk-Andreaggi, K.; Cady, J.; McMahon, T.P.; Armentrout, S.; Marshall, C. Ex-tended kinship analysis of historical remains using SNP capture. *Forensic Sci. Int. Genet.* **2021**, *57*, 102636. **[Google Scholar] [CrossRef]**

15. Turchi, C.; Previdere, C.; Bini, C.; Carnevali, E.; Grignani, P.; Manfredi, A.; Melchionda, F.; Onofri, V.; Pelotti, S.; Robino, C.; et al. Assessment of the Precision ID Identity Panel kit on challenging forensic samples. *Forensic Sci. Int. Genet.* **2020**, *49*, 102400. **[Google Scholar] [CrossRef] [PubMed]**

16. Ralf, A.; van Oven, M.; Montiel Gonzalez, D.; de Knijff, P.; van der Beek, K.; Wootton, S.; Lagace, R.; Kayser, M. Forensic Y-SNP analysis beyond SNaPshot: High-resolution Y-chromosomal haplogrouping from low quality and quantity DNA using Ion AmpliSeq and targeted massively parallel sequencing. *Forensic Sci. Int. Genet.* **2019**, *41*, 93–106. **[Google Scholar] [CrossRef]**

17. Tillmar, A.O.; Kling, D.; Butler, J.M.; Parson, W.; Prinz, M.; Schneider, P.M.; Egeland, T.; Gusmao, L. DNA Commission of the International Society for Forensic Genetics (ISFG): Guidelines on the use of X-STRs in kinship analysis. *Forensic Sci. Int. Genet.* **2017**, *29*, 269–275. **[Google Scholar] [CrossRef] [PubMed]**

18. Genomes Project, C.; Auton, A.; Brooks, L.D.; Durbin, R.M.; Garrison, E.P.; Kang, H.M.; Korbel, J.O.; Marchini, J.L.; McCarthy, S.; McVean, G.A.; et al. A global reference for human genetic variation. *Nature* **2015**, *526*, 68–74. **[Google Scholar] [CrossRef] [Green Version]**

19. Grandell, I.; Samara, R.; Tillmar, A.O. A SNP panel for identity and kinship testing using massive parallel sequencing. *Int. J. Legal Med.* **2016**, *130*, 905–914. **[Google Scholar] [CrossRef]**

20. Xavier, C.; de la Puente, M.; Mosquera-Miguel, A.; Freire-Aradas, A.; Kalamara, V.; Vidaki, A.; Gross, T.E.; Revoir, A.; Pospiech, E.; Kartasinska, E.; et al. Development and validation of the VISAGE AmpliSeq basic tool to predict appearance and ancestry from DNA. *Forensic Sci. Int. Genet.* **2020**, *48*, 102336. **[Google Scholar] [CrossRef] [PubMed]**

21. Kalia, S.S.; Adelman, K.; Bale, S.J.; Chung, W.K.; Eng, C.; Evans, J.P.; Herman, G.E.; Hufnagel, S.B.; Klein, T.E.; Korf, B.R.; et al. Recommendations for reporting of secondary findings in clinical exome and genome sequencing, 2016 update (ACMG SF v2.0): A policy statement of the American College of Medical Genetics and Genomics. *Genet. Med. Off. J. Am. Coll. Med. Genet.* **2017**, *19*, 249–255. **[Google Scholar] [CrossRef] [Green Version]**

22. Altschul, S.F.; Gish, W.; Miller, W.; Myers, E.W.; Lipman, D.J. Basic local alignment search tool. *J. Mol. Biol.* **1990**, *215*, 403–410. **[Google Scholar] [CrossRef]**

23. Abecasis, G.R.; Cherny, S.S.; Cookson, W.O.; Cardon, L.R. Merlin--rapid analysis of dense genetic maps using sparse gene flow trees. *Nat. Genet.* **2002**, *30*, 97–101. **[Google Scholar] [CrossRef]**

24. Matise, T.C.; Chen, F.; Chen, W.; De La Vega, F.M.; Hansen, M.; He, C.; Hyland, F.C.; Kennedy, G.C.; Kong, X.; Murray, S.S.; et al. A second-generation combined linkage physical map of the human genome. *Genome Res.* **2007**, *17*, 1783–1786. **[Google Scholar] [CrossRef] [Green Version]**

25. R Core Team. *R: A Language and Environment for Statistical Computing*; R Foundation for Statistical Computing: Vienna, Austria, 2013. **[Google Scholar]**

26. Edson, S.M. Extraction of DNA from Skeletonized Postcranial Remains: A Discussion of Protocols and Testing Modalities. *J. Forensic Sci.* **2019**, *64*, 1312–1323. **[Google Scholar] [CrossRef] [PubMed]**

27. Rohland, N.; Glocke, I.; Aximu-Petri, A.; Meyer, M. Extraction of highly degraded DNA from ancient bones, teeth and sediments for high-throughput sequencing. *Nat. Protoc.* **2018**, *13*, 2447–2461. **[Google Scholar] [CrossRef] [PubMed]**

28. *ForenSeq DNA Signature Prep Reference Guide, Rev C*; Verogen: San Diego, CA, USA, 2020.

29. Zhou, B.; Ho, S.S.; Greer, S.U.; Zhu, X.; Bell, J.M.; Arthur, J.G.; Spies, N.; Zhang, X.; Byeon, S.; Pattni, R.; et al. Comprehensive, integrated, and phased whole-genome analysis of the primary ENCODE cell line K562. *Genome Res.* **2019**, *29*, 472–484. **[Google Scholar] [CrossRef] [PubMed] [Green Version]**

30. Illumina Platinum Genomes. Available online: **https://www.illumina.com/platinumgenomes.html** (accessed on 13 November 2021).

31. Eberle, M.A.; Fritzilas, E.; Krusche, P.; Kallberg, M.; Moore, B.L.; Bekritsky, M.A.; Iqbal, Z.; Chuang, H.Y.; Humphray, S.J.; Halpern, A.L.; et al. A reference data set of 5.4 million phased human variants validated by genetic inheritance from sequencing a three-generation 17-member pedigree. *Genome Res.* **2017**, *27*, 157–164. **[Google Scholar] [CrossRef] [PubMed] [Green Version]**
32. Prcomp. Available online: **https://stat.ethz.ch/R-manual/R-devel/library/stats/html/prcomp.html** (accessed on 13 November 2021).
33. Santos, C.; Phillips, C.; Gomez-Tato, A.; Alvarez-Dios, J.; Carracedo, A.; Lareu, M.V. Inference of Ancestry in Forensic Analysis II: Analysis of Genetic Data. *Methods Mol. Biol.* **2016**, *1420*, 255–285. **[Google Scholar] [CrossRef]**
34. de la Puente, M.; Ruiz-Ramirez, J.; Ambroa-Conde, A.; Xavier, C.; Pardo-Seco, J.; Alvarez-Dios, J.; Freire-Aradas, A.; Mosquera-Miguel, A.; Gross, T.E.; Cheung, E.Y.Y.; et al. Development and Evaluation of the Ancestry Informative Marker Panel of the VISAGE Basic Tool. *Genes* **2021**, *12*, 1284. **[Google Scholar] [CrossRef] [PubMed]**
35. Chaitanya, L.; Breslin, K.; Zuniga, S.; Wirken, L.; Pospiech, E.; Kukla-Bartoszek, M.; Sijen, T.; Knijff, P.; Liu, F.; Branicki, W.; et al. The HIrisPlex-S system for eye, hair and skin colour prediction from DNA: Introduction and forensic developmental validation. *Forensic Sci. Int. Genet.* **2018**, *35*, 123–135. **[Google Scholar] [CrossRef] [Green Version]**
36. Hpsconvertsonline. Available online: **https://walshlab.sitehost.iu.edu/hpsconvertsonline.R** (accessed on 13 November 2021).
37. HirisPlex-S System Webtool. Available online: **https://hirisplex.erasmusmc.nl/** (accessed on 13 November 2021).
38. Ralf, A.; Gonzalez, D.M.; Zhong, K.; Kayser, M. Yleaf: Software for Human Y-Chromosomal Haplogroup Inference from Next-Generation Sequencing Data. *Mol. Biol. Evol.* **2018**, *35*, 1820. **[Google Scholar] [CrossRef] [PubMed] [Green Version]**
39. Yleaf Position File. Available online: **https://github.com/genid/Yleaf/blob/master/Position_files/WGS_hg38.txt** (accessed on 13 November 2021).
40. Dimery, I.W.; Ross, D.D.; Testa, J.R.; Gupta, S.K.; Felsted, R.L.; Pollak, A.; Bachur, N.R. Variation amongst K562 cell cultures. *Exp. Hematol.* **1983**, *11*, 601–610. **[Google Scholar] [PubMed]**
41. SWGDAM. *Recommendations of the SWGDAM ad hoc Working Group on Genotyping Results Reported as Likelihood Ratios*; Federal Bureau of Investigation's Scientific Working Group on DNA Analysis Methods (SWGDAM): Quantico, VA, USA, 2018. **[Google Scholar]**

25 ANCIENT DNA TECHNIQUES ENHANCE THE ACCURACY OF FORENSIC DNA PROFILING FOR UNIDENTIFIED INDIVIDUALS FROM THE KOREAN WAR AND WORLD WAR II

1. INTRODUCTION: AN OVERVIEW

Massively parallel sequencing (MPS), commonly referred to as next-generation sequencing (NGS), has lately started being incorporated into the forensic domain. The admissibility of MPS was initially established in a legal proceeding in the Netherlands in 2019 [1]. The method has received approval for utilization in the U.S. criminal justice system for both STR profiling and mitochondrial (mt) DNA sequencing [2]. In addition, the technique of whole genome (shotgun) sequencing and the usage of genome-wide SNP arrays are commonly employed in the field of investigative genetic genealogy to gather information that might be helpful in both ongoing and unsolved cases [3]. Although there have been significant advancements in DNA profiling using MPS for complex instances involving historical skeletal remains, such as those of historical figures, unmarked graves, and unidentified war victims, the success rates are still rather low [4,5,6]. The reason for this could be attributed to the fact that effective forensic DNA profiling usually necessitates the presence of very sizable DNA fragments (exceeding ~100 base pairs) using contemporary techniques, which may not always be adequately abundant in deteriorated samples.

One set of historical remains that exemplifies the challenges faced by historical identification cases is Unknowns from the Korean War buried in the National Memorial Cemetery of the Pacific (NMCP) in Honolulu, Hawaii [7]. These remains contain poorly preserved DNA due to extensive chemical treatment that was standard postmortem military practice in the 1940s and 1950s. The war dead were initially collected in temporary graves (**Figure 1**a) by the Army Graves Registration Service. They were then transferred to the United Nations Cemetery in Tanggok,

Republic of Korea [**8,9**] where they were treated with insecticide and deodorant. The remains were subsequently shipped to the Army's Central Identification Unit in Kokura, Japan [7,**10**]. At this location, the remains were soaked in 40–50% formaldehyde solution for three to five days (**Figure 1**b). Formaldehyde is known to lead to the formation of cross-linked complexes between DNA molecules or between DNA and proteins, complicating DNA extraction [**11**]. In addition, Hawaii has a warm and wet environment, while Korea is climatically diverse with temperatures fluctuating from freezing to extremely hot. Such environmental conditions involving high heat have been found to increase DNA degradation, specifically deamination and fragmentation rates [**12**]. After further treatment with fungicide and a powdered hardening compound (**Figure 1**c), bodies were wrapped and casketed. In total, the remains of 859 unidentified Korean War service members were shipped to Hawaii in 1956 and buried as Unknowns at the NMCP, also known as the Punchbowl [7].

Figure 1. These historical photographs show some of the important events that occurred after service members were killed in action during the Korean War. (A) In this photograph, taken on 19 August 1950, a battlefield cemetery in Daegu (formerly Taegu), Korea, contains dozens of temporary graves where the bodies of soldiers killed in action were laid to rest. This image was reprinted with permission of the U.S. Army Quartermaster Museum in Fort Lee, Virginia. (B) This photograph of the Kokura embalming laboratory was originally printed in the American Graves Registration Service Group 8204th Army Unit APO 3 Brochure, April 1955. Examination tables were equipped with exhaust systems for formaldehyde treatment. (C) These remains from a Korean War service member were covered with a powdered hardening compound prior to shipment to the United States. This photograph was reprinted courtesy of the DPAA. (D) In Operation GLORY, the first of the United Nations war dead from North Korea were received at Moji Port near Kokura, Japan (ca. fall of 1954). This photograph was originally printed in the American Graves Registration Service Group 8204th Army Unit APO 3 Brochure, April 1955.

In 1999, the Central Identification Laboratory, Hawaii (CILHI), which is now known as the Defense POW/MIA Accounting Agency (DPAA), initiated a new exhumation campaign with the aim of identifying Korean War Unknowns at NMCP. The Armed Forces DNA Identification Laboratory, a component of the Armed Forces Medical Examiner System, collaborated with the CILHI to conduct DNA testing on the first five excavated cases. This testing involved the use of standard forensic DNA profiling procedures [14]. Nevertheless, the outcomes were not able to be replicated, therefore rendering them untrustworthy.

In 2016 the AFMES-AFDIL validated a novel MPS protocol for mitochondrial genome (mitogenome) sequencing using a probe capture and Illumina sequencing approach. This method was shown to be more amenable to the crosslinked and severely fragmented DNA from NMCP Unknowns, which was detailed in a 2017 performance evaluation [**4**]. As a result of this MPS mitogenome sequencing validation, the AFMES-AFDIL became the first accredited laboratory to adopt MPS for forensic casework worldwide. The study showed severe degradation of the DNA from Korean War NMCP Unknowns, which averaged <70 base pairs (bp) in length. The validated MPS methods are now routinely applied to all chemically treated or severely degraded DNA samples at the AFMES-AFDIL, regardless of the military conflict. These include remains from U.S. military WWII cemetery exhumations with similar preservation problems to those encountered with the Korean War unknowns from the NMCP. To date, over 3000 samples have been tested using the validated mitogenome MPS protocol with a success rate of ~60%. While this is a marked improvement from before the integration of MPS into the casework workflow (6% success rate with PCR and Sanger sequencing), the sample failure rate is still substantial, even when targeting mtDNA, which is high in copy number relative to nuclear DNA. Additional advancements to improve MPS sensitivity could elevate success rates in reportable mitogenome profile generation and may furthermore enable SNP profiling for extended kinship analysis [**15**].

While the forensic field has faced obstacles in generating DNA profiles from historical remains, the ancient DNA field has been successful in recovering authentic DNA from very old and degraded samples from a variety of substrates and environments. Advancements in the extraction [**16,17**], library preparation [**18**], and enrichment of target DNA [**19,20**] have enabled the generation of hominin nuclear and mtDNA sequence profiles from skeletal elements as well as sediment over 300,000 years old [**21,22,23**]. Of particular importance in these developments was an extraction protocol that can recover DNA fragments <50 bp in length [**16**] and transitioning from double- to single-stranded DNA library preparation [**24**], a method that has also been shown to improve the recovery of DNA molecules from formalin-fixed samples [**25,26,27**]. However, the focus for method development for forensic casework and ancient DNA studies is different. This is due to the consistent, degraded nature of ancient DNA samples that motivates the foundation for methodological advancements in the field. Forensic samples, by contrast, are much more heterogeneous in terms of DNA quality. Novel forensic methods are largely aimed to improve DNA profiling of traditional forensic targets that exceed 100 bp in size. Furthermore, new forensic DNA methods must meet rigorous validation standards, making it difficult to test and adopt new protocols. These differing forces that drive the two fields have positioned ancient DNA considerably ahead of forensics in terms of method development. Yet both fields share the common problem of analyzing DNA that is (or can be) highly degraded. This commonality prompted the question if methods from the ancient DNA field could improve success rates for generating mtDNA profiles for historical identifications, which has been suggested in recent studies and reviews [**28,29,30**]. Hence, we conducted an assessment to examine the practicality of incorporating ancient DNA techniques into a forensic investigation process. This involved evaluating various methods for extracting DNA and preparing libraries, utilizing fifteen skeletal remains with DNA of different levels of quality, as indicated by prior analysis. Subsequently, we assessed the efficacy of each technique by considering the quantity of informative sequences, the comprehensiveness of mtDNA profiles, and the agreement with predicted profiles, if applicable. This study establishes the foundation for future advancements that will enhance DNA profiling and identification of historical remains samples of extremely low quality.

2. MATERIALS AND METHODS

2.1. Skeletal Samples

This study employed fifteen bone samples that were not suitable for providing conclusive evidence, for the purpose of conducting DNA testing (Table 1). The samples were obtained from exhumed corpses of World War II and Korean War soldiers who were probably subjected to chemical treatment, either through the use of formalin or the application of chemical powders (such as deodorant or pesticide). The fifteen samples underwent prior processing in ordinary casework at the AFMES-AFDIL, utilizing the validated mitogenome sequencing methods described in [4].. Samples were selected based on DNA quality in order to cover a range in quality types and not based on skeletal element. The samples ranged in DNA quality based on casework sequencing success: seven failed with <10-fold average coverage, four provided mtDNA sequences with 10-35-fold (low) average coverage, and four yielded high coverage mtDNA data (100-360-fold).

Table 1. Sample descriptions and case contexts. World War II (WWII) conflict locations are shown, as well as the cemetery of disinterment. Formalin treatment was not documented, but was noted as being likely in certain cases. Powder application was evidenced by visible powder on the remains. Casework results were based on the coverage of the revised Cambridge Reference Sequence (rCRS) [31] produced from the same skeletal sample using the methods outlined in [4].

Sample ID	Skeletal Element	Context	Cemetery of Disinterment	Formalin Treatment	Powder Application	Casework Results
2224	Tibia	WWII-Tarawa	NMCP [1]	Unknown	Not visible	High Coverage
0858	Temporal	WWII-USS Oklahoma	NMCP [1]	No, Oil Soaked	Likely	High Coverage
0994	Femur	WWII-Buna	MAC [2]	Unknown	Not visible	High Coverage
2255	Humerus	WWII-Cabanatuan	MAC [2]	Unknown	Not visible	High Coverage
0899	Humerus	WWII-Tarawa	NMCP [1]	Unknown	Yes	Low Coverage
0100	Femur	Korea	NMCP [1]	Likely	Yes	Low Coverage
1001	Os coxa	WWII-Yugoslavia	SRAC [3]	Unknown	Not visible	Low Coverage
3140	Radius	WWII-Burma	NMCP [1]	Unknown	Yes	Low Coverage
0378	Femur	WWII-Tarawa	NMCP [1]	Unknown	Yes	Failed
1837	Tibia	WWII-Tarawa	NMCP [1]	Unknown	Yes	Failed
1198	Tibia	Korea	NMCP [1]	Likely	Yes	Failed
1964	Humerus	Korea	NMCP [1]	Likely	Yes	Failed
0917	Tibia	Korea	NMCP [1]	Likely	Yes	Failed
3825	Radius	Korea	NMCP [1]	Likely	Yes	Failed
1487	Femur	WWII-Italy	SRAC [3]	Unknown	Yes	Failed

[1] National Memorial Cemetery of the Pacific, Honolulu, Hawaii, USA; [2] Manila American Cemetery, Philippines; [3] Sicily-Rome American Cemetery, Italy.

2.2. DNA Extraction

All DNA extractions were performed at the AFMES-AFDIL in a clean, pre-PCR room with positive pressure, per standard ancient and forensic DNA practices. Under sterile conditions, each bone was sanded to remove the external surface, washed in ethanol, and then powdered using a blender cup. Bone powder samples were digested and a reagent blank (RB) was initiated for each extraction set. DNA was extracted using one of the conditions outlined in **Table 2.** As described in Edson et al. [32], the AFDIL method utilized 1.0 g of bone powder incubated overnight at 56 °C in 7.5 mL digestion buffer (0.5 M EDTA, 1% lauroylsarcosine) with 200 μL proteinase K (20 mg/mL). DNA was then isolated using an organic DNA extraction protocol with phenol chloroform isoamyl alcohol (PCIA), followed by concentration and buffer exchange using an Amicon Ultra-4 (molecular weight cut-off 30 kDa) (MilliporeSigma, Burlington, MA, USA). The resulting DNA extracts were divided into three aliquots to assess the effect of DNA repair on the sequence data. The following repair conditions were evaluated: (1) no repair, (2) USER (New England Biolabs (NEB), Ipswich, MA, USA) treatment to remove uracil residues resulting from cytosine deamination, and (3) NEBNext FFPE DNA Repair Mix (NEB) to remove uracils and also repair nicks, gaps, oxidized bases, and blocked 3′ ends [33]. Repair treatments followed their respective manufacturer protocols. All three DNA extracts (no repair, USER, and FFPE DNA Repair) underwent MinElute (QIAGEN, Hilden, Germany) purification after repair. This resulted in a total of three different PCIA-based protocols that were tested (PM: no repair, PUM: USER, and PFM: FFPE DNA Repair).

The ancient DNA extraction protocol followed was the Dabney method [16] as described in Rohland et al. [17], substituting the binding buffer 'D' with QIAGEN PB buffer. This method utilized 0.2 g bone powder digested at 37 °C in 1 mL 0.45 M EDTA with 0.05% Tween 20 and 25 μL proteinase K (20 mg/mL). Unlike the AFDIL digestion, there was not a complete digestion of bone powder. Therefore, the bone powder pellets remaining after the initial 37 °C Dabney extraction were re-digested overnight with 1.0 mL buffer at 56 °C, which resulted in the powder going into solution. Purification was then performed with the Roche High Pure Spin Column (a large volume silica column) (Pleasanton, CA, USA). This resulted in two ancient DNA extractions, although each were from the same bone powder aliquot (first digestion: aDNA37 and second digestion: aDNA56).

Finally, a hybrid "AFDIL-Dabney" DNA extraction method was performed following the AFMES-AFDIL digestion procedures (i.e., 1.0 g of bone powder incubated overnight at 56 °C in 7.5 mL digestion buffer and 200 μL proteinase K) and Dabney purification (i.e., High Pure Spin Column).

2.3. Library Preparation

2.3.1. In House MPI Method

The process of preparing a library of single-stranded DNA was carried out according to the method given in reference [19]. This was done using a Bravo NGS Workstation (Agilent Technologies) at the Max Planck Institute for Evolutionary Anthropology (MPI-EVA) in Leipzig, Germany.

Table 2. DNA extraction methods tested.

Method	Bone Powder (g)	Digestion Buffer	Digestion Buffer Volume (mL)	Proteinase K (20 mg/mL) (μL)	Incubation Temperature (°C)	DNA Purification	Repair Protocol
AFDIL (PM)	1.0	Demin buffer [1]	7.5	200	56	PCIA with buffer exchange	NA (MinElute purification)
AFDIL-USER (PUM)	1.0	Demin buffer [1]	7.5	200	56	PCIA with buffer exchange	USER (NEB, Ipswich, MA, USA)
AFDIL-FFPE (PFM)	1.0	Demin buffer [1]	7.5	200	56	PCIA with buffer exchange	NEBNext FFPE DNA Repair Mix (NEB)
Dabney—37 (aDNA37)	0.2	Dabney buffer [2]	1.0	25	37	Silica column and PB Buffer	NA
Dabney—56 (aDNA56)	Dabney 37 remaining pellet	Dabney buffer [2]	1.0	25	56	Silica column and PB Buffer	NA
AFDIL-Dabney (A_D)	0.2	Demin buffer [1]	4.0	200	56	Silica column and PB Buffer	NA

[1] Demin buffer: 0.5 M EDTA, %1 lauroyl sarcosine; [2] Dabney buffer: 0.45 M, %0.5 Tween 20.

Each sample extract and RB were used in a volume of 5µL. Essentially, the DNA was subjected to heat to cause denaturation, after which the phosphate groups at the 5′ and 3′ ends were eliminated. T4 DNA ligase was used to ligate an adapter oligonucleotide with a 3′ biotin to the 3′ end of each DNA fragment, aided by a splinter oligonucleotide. This product was then immobilized on streptavidin magnetic beads and a wash was performed at 45 °C to remove the splinter. Primer and the *E.coli* DNA polymerase I Klenow fragment was added to make a copy of the template strand from the attached adapter. A bead wash was then performed to remove excess primers and prevent adapter dimer formation. Next, T4 DNA ligase was used for a blunt-end ligation of a second, double-stranded adapter. Heat denaturation at 95 °C was then used to release the library products from the beads. Library amplification and indexing was performed with AccuPrimePfx DNA polymerase with the following cycling parameters: 95 °C for 2 min, then 35 cycles of 95 °C for 20 s, 60 °C for 30 s and 68 °C for 1 min, with a final additional 5 min at 68 °C. Post-amplification purification was performed using the MinElute PCR purification kit. Libraries were generated from all 15 samples, one RB, one library positive control and one library negative control.

2.3.2. SRSLY Kit

The SRSLY single-stranded DNA library preparations were performed manually at the AFMES-AFDIL using Claret's SRSLY PicoPlus Library Prep Kit for Illumina (Santa Cruz, CA, USA). Unless otherwise noted, library preparation was carried out according to the manufacturer protocol for ancient and highly degraded DNA samples, including purification steps with AMPure XP (Beckman Coulter, Brea, CA, USA) for maximum short fragment retention [**34**]. In summary, 5 µL of extract was used for library preparation, and a positive control and negative control were included with each sample set. The positive control was enzymatically fragmented K562 cell line with 1 ng library input, as described in [**4**]. An initial denaturation step rendered all DNA single-stranded and then stabilized by a single-stranded binding protein, followed by adapter ligation at 37 °C (lid 45 °C) for 1 h with splinted adapters that retain the original strand orientation. AMPure XP purification after adapter ligation included the addition of isopropanol and 10 mM Tris-HCl pH 8.5 with a final ratio of 0.54× (0.99× ratio to library prior to isopropanol and Tris-HCl addition). The full library volume was amplified in a single reaction using Claret unique dual indexes (undiluted) and KAPA HiFi HotStart Uracil+ ReadyMix (Roche) according to the manufacturer recommendations. Uracil+ was selected in order to tolerate uracil bases present in the DNA due to cytosine deamination. Library amplification was completed with the following parameters: 95 °C for 3 min, then 35 cycles of 95 °C for 20 s, 60 °C for 15 s, 72 °C for 1 min, followed by 72 °C for 7 min. Post-amplification purification utilized a 1.2× AMPure XP ratio. Libraries were generated from a total of 15 samples, one RB, two positive controls and two negative controls.

2.3.3. KAPA Hyper Prep Kit

The KAPA double-stranded DNA library preparations were carried out manually at the AFMES-AFDIL facility using the Roche KAPA Hyper Prep Kit. Unless stated otherwise, library preparation was conducted in accordance with the manufacturer's protocol. Again, 5 μL of extract was used for library preparation, and a positive control and negative control were included with each sample set. The positive control was fragmented DNA from a K562 cell line with 1 ng library input. End repair and A-tailing took place in a single step with incubation at 20 °C for 30 min, then 65 °C for 30 min (lid 85 °C), followed by adapter ligation using 15 μM KAPA dual indexed adapters for all libraries at 20 °C for 15 min. Post-ligation clean-up used an increased 2× AMPure XP ratio in order to retain smaller fragments. The full library volume was amplified using KAPA HiFi HotStart Uracil+ ReadyMix and KAPA Library Amplification Primer mix, according to the manufacturer's recommendations with the following parameters: 95 °C for 3 min, then 35 cycles of 98 °C for 20 s, 60 °C for 15 s, 72 °C for 1 min, followed by 72 °C for 7 min. The same post-amplification AMPure XP purification was utilized as with the SRSLY libraries. Libraries were generated from a total of 15 samples, one RB, two positive controls and two negative controls.

2.3.4. Quantification

In order to assess the quantity of DNA molecules transformed into library molecules for each of the three library production procedures, a quantitative (q) PCR analysis was conducted on the unamplified library products of each method. The quantitative polymerase chain reaction (qPCR) was conducted following the methodology outlined in Gansauge et al. [18] at the Max Planck Institute for Evolutionary Anthropology (MPI-EVA). All the libraries were processed on a single plate to eliminate any potential difference across plates.

2.4. MtDNA Hybridization Capture

In-solution hybridization capture was used for the enrichment of human mtDNA for the three different types of library preparation methods. Two rounds of capture were performed following the protocol outlined in Fu et al. [19] in 384-well format on a Bravo NGS Workstation B (Agilent, Santa Clara, CA, USA) with adjustments to the number of amplification cycles per library to prevent each from reaching plateau.

2.5. Sequencing and Data Processing

For the comparison of extraction protocols, all resulting single-stranded DNA libraries were pooled for shotgun sequencing. For the comparison of library preparation protocols, each library type was pooled separately (three total pools) for shotgun sequencing and the human mtDNA enriched SRSLY and MPI library sets (two pools). In order to avoid captured libraries with identical indices being sequenced together, the KAPA Hyper Prep human mtDNA enriched libraries were split between nine pools and sequenced with other libraries not included in this study. All sequencing pools were created by combining equal amounts of sample libraries and 1:10 dilutions of controls (RBs and library negative controls). Library positive controls were not sequenced. Sequencing was performed on a MiSeq or NextSeq (Illumina technology) in 2× 75 bp paired-end configuration with 8-bp dual-indexed reads. Base calling was completed with bustard (Illumina) and overlapping paired-end reads were merged using leeHom [**35**]. For shotgun sequencing, reads were then mapped to the human reference genome GRCh37/1000 (Genomesrelease; **ftp://ftp.1000genomes.ebi.ac.uk/vol1/ftp/technical/reference/phase2_reference_assembly_sequence/**, accessed on 12 November 2021) and captured library reads were mapped to the revised Cambridge reference sequence [**31**]. PCR duplicates were then removed using bam-rmdup (**https://github.com/mpieva/biohazard-tools**, accessed on 15 October 2021).

For the processing of the shotgun results, all sequences less than 35 bp and unmapped sequences were removed. The remaining sequences were then used to calculate the average fragment size, rates of C-to-T substitutions, and number of informative sequences [**26**] in each sample. The 35 bp cut-off was used to allow comparisons to previous published ancient DNA data.

For the processing of the capture enriched data, the resulting de-duplicated sequences from each library preparation type continued either through the MPI workflow or a CLC Genomic Workbench (QIAGEN, Hilden, Germany) workflow. For the MPI workflow, unmapped sequences, sequences shorter than 30 bp or with a mapping quality less than 25 were removed. The resulting sequences were used to calculate the coverage of the mtDNA genome. For the CLC workflow, the de-duplicated bam files were ingested into the CLC at the AFMES-AFDIL for variant calling, mtDNA haplogroup prediction, and forensic profile reporting using AQME [**36**], in addition to determining the mtDNA coverage. A minimum read depth of 5 was required for bases to be included in the reported sequence range. Variant detection required a minimum of 4 reads supporting the variant as well as a minimum frequency of 10%. Point heteroplasmy was documented when variants exhibited 10–90% frequency. Length heteroplasmy was reported based on the major molecule; however, this was ignored in concordance assessments due to the impact of differing polymerases on length variants [**37**]. Additionally, variants consistent with cytosine deamination (C-to-T or G-to-A substitutions) that were low in frequency (<30%) and/or lacking in forward/reverse read balance (<0.30) were excluded. These variant detection parameters minimized the reporting of stochastic error while allowing for the mixture and contamination detection necessary in forensic casework. When processing the SRSLY libraries, contaminant pig DNA was identified as mismapping to the human mtDNA reference genome (rCRS), resulting in spikes of coverage and unexpected variants. The presence of pig sequences in these libraries was confirmed using a pipeline [**38**] that uses BLAST [**39**] and MEGAN [**40**] to assign sequences from sediment samples to different mammalian families. To remove these sequences a "pig-out" pipeline was added to both the CLC and MPI workflows.

3. RESULTS

3.1. Evaluation of Extraction Methods

To assess the viability of incorporating ancient DNA techniques into a forensic casework process, we initially examined the effectiveness of several extraction strategies in terms of their ability to successfully retrieve DNA fragments. Fifteen skeletal remains, which had already undergone testing at the AFMES-AFDIL and were thus proven to be of reliable quality, are included in Table 1. We selected five different extraction/repair protocols to test: three PCIA-based from the forensic field [**32**], one from the ancient DNA field ([**17**], with PB buffer), and one that combined elements of each the forensic and ancient DNA protocols (see **Section 2.2**). As the bone powder pellet did not go into solution in the initial ancient DNA digestion at 37 °C, a redigestion was performed at 56 °C of the same pellet and each of the fractions was purified. Therefore, six extraction/repair conditions were tested in total (**Table 2**); however, the two ancient DNA extractions were from the same original powder sample.

DNA extraction performance was assessed using shotgun sequencing data produced from single-stranded libraries constructed following the method outlined in Gansauge et al. [**18**]. To determine which extraction protocol was most effective, we estimated the number of unique DNA fragments present in each library that mapped to the human DNA reference genome (informative sequences) in each library. For each sample we then determined which extraction protocol resulted in the most recovered informative sequences and took the ratio of the informative sequences from each other extraction protocol to this "best" protocol. This resulted in the relative informative sequences for each sample and extraction combination This calculation allowed us to compare the relative informative sequences across extraction types for the different samples (**Figure** 2A). We found that the combined forensic ancient DNA extraction protocol (AFDIL-Dabney) was significantly worse than all other protocols (n = 15, Wilcoxon test with correction for multiple testing, p-values: 6.5×10^{-4} to $^{4-}10 \times 1.1$). The redigestion at 56 °C with the ancient DNA protocol (aDNA56) recovered significantly more informative fragments than the two forensic extraction protocols that included extra DNA repair steps intended to improve recovery of damaged DNA (n = 15, Wilcoxon test with correction for multiple testing, p-values both 2.8×10^{-2}). This loss of recovered DNA is likely, at least in part, due to the addition of extra purification steps. No significant difference in recovered informative sequences was observed between the two ancient DNA protocols and the forensic protocol with no DNA repair (p-values: 6.4×10^{-1} to 1).

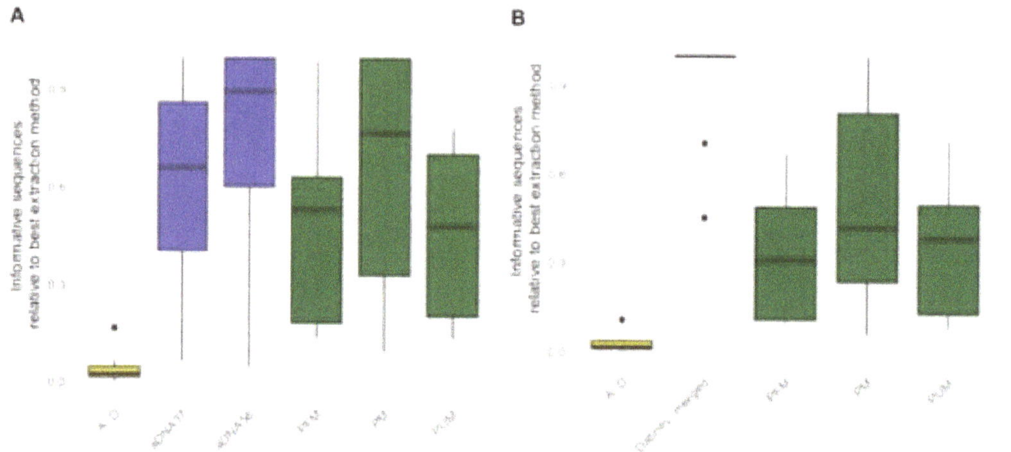

Figure 2. Relative informative sequences recovered per extraction protocol across 15 historical skeletal remains with the ancient DNA extraction fractions (A) separate and (B) merged. The total informative sequences per library was calculated by multiplying the number of DNA molecules in each library as determined by qPCR by the proportion of sequences longer than 35 bp that mapped to the human reference genome. The relative informative sequences were then calculated by taking the ratio of total informative sequences for each extraction/sample combination to the total number of informative sequences from the best performing extraction protocol per sample. The box plots follow the standard Tukey representation to show the distribution of the relative informative sequences per extraction type (*n* = 15). The significance of different relative amounts recovered per extraction was evaluated with a Wilcoxon test with a correction for multiple testing. The ancient DNA protocols (aDNA37 :37 °C digestion; aDNA56 :56 °C digestion; Dabney merged: merging of total informative sequences per extraction/sample from aDNA37 and aDNA56) are colored in blue; forensic protocols (PM: PCIA with Min Elute; PUM: PCIA with Min Elute and USER treatment; PFM: PCIA with Min Elute and FFPE treatment) in green; and the combined forensic ancient DNA protocol (A_D: AFDIL digestion with Dabney purification) in yellow.

However, as the two ancient DNA protocols originate from the same bone powder pellet, theoretically if the initial lysis had been performed at 56 °C the DNA recovered in the 37 °C and 56 °C fractions would have been extracted together. If we combine the recovered informative sequences from both of these fractions, the ancient DNA protocol results in significantly more informative sequences than the forensic extraction protocols (n = 15, Wilcoxon test with correction for multiple testing, p-values all $<4.8 \times 10^{-4}$, **Figure** 2B). These results indicate that the ancient

DNA extraction protocol does improve the recovery of target DNA from historical skeletal remains that are of interest for forensic casework particularly in a MPS workflow.

3.2. Characteristics of Recovered DNA

Single-stranded DNA library preparation coupled with shotgun sequencing allows for the characterization of nuclear DNA recovered from each of these samples. Specifically, we examined the percentage of sequences longer than 35 bp that mapped to the human genome, average fragment size for each sample, and deamination as represented by cytosine to thymine (C-to-T) substitutions. The latter two analyses were performed after removing unmapped reads, PCR duplicates, and sequences shorter than 35 bp or with mapping qualities less than 25. We used the data from the ancient DNA extraction protocol with a 56 °C incubation to investigate each of these factors.

The percent mapped for the samples ranged from 0.13–72.5% (mean 7.88%) and the average fragment size for each of the samples ranged from 44 to 64 bp (mean 49 bp). The observed percent of C-to-T substitutions on the 5′ ends ranged from 8.8% to 30.7% (mean 18.3%) and 8% to 22.3% (mean 13.3%) on the 3′ end. These values are similar to those observed in ancient DNA samples, where the presence of short DNA fragments and deamination are used for the authentication of ancient DNA [**41**] as well as to disentangle endogenous ancient DNA from modern human contamination [**42**]. While there have been multiple studies on the impacts of various factors on the DNA degradation patterns from ancient samples [**12,22,43**] and modern samples [**44,45**], there is limited data on DNA degradation in the more recent past. Therefore, we compared the average fragment size and deamination patterns of this set of forensic samples to published shotgun data from seven Late Pleistocene modern human specimens (~37 to 46,000 years ago) [**46**]. We found that the percent of sequences that mapped to the human reference genome (fragment length ≥ 35 bp and mapping quality ≥ 25) for about half of the historical remains fell within two standard deviations of the mean for the ancient samples (**Figure 3**A). Additionally, the average sizes of mapped human DNA fragments recovered from all of the historical remains were within the range for the ancient samples (**Figure 3**B). As the library preparation method used is known to impact the observed 3′ deamination [**26**], we compared the values on the 5′ end only. In this comparison, many of the historical remains had lower deamination rates than the average ancient sample (**Figure 3**C). In ancient DNA studies, a cut-off of 10% deamination is often used for the identification of authentic ancient DNA [**38,43,47**], and the majority of the historical samples exceeded this threshold. When looking at the rate of deamination along the DNA fragments, while the terminal base has the highest rate, elevated rates of deamination continue until the third terminal base for the historical samples (**Figure 3**D). This trend is known for ancient samples, but this observation indicates that it may be important to consider deamination during the downstream analyses of historical DNA samples, particularly when no DNA repair step is included. In addition, the similarity between the DNA recovered from the historical and ancient samples emphasizes the difficulty of performing identifications for forensic casework when remains may have been formalin treated or are poorly preserved.

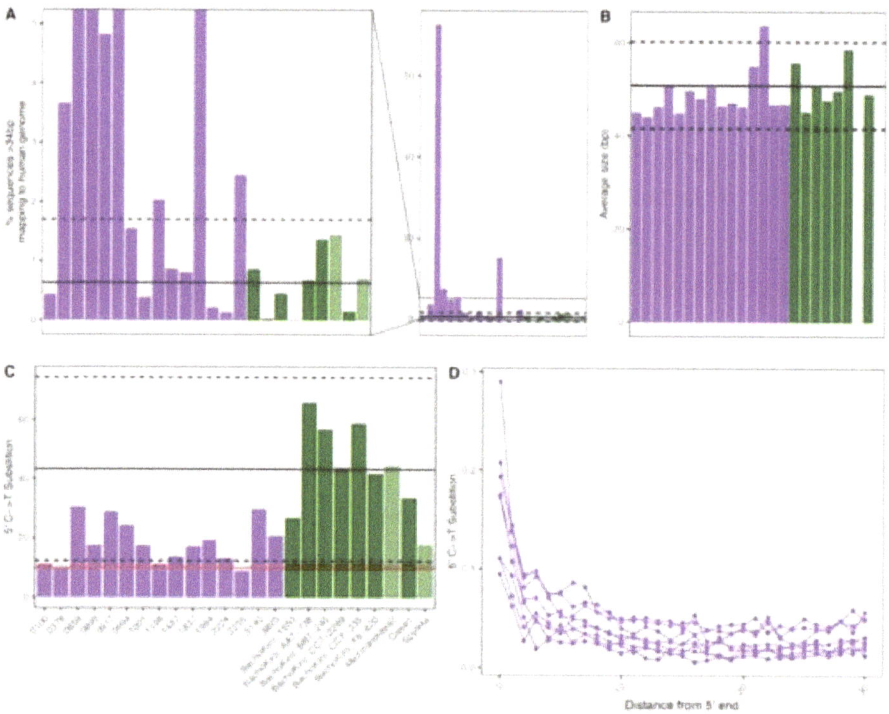

Figure 3. The (A) percent mapped (B) average fragment size (bp) and (C,D) observed 5′ C to T substitution frequencies of DNA recovered from historical and ancient DNA samples. Each bar represents a single sample for fifteen historical samples (purple) and seven Late Pleistocene modern human skeletal samples from ~37,000–46,000 years ago (dark green). In (A–C) the *x*-axis is the same for each and the black line represents the average for the ancient skeletal (dark green) samples and the dashed lines show two standard deviations above and below this average. In (C) the red line represents the 10% threshold typically used to determine if a sample contains ancient DNA. In (D) each point represents the 5′ C to T substitution frequency at a position from the 5′ end of DNA fragments recovered from a single sample. Note that only historical samples are shown in (D).

3.3. Evaluation of Library Preparartion Protocols

The analysis of the DNA obtained from the historical remains corroborated the results of prior investigations, indicating that the DNA extracted from these samples has undergone significant degradation [4,15,48]. Due to the utilization of a double-stranded library preparation approach in the present approved MPS workflow at AFMES-AFDIL, it is probable that a significant number of these DNA fragments are not retained and fail to be turned into libraries. Consequently, the utilization of a single-stranded library preparation method could enhance success rates.. We tested this hypothesis by taking three 5 μL aliquots from the ancient DNA extracts with a 56 °C incubation and preparing libraries with the KAPA Hyper Prep double-stranded kit (currently being validated for casework by AFMES-AFDIL) [**15**], the SRSLY single-stranded kit [**49**], and the single-stranded DNA library preparation method described in [**18**], which we refer to as the "MPI" method. For the double-stranded KAPA libraries, shotgun sequencing resulted in between 4 and 6997 unique nuclear DNA fragments recovered (two libraries failed during library prep).

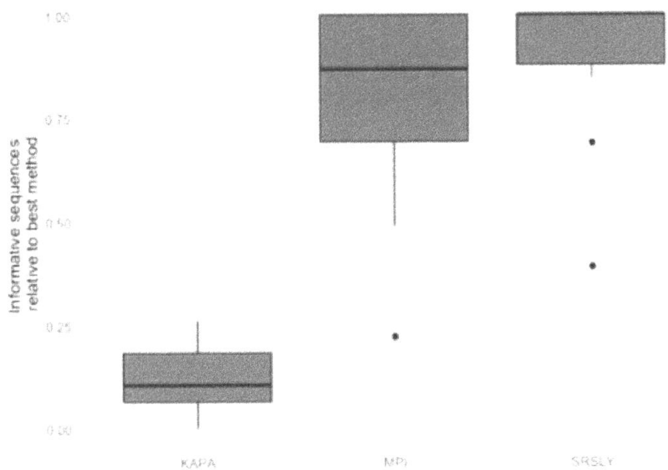

Figure 4. Relative informative nuclear DNA sequences recovered per library preparation protocol across 15 historical skeletal remains. The relative informative sequences per library was calculated as described in Supporting Information Text S2. The box plots follow the standard Tukey representation to show the distribution of the relative informative sequences per extraction type (*n* = 15). The significance of different relative amounts recovered per library prep type was evaluated with a Wilcoxon test with a correction for multiple testing.

The two single-stranded DNA libraries resulted in between 23 and 143,388 unique DNA fragments for the SRSLY kit and 908 to 407,183 for the MPI method. It is important to note, however, that none of these libraries were sequenced to exhaustion. As with the extraction method comparison, we determined the relative informative sequences per sample for the three different library preparation methods based on the shotgun data (**Figure 4**). This confirmed that single-stranded library preparation methods were more effective at recovering informative sequences (*n* = 15 for the number of samples used for comparisons across the three different library preparation methods, Wilcoxon test with correction for multiple testing, both p-values $< 2.0 \times 10^{-5}$), although there was no significant difference between the two single-stranded methods (p-value = 0.29).

Next, we wanted to evaluate the mtDNA variant calling success rates of the two single-stranded methods for each sample to determine if there was a difference from previous casework attempts. In order to test this, each library was enriched for human mtDNA using hybridization capture. The resulting sequencing data was then processed both at the MPI and using a CLC workflow at the AFMES-AFDIL (see **Section 2.5**). No significant difference was observed in the percentage of sequences that mapped between the two library preparation methods (**Figure 5**A, Wilcox test with correction for multiple testing, $n = 15$, $p = 0.57$).

However, during the processing of the SRSLY libraries, pig mtDNA was identified as mismapping to the human mtDNA genome (0 to 17 fragments, contributing up to %6.5 of sequences per sample library). Therefore, a "pig-out" alternative workflow was used to remove these sequences via competitive mapping for all the SRSLY libraries. The MPI method was found to result in more mtDNA sequences (**Figure 5**B, p-values adjusted for multiple testing 5.5×10^{-4} for MPI workflow and $^{6-}10 \times 7.6$ for CLC workflow, $n = 12$ samples compared between library types, Wilcoxon test) with no significant difference in number of unique mtDNA sequences ($n = 40$ sample, library and workflow comparisons, p-value adjusted for multiple testing: 0.24, Wilcoxon test). Importantly, the maximum coverage for the negative controls was 0.3-fold for the CLC workflow, which is lower than typically observed in MPS data using the validated AFMES-AFDIL double-stranded DNA workflow [**4**], allowing for lowering thresholds below 10-fold for variant detection. For the CLC processing, two standard deviations above mean coverage of the controls was 0.38-fold. Thus the analytical threshold, above which sequence data can be distinguished from background noise, could be set as low as 2-fold (2×) average coverage.

The SRSLY library preparation method resulted in 13 samples with an average coverage of at least 10-fold for the mtDNA genome and the MPI library preparation method resulted in 14 samples with at least 10-fold average coverage. While 10-fold was the previous coverage threshold, the low coverages observed in the negative controls supported decreasing this threshold comfortably to 5-fold, well above the 2-fold level of background noise. All samples resulted in at least 5-fold average coverage using either single-stranded library preparation technique (**Figure 6**).

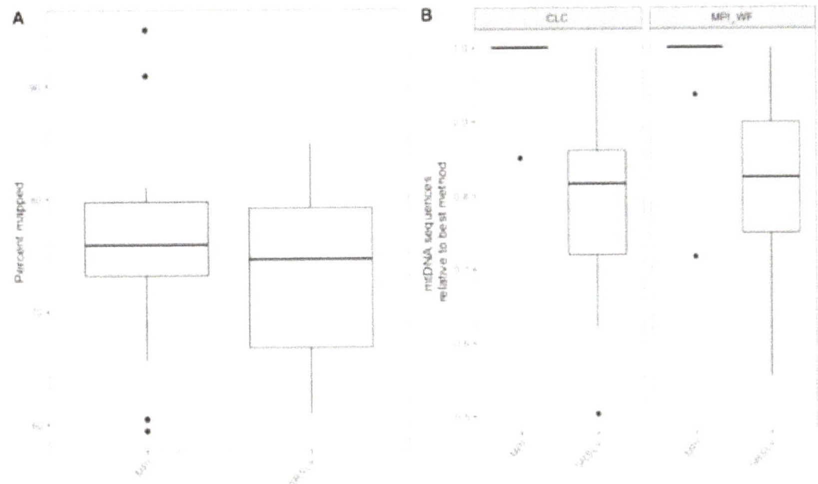

Figure 5. (A) The distribution of the percentage of sequences at least 30 bp long that mapped to the human mtDNA reference genome after capture using the MPI workflow. (B) The relative number of mapped mtDNA sequences across library preparation methods for both the CLC and MPI workflows. The box plots follow the standard Tukey representation to show the distribution of the relative informative sequences per extraction type (n = 15).

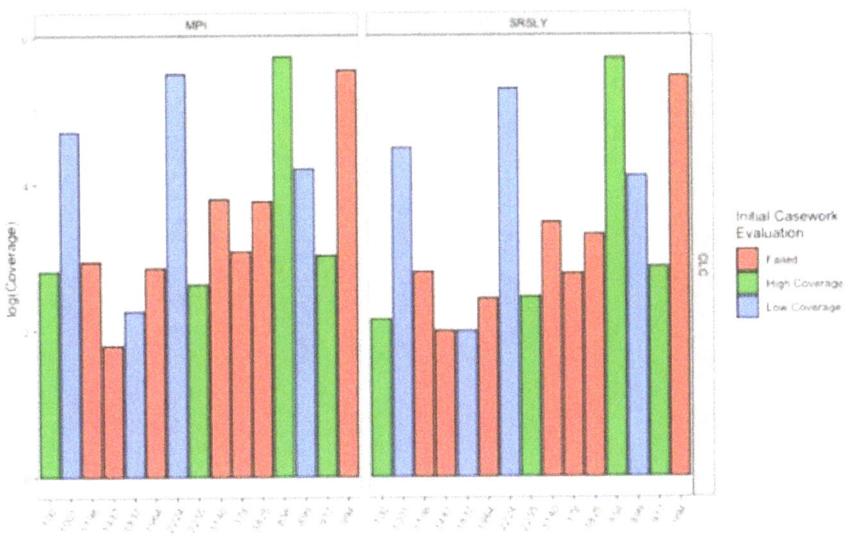

Figure 6. Average coverage of different qualities of samples with MPI and SRSLY library preparations after enrichment for human mtDNA using the CLC workflow. Samples are colored based on the quality as determined from the initial casework evaluation before changing the extraction and library preparation methods (Table 1). The left plots underwent MPI library preparation and the right plots SRSLY library preparation.

For variant calling, only the CLC workflow was used as it applies forensic standards to haplotype reporting that have been previously validated for forensic casework use. When compared with previously generated casework results, the low-coverage and high-coverage sample data were nearly 100% concordant (**Table 3**). It is important to note that the previous casework results were generated from two DNA libraries comprised of roughly 1.0 g of bone powder each; whereas the data in this study utilized DNA libraries comprised of 0.01 g of bone powder. The only discordance was observed in the high coverage sample 2255 at nucleotide position (np) 6755. Heteroplasmy (G6755R) was called in the 473-fold casework data due to a minor guanine (G) variant at 21.1% frequency. However, heteroplasmy was not observed above the set thresholds in either the MPI or the SRSLY data, which both produced a G6755A variant. It is noted that the MPI data exhibited 15-fold coverage at this position, with 13 (87%) of reads exhibiting adenine (A) at this position and two reads (13%) producing a guanine (G) base. The SRSLY data had 11× coverage at np 6755 and produced 100% adenine. This decreased observation of guanine at this position may be due to a combination of lower coverage and cytosine deamination. It is possible that higher coverage, gained through either higher DNA input into library or through combining data produced from multiple libraries and/or sequencing events, would result in improved heteroplasmy detection at this position in either of the single-stranded DNA libraries.

Table 3. Haplotype comparison between previously generated forensic casework data using the methods described in [4] and two single-stranded DNA library preparations (the Max Planck Institute (MPI) [18] and SRSLY [49]) after enrichment for human mtDNA. The AFMES-AFDIL casework minimum coverage threshold was 10-fold and the maximum number of reportable bases was 16,507, as regions of length heteroplasmy are not reported in routine casework. The MPI and SRSLY coverage thresholds were 5-fold. The SRSLY bioinformatics workflow required removal of contaminating *Sus scrofa* mitochondrial DNA (mtDNA) sequences prior to mapping to the rCRS [31].

Sample ID	Casework		MPI			SRSLY (Pig Out)		
	Reported Bases	Predicted Haplogroup	Reported Bases	Predicted Haplogroup	Discordant Sites (Compared to Casework)	Reported Bases	Predicted Haplogroup	Discordant Sites (Compared to Casework)
High Coverage								
2224	16507	H18	16569	H18	0	16568	H18	0
0858	16507	V3c	16568	V3c	0	16569	V3c	0
0994	16507	J1c4b	16569	J1c4b	0	16569	J1c4b	0
2255	16507	J1c3	16268	J1c3/J1c3h/J1c3k	1	16144	J1c3	1
Low Coverage								
0899	16507	H4a1a4b	16565	H4a1a4b	0	16563	H4a1a4b	0
0100	16183	H24a	16181	H24a	0	13585	H24/H24a	0
1001	15765	H1c6	16566	H1c6	0	16567	H1c6	0
3140	16505	U5b1b1-16192	16555	U5b1b1- 16192	0	16515	U5b1b1-16192	0
Failed in Casework								
0378	NA	NA	15951	U5a1a1+16362/U5a1a1d/ U5a1a1d1	NA	15681	U5a1a1+16362/ U5a1a1d/U5a1a1d1	NA
1837	NA	NA	13504	K1c2	NA	11526	K1c2	NA

Sample ID	Casework		MPI			SRSLY (Pig Out)		
	Reported Bases	Predicted Haplogroup	Reported Bases	Predicted Haplogroup	Discordant Sites (Compared to Casework)	Reported Bases	Predicted Haplogroup	Discordant Sites (Compared to Casework)
1198	NA	NA	16091	H1c3b	NA	15821	H1c3b	NA
1964	NA	NA	16010	L2a1c+16086/L2a1c3b	NA	14871	L2a1c+16086	NA
0917	NA	NA	16386	K1a1b2b	NA	16004	K1a1b2b	NA
3825	NA	NA	16550	H11a	NA	16319	H11a	NA
1487	NA	NA	10715	K1a (and subhaplogroups of K1a)	NA	12009	K1a9	NA

The seven samples that did not generate mtDNA profiles that could be included in the report during regular casework, using a threshold of 10-fold coverage, showed consistent and partially matching profiles when comparing the 5-fold MPI and SRSLY data. The estimated haplogroups for all seven samples were nearly same between the MPI and SRSLY data, despite some variations in the ranges. The lack of complete similarity between them can be attributed to the absence of some variations in regions with insufficient coverage. These regions had a minimum coverage threshold of 5-fold, but did not meet the minimum variant count of 4 due to factors such as cytosine deamination or low-level background noise. The merging of sequence data from a second/replicate library is expected to enhance the coverage and accuracy of variant detection. The integration of data from two separate libraries, a common practice in AFMES-AFDIL casework, would probably result in these profiles being eligible for reporting. Significantly, out of the seven samples that were unsuccessful, only samples 1837 and 1964 had a coverage that exceeded five times the average in routine DNA casework. Therefore, the MPI and SRSLY techniques enhanced the success rate of mtDNA profiling for the lowest quality AFMES-AFDIL samples, specifically the bottom third (n = 5), from 0% to 100%. Put simply, the previous approach failed to provide even partially complete mtDNA profiles with enough coverage to be reported, but the ssDNA methods provided enough coverage to determine haplogroups and differentiate mtDNA profiles.

4. DISCUSSION

Historical remains are typically analyzed by the forensic science community, especially when they originate from armed combat battles. The DNA obtained from the samples in this collection, which underwent diverse preservation circumstances, including formalin treatment, exhibited comparable characteristics to DNA retrieved from remains dating back more than 30,000 years. Prior studies have shown that ancient DNA extraction methods can enhance DNA yields in historical cases using artificially degraded DNA and forensic casework samples [28,30]. However, this study specifically highlights the advantages of implementing this change for the exhumed remains of WWII and Korean War service members. The DNA obtained from the examined remains has undergone degradation, perhaps due to the application of powder and/or treatment with formalin [25,26]. As a result, these remains are suitable for the production of single-stranded DNA libraries. Out of the seven samples that previously did not produce data meeting the necessary 10-fold average coverage for casework, six samples successfully achieved this criterion using at least one of the two single-stranded DNA library production methods. Significantly, these techniques also yielded more accurate negative controls, enabling the reduction of thresholds for data reporting. This is advantageous when dealing with degraded samples that have limited data quantities. By reducing the average coverage criteria to 5-fold, we were able to include two additional samples in our report. If we had kept the previous 10-fold threshold, these samples would have been excluded. As a result, all samples were able to be included in the report.

Typically, in forensic DNA analysis, analytical thresholds are established based on 2–3 standard deviations above the mean coverage in negative controls [**4**]. In this study that would place the average coverage threshold at 2-fold. However, we raised this threshold to 5-fold, in order to minimize the reporting of background noise, low-level variants, and contamination. This also raises an important discussion about the detection of contamination. Although both the ancient DNA and forensic fields follow similar clean room practices and maintain negative controls throughout processing, the methods for identification of contamination in sample data differs. Expected elevated rates of deamination in ancient samples allow for the examination of deamination patterns to determine the presence of endogenous DNA and estimate amounts of modern human contamination [**41,42**], which may result in decisions to restrict analyses to only fragments that have deamination. While the samples in this study showed elevated deamination rates, this is not the case for all forensic samples and cannot be relied upon for confirmation of endogenous DNA content. For this reason, the forensic field has established three pathways for detecting contamination. First, DNA profiles from all staff members and individuals known to have handled the remains are maintained allowing for source identification from any contamination that may arise from the handling of the remains during processing. Profiles among unidentified remains can also be compared this way in case cross-contamination occurs due to commingling of remains. Second, all results from degraded skeletal remains are replicated from a new/second bone powder aliquot that is taken through DNA extraction, downstream processing, and analysis. Finally, each profile is evaluated as being either single-source or a mixture using interpretation parameters established through comprehensive validation studies. For mtDNA, mixed/contaminated DNA samples are known to cause a reduction in variant frequency as well as an excess of heteroplasmy (>3 nucleotide positions across the mitogenome) [**4**]. Mixed DNA samples are not reportable, and the laboratory work will be repeated in an attempt to determine the source of the mixture. If the bone itself is contaminated, as evidenced by a repeat mixture after a second DNA extraction, the sample results will not be reported. Determining contamination within the bone requires having multi-fold coverage across the mitogenome, which is why we decided to use a threshold of 5-fold. Attempts to lower this threshold further would require the development of other bioinformatic tools for the identification of mixtures from low-coverage data.

Although exogenous human DNA is highly scrutinized, as described above, low-level contaminants from non-human sources can also impact the accuracy of variant calling. We were able to develop a pipeline to remove the identified pig mtDNA contaminant, but currently validated workflows at the AFMES-AFDIL do not include these steps or a way to easily screen for the presence of non-human DNA. There are different taxonomic classifiers and methods used in other fields to identify different species and families in sequence data [**39,40,50**]. Based on the pig DNA observed in the SRSLY data, it may be worthwhile considering integrating one of these into workflows when working with historical casework. Additionally, the presence of deamination in all samples raises the question of how to ensure this does not impact downstream variant calling. In the presented analysis, variants that appeared to be due to deaminated bases were removed from the profile for concordance purposes. For the typical casework workflow, the impacts of deamination would likely be mitigated by merging the data from multiple extractions and/or libraries. In ancient DNA studies, different methods are used to minimize the impact of deamination on downstream analyses. For example, the terminal three positions of all fragments are often masked or, in extreme cases, only variants that are transversions are used [**51,52**]. A previous sensitivity study on the impact of DNA degradation on STR and SNP concordance with Verogen'sForenSeq kit also found differences in concordance and heterozygosity balance in comparison to studies with non-degraded DNA [**28**], indicating that considerations for the analysis of degraded DNA across platforms is of importance. Future studies on the impacts of low-input and degraded DNA for downstream forensics analyses will both help to increase the accuracy when working with more challenging data and open up the possibility of producing accurate, identification informative data from previously inaccessible samples.

5. AT A GLANCE

We have discovered that modifying the ancient DNA extraction methodology and combining it with single-stranded library preparation significantly enhances the quantity of mitochondrial DNA recovered from excavated skeletal remains of military service members. These findings highlight the importance of sharing techniques between the fields of ancient DNA and forensics [29], which prove to be mutually advantageous, as demonstrated by the ancient DNA community's adoption of bleach treatment to eliminate contamination from samples [53,54]. The use of the techniques employed in this research onto forensic evidence containing extensively deteriorated DNA holds the promise of enhancing the efficacy of identifying ancient remains.

6. CONCISE SUMMARY

The incorporation of massively parallel sequencing (MPS) technology into forensic casework has specifically aided in the identification of unidentified military personnel. Nevertheless, skeletal remains that have undergone significant degradation or chemical treatment frequently do not yield viable DNA profiles, despite the utilization of advanced mitochondrial (mt) DNA capture and MPS techniques. Simultaneously, the field of ancient DNA has created systematic procedures tailored for fragmented DNA, leading to the effective retrieval of nuclear DNA and mtDNA from skeletal remains and sediment that are more than 100,000 years old. This study utilizes a collection of exhumed skeleton remains from the Korean War and World War II to examine if the application of ancient DNA extraction and library preparation techniques enhances the accuracy of forensic DNA profiling. We have discovered a method for extracting ancient DNA that has yielded a greater number of human mtDNA fragments compared to previously employed techniques in forensic investigations. Furthermore, employing single-stranded library preparation instead of double-stranded library preparation led to a higher achievement of detectable mtDNA profiles. This study highlights that the integration of ancient DNA extraction and library preparation methods assessed in this research enhances the efficacy of DNA profiling and the probability of recognizing historical remnants.

REFERENCES

1. First Criminal Conviction Secured with Next-Gen Forensic DNA Technology. 2019. Available online: **https://verogen.com/first-criminal-conviction-with-next-gen-forensic-dna/** (accessed on 2 December 2021).
2. *National DNA Index System (NDIS) Operational Procedures Manual*; Version 10; FBI Laboratory: Quantico, VA, USA, 2021.
3. Greytak, E.M.; Moore, C.; Armentrout, S.L. Genetic genealogy for cold case and active investigations. *Forensic Sci. Int.* **2019**, *299*, 103–113. **[Google Scholar]** **[CrossRef]**
4. Marshall, C.; Andreaggi, K.; Daniels-Higginbotham, J.; Oliver, R.S.; Barritt-Ross, S.; McMahon, T.P. Performance evaluation of a mitogenome capture and Illumina sequencing protocol using non-probative, case-type skeletal samples: Implications for the use of a positive control in a next-generation sequencing procedure. *Forensic Sci. Int. Genet.* **2017**, *31*, 198–206. **[Google Scholar]** **[CrossRef]** **[PubMed]** **[Green Version]**
5. Amorim, A.; Fernandes, T.; Taveira, N. Mitochondrial DNA in human identification: A review. *PeerJ* **2019**, *7*, e7314. **[Google Scholar]** **[CrossRef]**

6. Ambers, A.; Bus, M.M.; King, J.L.; Jones, B.; Durst, J.; Bruseth, J.E.; Gill-King, H.; Budowle, B. Forensic genetic investigation of human skeletal remains recovered from the La Belle shipwreck. *Forensic Sci. Int.* **2020**, *306*, 110050. **[Google Scholar]** **[CrossRef]**

7. Cole, P.M. *POW/MIA Issues: The Korean War*; NDRI: Santa Monica, CA, USA, 1994; Volume 1. **[Google Scholar]**

8. Keene, J. Bodily matters above and below ground: The treatment of American remains from the Korean war. *Public Hist.* **2010**, *32*, 59–78. **[Google Scholar]** **[CrossRef]** **[PubMed]**

9. Morris-Suzuki, T. Lavish are the dead: Re-envisioning Japan's Korean war. *Asia Pac. J. Jpn. Focus* **2013**, *11*, 1–8. **[Google Scholar]**

10. Coleman, B.L. Recovering the Korean war dead, 1950–1958: Graves registration, forensic anthropology, and wartime memorialization. *J. Mil. Hist.* **2008**, *72*, 179–222. **[Google Scholar]** **[CrossRef]**

11. Gilbert, M.T.P.; Haselkorn, T.; Bunce, M.; Sanchez, J.J.; Lucas, S.B.; Jewell, L.D.; Van Marck, E.; Worobey, M. The isolation of nucleic acids from fixed, paraffin-embedded tissues-which methods are useful when? *PLoS ONE* **2007**, *2*, e537. **[Google Scholar]** **[CrossRef]**

12. Kistler, L.; Ware, R.; Smith, O.; Collins, M.; Allaby, R.G. A new model for ancient DNA decay based on paleogenomic meta-analysis. *Nucleic Acids Res.* **2017**, *45*, 6310–6320. **[Google Scholar]** **[CrossRef]** **[PubMed]**

13. Holland, T.; Byrd, J.; Sava, V. Joint POW/MIA Accounting Command's Central Identification. *Forensic Anthropol. Lab.* **2008**, *47*, 63–80. **[Google Scholar]**

14. Edson, S.; Ross, J.P.; Coble, M.D.; Parsons, T.J.; Barritt, S.M. Naming the dead—Confronting the realities of rapid identification of degraded skeletal remains. *Forensic Sci. Rev.* **2004**, *16*, 63–90. **[Google Scholar]** **[PubMed]**

15. Gorden, E.M.; Greytak, E.M.; Sturk-Andreaggi, K.; Cady, J.; McMahon, T.P.; Armentrout, S.; Marshall, C. Extended kinship analysis of historical remains using SNP capture. *Forensic Sci. Int. Genet.* **2022**, *57*, 102636. **[Google Scholar]** **[CrossRef]**

16. Dabney, J.; Knapp, M.; Glocke, I.; Gansauge, M.-T.; Weihmann, A.; Nickel, B.; Valdiosera, C.; García, N.; Pääbo, S.; Arsuaga, J.-L.; et al. Complete mitochondrial genome sequence of a Middle Pleistocene cave bear reconstructed from ultrashort DNA fragments. *Proc. Natl. Acad. Sci. USA* **2013**, *110*, 15758–15763. **[Google Scholar]** **[CrossRef]** **[Green Version]**

17. Rohland, N.; Glocke, I.; Aximu-Petri, A.; Meyer, M. Extraction of highly degraded DNA from ancient bones, teeth and sediments for high-throughput sequencing. *Nat. Protoc.* **2018**, *13*, 2447–2461. **[Google Scholar]** **[CrossRef]**

18. Gansauge, M.-T.; Aximu-Petri, A.; Nagel, S.; Meyer, M. Manual and automated preparation of single-stranded DNA libraries for the sequencing of DNA from ancient biological remains and other sources of highly degraded DNA. *Nat. Protoc.* **2020**, *15*, 2279–2300. **[Google Scholar] [CrossRef]**

19. Fu, Q.; Meyer, M.; Gao, X.; Stenzel, U.; Burbano, H.A.; Kelso, J.; Pääbo, S. DNA analysis of an early modern human from Tianyuan Cave, China. *Proc. Natl. Acad. Sci. USA* **2013**, *110*, 2223–2227. **[Google Scholar] [CrossRef] [PubMed] [Green Version]**

20. Suchan, T.; Kusliy, M.A.; Khan, N.; Chauvey, L.; Tonasso-Calvière, L.; Schiavinato, S.; Southon, J.; Keller, M.; Kitagawa, K.; Krause, J.; et al. Performance and automation of ancient DNA capture with RNA hyRAD probes. *Mol. Ecol. Resour.* **2021**, 1–17. **[Google Scholar] [CrossRef] [PubMed]**

21. Vernot, B.; Zavala, E.I.; Gómez-Olivencia, A.; Jacobs, Z.; Slon, V.; Mafessoni, F.; Romagné, F.; Pearson, A.; Petr, M.; Sala, N.; et al. Unearthing Neanderthal population history using nuclear and mitochondrial DNA from cave sediments. *Science* **2021**, *372*, eabf1667. **[Google Scholar] [CrossRef] [PubMed]**

22. Zavala, E.I.; Jacobs, Z.; Vernot, B.; Shunkov, M.V.; Kozlikin, M.B.; Derevianko, A.P.; Essel, E.; de Fillipo, C.; Nagel, S.; Richter, J.; et al. Pleistocene sediment DNA reveals hominin and faunal turnovers at Denisova Cave. *Nature* **2021**, *595*, 399–403. **[Google Scholar] [CrossRef]**

23. Meyer, M.; Arsuaga, J.L.; de Filippo, C.; Nagel, S.; Aximu-Petri, A.; Nickel, B.; Martinez, I.; Gracia, A.; de Castro, J.M.B.; Carbonell, E.; et al. Nuclear DNA sequences from the middle pleistocene sima de loshuesos hominins. *Nature* **2016**, *531*, 504–507. **[Google Scholar] [CrossRef]**

24. Meyer, M.; Kircher, M.; Gansauge, M.-T.; Li, H.; Racimo, F.; Mallick, S.; Schraiber, J.G.; Jay, F.; Prüfer, K.; De Filippo, C.; et al. A high-coverage genome sequence from an archaic Denisovan individual. *Science* **2012**, *338*, 222–226. **[Google Scholar] [CrossRef] [Green Version]**

25. Stiller, M.; Sucker, A.; Griewank, K.; Aust, D.; Baretton, G.B.; Schadendorf, D.; Horn, S. Single-strand DNA library preparation improves sequencing of formalin-fixed and paraffin-embedded (FFPE) cancer DNA. *Oncotarget* **2016**, *7*, 59115–59128. **[Google Scholar] [CrossRef] [Green Version]**

26. Gansauge, M.-T.; Gerber, T.; Glocke, I.; Korlević, P.; Lippik, L.; Nagel, S.; Riehl, L.M.; Schmidt, A.; Meyer, M. Single-stranded DNA library preparation from highly degraded DNA using T4 DNA ligase. *Nucleic Acids Res.* **2017**, *45*, e79. **[Google Scholar]**

27. Straube, N.; Lyra, M.L.; Paijmans, J.L.A.; Preick, M.; Basler, N.; Penner, J.; Rödel, M.; Westbury, M.V.; Haddad, C.F.B.; Barlow, A.; et al. Successful application of ancient DNA extraction and library construction protocols to museum wet collection specimens. *Mol. Ecol. Resour.* **2021**, *21*, 2299–2315. **[Google Scholar] [CrossRef]**

28. Zavala, E.I.; Rajagopal, S.; Perry, G.H.; Kruzic, I.; Bašić, Ž.; Parsons, T.J.; Holland, M.M. Impact of DNA degradation on massively parallel sequencing-based autosomal STR, iiSNP, and mitochondrial DNA typing systems. *Int. J. Legal. Med.* **2019**, *133*, 1369–1380. **[Google Scholar] [CrossRef]**

29. Hofreiter, M.; Sneberger, J.; Pospisek, M.; Vanek, D. Progress in forensic bone DNA analysis: Lessons learned from ancient DNA. *Forensic Sci. Int. Genet.* **2021**, *54*, 102538. **[Google Scholar] [CrossRef]**

30. Xavier, C.; Eduardoff, M.; Bertoglio, B.; Amory, C.; Berger, C.; Casas-Vargas, A.; Pallua, J.; Parson, W. Evaluation of DNA Extraction Methods Developed for Forensic and Ancient DNA Applications Using Bone Samples of Different Age. *Genes* **2021**, *12*, 146. **[Google Scholar] [CrossRef] [PubMed]**

31. Andrews, R.M.; Kubacka, I.; Chinnery, P.F.; Lightowlers, R.N.; Turnbull, D.M.; Howell, N. Reanalysis and revision of the Cambridge reference sequence for human mitochondrial DNA. *Nat. Genet.* **1999**, *23*, 147. **[Google Scholar] [CrossRef] [PubMed]**

32. Edson, S.M. Extraction of DNA from Skeletonized Postcranial Remains: A Discussion of Protocols and Testing Modalities. *J. Forensic Sci.* **2019**, *64*, 1312–1323. **[Google Scholar] [CrossRef] [PubMed]**

33. Gorden, E.M.; Sturk-Andreaggi, K.; Marshall, C. Repair of DNA damage caused by cytosine deamination in mitochondrial DNA of forensic case samples. *Forensic Sci. Int. Genet.* **2018**, *34*, 257–264. **[Google Scholar] [CrossRef] [Green Version]**

34. *SRSLY NGS Library Prep Kit for Illumina PicoPlus and NanoPlus UDIs and UMIs/UDIs: Unabridged Instruction Manual*; ClarertBiosience, LLC.: Santa Cruz, CA, USA, 2020.

35. Renaud, G.; Stenzel, U.; Kelso, J. leeHom: Adaptor trimming and merging for Illumina sequencing reads. *Nucleic Acids Res.* **2014**, *42*, e141. **[Google Scholar] [CrossRef]**

36. Sturk-Andreaggi, K.; Peck, M.A.; Boysen, C.; Dekker, P.; McMahon, T.P.; Marshall, C.K. AQME: A forensic mitochondrial DNA analysis tool for next-generation sequencing data. *Forensic Sci. Int. Genet.* **2017**, *31*, 189–197. **[Google Scholar] [CrossRef]**

37. Sturk-Andreaggi, K.; Parson, W.; Allen, M.; Marshall, C. Impact of the sequencing method on the detection and interpretation of mitochondrial DNA length heteroplasmy. *Forensic Sci. Int. Genet.* **2020**, *44*, 102205. **[Google Scholar]** **[CrossRef]** **[PubMed]** **[Green Version]**

38. Slon, V.; Hopfe, C.; Weiß, C.L.; Mafessoni, F.; de la Rasilla, M.; Lalueza-Fox, C.; Rosas, A.; Soressi, M.; Knul, M.V.; Miller, R.; et al. Neandertal and Denisovan DNA from Pleistocene sediments. *Science* **2017**, *356*, 605–608. **[Google Scholar]** **[CrossRef]** **[Green Version]**

39. Altschul, S.F.; Gish, W.; Miller, W.; Myers, E.W.; Lipman, D.J. Basic local alignment search tool. *J. Mol. Biol.* **1990**, *215*, 403–410. **[Google Scholar]** **[CrossRef]**

40. Huson, D.H.; Auch, A.F.; Qi, J.; Schuster, S.C. MEGAN analysis of metagenomic data. *Genome Res.* **2007**, *17*, 377–386. **[Google Scholar]** **[CrossRef]** **[Green Version]**

41. Briggs, A.W.; Stenzel, U.; Johnson, P.L.F.; Green, R.; Kelso, J.; Prufer, K.; Meyer, M.; Krause, J.; Ronan, M.T.; Lachmann, M.; et al. Patterns of damage in genomic DNA sequences from a neandertal. *Proc. Natl. Acad. Sci. USA* **2007**, *104*, 14616–14621. **[Google Scholar]** **[CrossRef]** **[PubMed]** **[Green Version]**

42. Peyregne, S.; Peter, B.M. AuthentiCT: A model of ancient DNA damage to estimate the proportion of present-day DNA contamination. *Genome Biol.* **2020**, *21*, 246. **[Google Scholar]** **[CrossRef]**

43. Sawyer, S.; Krause, J.; Guschanski, K.; Savolainen, V.; Pääbo, S. Temporal patterns of nucleotide misincorporations and DNA fragmentation in ancient DNA. *PLoS ONE* **2012**, *7*, e34131. **[Google Scholar]**

44. Rathbun, M.M.; McElhoe, J.A.; Parson, W.; Holland, M.M. Considering DNA damage when interpreting mtDNAheteroplasmy in deep sequencing data. *Forensic Sci. Int. Genet.* **2017**, *26*, 1–11. **[Google Scholar]** **[CrossRef]**

45. Holland, C.A.; McElhoe, J.A.; Gaston-Sanchez, S.; Holland, M.M. Damage patterns observed in mtDNA control region MPS data for a range of template concentrations and when using different amplification approaches. *Int. J. Leg. Med.* **2021**, *135*, 91–106. **[Google Scholar]** **[CrossRef]**

46. Hajdinjak, M.; Mafessoni, F.; Skov, L.; Vernot, B.; Hübner, A.; Fu, Q.; Essel, E.; Nagel, S.; Nickel, B.; Richter, J.; et al. Initial Upper Palaeolithic humans in Europe had recent Neanderthal ancestry. *Nature* **2021**, *592*, 253–257. **[Google Scholar]** **[CrossRef]**

47. Fu, Q.; Posth, C.; Hajdinjak, M.; Petr, M.; Mallick, S.; Fernandes, D.; Furtwängler, A.; Haak, W.; Meyer, M.; Mittnik, A.; et al. The genetic history of Ice Age Europe. *Nature* **2016**, *534*, 200–205. **[Google Scholar]** **[CrossRef]** **[PubMed]** **[Green Version]**

48. Marshall, C.; Taylor, R.; Sturk-Andreaggi, K.; Barritt-Ross, S.; Berg, G.E.; McMahon, T.P. Mitochondrial DNA haplogrouping to assist with the identification of unknown service members from the World War II Battle of Tarawa. *Forensic Sci. Int. Genet.* **2020**, *47*, 102291. **[Google Scholar] [CrossRef] [PubMed]**

49. Troll, C.J.; Kapp, J.; Rao, V.; Harkins, K.; Cole, C.; Naughton, C.; Morgan, J.M.; Shapiro, B.; Green, R.E. A ligation-based single-stranded library preparation method to analyze cell-free DNA and synthetic oligos. *BMC Genom.* **2019**, *2*, 1023. **[Google Scholar] [CrossRef] [PubMed]**

50. Wood, D.E.; Lu, J.; Langmead, B. Improved metagenomic analysis with Kraken 2. *Genome Biol.* **2019**, *20*, 257. **[Google Scholar] [CrossRef] [Green Version]**

51. Green, R.E.; Krause, J.; Briggs, A.W.; Maricic, T.; Stenzel, U.; Kircher, M.; Patterson, N.; Li, H.; Zhai, W.; Fritz, M.H.-Y.; et al. A draft sequence of the Neandertal genome. *Science* **2010**, *328*, 710–722. **[Google Scholar] [CrossRef] [Green Version]**

52. Meyer, M.; Fu, Q.; Aximu-Petri, A.; Glocke, I.; Nickel, B.; Arsuaga, J.L.; Martinez, I.; Gracia, A.; De Castro, J.M.B.; Carbonell, E.; et al. A mitochondrial genome sequence of a hominin from Sima de losHuesos. *Nature* **2014**, *505*, 403–406. **[Google Scholar] [CrossRef]**

53. Hajdinjak, M.; Fu, Q.; Hübner, A.; Petr, M.; Mafessoni, F.; Grote, S.; Skoglund, P.; Narasimham, V.; Rougier, H.; Crevecoeur, I.; et al. Reconstructing the genetic history of late Neanderthals. *Nature* **2018**, *555*, 652–656. **[Google Scholar] [CrossRef] [Green Version]**

54. Korlević, P.; Gerber, T.; Gansauge, M.-T.; Hajdinjak, M.; Nagel, S.; Aximu-Petri, A.; Meyer, M. Reducing microbial and human contamination in DNA extractions from ancient bones and teeth. *Biotechniques* **2015**, *59*, 87–93. **[Google Scholar] [CrossRef]**

Index

www.ingramcontent.com/pod-product-compliance
Lightning Source LLC
LaVergne TN
LVHW062217200926
848484LV00006B/1418